Contributors

B. ANDERHUB, *Hygien.-mikrobiol. Institut, Kantonsspital, CH-6004, Lucerne, Switzerland*

F. AYLWARD, *Department of Food Science, The University of Reading, London Road, Reading RG1 5AQ, England*

A. C. BAIRD-PARKER, *Unilever Ltd. Colworth House, Sharnbrook, Bedford, England*

ANN BAILLIE, *Unilever Ltd. Colworth House, Sharnbrook, Bedford, England*

W. H. BARKER, *Epidemiology Program, Center for Disease Control, Atlanta, Georgia 30333, U.S.A.*

G. I. BARROW, *Public Health Laboratory, Royal Cornwall Hospital (City), Infirmary Hill, Truro, Cornwall, England*

V. BARTL, *Hygiene Laboratories, Safarikova 14, Prague 2, Czechoslovakia*

H. BEERENS, *C.E.R.T.I.A. 59 Villeneuve-d'Ascq, Rue Jules-Guesde, France*

M. S. BERGDOLL, *University of Wisconsin, 2115 Herrick Drive, Madison, Wisconsin 53706, U.S.A.*

P. S. BRACHMAN, *Epidemiology Program, Center for Disease Control, Atlanta, Georgia 30333, U.S.A.*

H. BRODHAGE, *Hygien.-mikrobiol. Institut, Kantonsspital, CH-6004 Lucerne, Switzerland*

F. L. BRYAN, *Health Agencies Branch, Center for Disease Control, Atlanta, Georgia 30333, U.S.A.*

J. H. B. CHRISTIAN, *Commonwealth Scientific & Industrial Research Organization, Division of Food Research, Box 52, North Ryde, New South Wales 2113, Australia*

J. S. CROWTHER, *Unilever Ltd. Colworth House, Sharnbrook, Bedford, England*

F. L. DAVIES, *National Institute for Research in Dairying, Shinfield, Reading RG2 9AT, England*

W. EDEL, *Rijks Institut Voor de Volksgezondheid, Sterrenbos 1, Utrecht, The Netherlands*

M. S. EDMONDS, *Marks & Spencer Ltd. Michael House, Baker Street, London W1A 1DN, England*

M. W. EKLUND, *National Marine Fisheries Service, Pacific Fishery Products Center, 2725 Montlake Blvd. East, Seattle, Washington 98102, U.S.A.*

D. W. ELLIOTT, *Marks & Spencer, Michael House, Baker Street, London W1A 1DN, England*

L. FIEVEZ, *C.E.R.T.I.A. Rue Jules Guesde–Flers-Bourg, 59-Villeneuve D'Ascq, France*

E. J. GANGAROSA, *Epidemiology Program, Center for Disease Control, Atlanta, Georgia 30333, U.S.A.*

v

R. J. GILBERT, *Food Hygiene Laboratory, Central Public Health Laboratory, Colindale Avenue, London NW9 5HT, England*

B. A. GLATZ, *Food Research Institute, University of Wisconsin, Madison, Wisconsin, U.S.A.*

J. M. GOEPFERT, *Food Research Institute, University of Wisconsin, Madison, Wisconsin, U.S.A.*

N. GOLDENBERG, *Marks & Spencer, Michael House, Baker Street, London W1A 1DN, England*

P. A. M. GUINEE, *Rijks Institut Voor de Volksgezondheid, Sterrenbos 1, Utrecht, The Netherlands*

G. A. HARRWEIJN, *Central Institute for Nutrition and Food Research T.N.O. Utrechtseweg 48, Zeist, The Netherlands*

R. W. S. HARVEY, *Public Health Laboratory Service, University Hospital of Wales, Heath Park, Cardiff CF4 4XW, Wales*

W. J. HAUSLER, Jr., *State Hygienic Laboratory, University of Iowa, Iowa City, Iowa 52240, U.S.A.*

E. HESS, *Veterinar-Bakteriologisches Institut der Universitat Zurich, Winterthurerstrasse 270, Zurich 11, Switzerland*

BETTY C. HOBBS, *Food Hygiene Laboratory, Central Public Health Laboratory, Colindale Avenue, London NW9 5HT, England*

C. IENISTEA, *Institute of Hygiene and Public Health, Str. Dr. Leonte 1-3 Bucharest, Romania*

A Chairman M. INGRAM, *Meat Research Institute, Langford, Nr. Bristol, England*

B. JARVIS, *British Food Manufacturing Institute Research Association, Randalls Road, Leatherhead, Surrey, England*

E. H. KAMPELMACHER, *The Agricultural University, Wageningen, The Netherlands*

H. U. KIM, *Food Research Institute, University of Wisconsin, Madison, Wisconsin, U.S.A.*

E. P. LARKIN, *Virology Branch, Department of Health, Education & Welfare, 1090 Tusculum Avenue, Cincinnati, Ohio 45226, U.S.A.*

J. A. LEE, *Epidemiology Research Laboratory, Central Public Health Laboratory, Colindale Avenue, London NW9 5HT, England*

H. E. MARTHEDAL, *Royal Veterinary & Agricultural University, 13 Bulowsvej, 1870 Copenhagen V, Denmark*

M. H. MERSON, *Epidemiology Program, Center for Disease Control, Atlanta, Georgia 30333, U.S.A.*

M. O. MOSS, *Department of Biological Science, University of Surrey, Guildford, Surrey, England*

D. A. A. MOSSEL, *The Catholic University, Louvain, Belgium*

F. T. POYSKY, *National Marine Fisheries Service, Pacific Fishery Products Center, 2725 Montlake Blvd. East, Seattle, Washington 98102, U.S.A.*

H. J. REHM, *Institut für Mikrobiologie, Piusallee 7, 44 Munster (Westf), Germany*

H. RIEMANN, *University of California, 2046 Haring Hall, Davis, California 95616, U.S.A.*

R. K. ROBINSON, *Department of Food Science, The University of Reading, London Road, Reading RG1 5AQ, England*

C. ROMOND, *C.E.R.T.I.A. 59 Villeneuve-d'Ascq, Rue Jules-Guesde, France*

B. ROWE, *Salmonella & Shigella Reference Laboratory, Central Public Health Laboratory, Colindale Avenue, London NW9 5HT, England*

HALINA SADOWSKA, *Sanit-Epidemiological Department, Ministry of Health & Social Welfare, 15 Miodows Street, Warszawa 1, Poland*

R. SAKAZAKI, *Division of Bacteriology 1, National Institute of Health, 10-35 Kamiosaki 2-chome, Shinagawa-ku, Tokyo, Japan*

M. VAN SCHOTHORST, *Rijks Institut Voor de Volksgezondheid, Sterrenbos 1, Utrecht, The Netherlands*

R. J. SINNELL, *Institut fur Lebensmittelhygiene, Bitterstrasse 8-12, 1 Berlin 33 (Dahlem) Germany*

N. SKOVGAARD, *Royal Veterinary & Agricultural University, Bulowsvej 13, 1870 Copenhagen V, Denmark*

R. SPENCER, *J. Sainsbury Ltd., Stamford House, Stamford Street, London S.E.1, England*

W. M. SPIRA, *Food Research Institute, University of Wisconsin, Madison, Wisconsin, U.S.A.*

F. J. VAN SPRANG, *Central Institute for Nutrition and Food Research T.N.O. Utrechtseweg 48, Zeist, The Netherlands*

BETTY J. STEWART, *Commonwealth Scientific & Industrial Research Organization, Division of Food Research, Box 52, North Ryde, New South Wales 2113, Australia*

R. G. A. SUTTON, *School of Public Health & Tropical Medicine, University of Sydney, Australia*

A. TAYLOR, *Epidemiology Program, Center for Disease Control, Atlanta, Georgia 30333, U.S.A.*

G. WILKINSON, *National Institute for Research in Dairying, Shinfield, Reading RG2 9AT, England*

E. F. WILLIAMS, *J. Sainsbury Ltd., Stamford House, Stamford Street, London S.E.1, England*

SIR GRAHAM WILSON, *London School of Hygiene & Tropical Medicine, Keppel Street, London WC1E 7HT, England*

ANTONNETTE A. WIENEKE, *Food Hygiene Laboratory, Central Public Health Laboratory, Colindale Avenue, London NW9 5HT, England*

A Chairman B. WEITZ, *National Institute for Research in Dairying, Shinfield, Reading RG2 9AT, England*

INTERNATIONAL ASSOCIATION OF MICROBIOLOGICAL SOCIETIES

Committee on Food Microbiology and Hygiene

President: Professor D. A. A. Mossel,
                  Zeist, The Netherlands
Secretary: Dr N. Skovgaard,
                  Copenhagen, Denmark
Vice-President: Dr Betty C. Hobbs,
                  London, England
Treasurer: Dr H. E. Bauman,
                  Minneapolis, U.S.A.

SYMPOSIUM ORGANIZING COMMITTEE

Sir Graham Wilson (Honorary President)
Professor F. Aylward (President)
Dr Betty C. Hobbs (Secretary)
Dr R. K. Robinson (Treasurer)
Miss Nancy Cockman (Assistant Secretary)

THE MICROBIOLOGICAL SAFETY OF FOOD

Proceedings of the Eighth International Symposium on Food Microbiology
Reading, England
September 1972
Organized by the Committee on Food Microbiology and Hygiene of The
International Association of Microbiological Societies

Edited by

BETTY C. HOBBS

AND

J. H. B. CHRISTIAN

1973

ACADEMIC PRESS · LONDON · NEW YORK

ACADEMIC PRESS INC. (LONDON) LTD
24-28 OVAL ROAD
LONDON N.W.1

U.S. Edition published by
ACADEMIC PRESS INC.
111 FIFTH AVENUE
NEW YORK, NEW YORK 10003

Library of Congress Catalog Card Number: LCCCN 73-9464
ISBN: 12-350750-2

Printed in Great Britain by
The Whitefriars Press Ltd., London and Tonbridge, England

Preface

INVESTIGATION, surveillance, research, Codes of Practice, legislation and education are all necessary parts of the process of learning to prevent food-borne infections and intoxications.

Many workers over the years and from many nations have contributed to the knowledge available today.

It seemed fitting to share and to discuss our present experience with the hope that we will lead each other into new pathways, clarify the older ones and from our combined efforts see more plainly the best way to proceed in the future.

With these objects in mind an International Symposium on the Microbiological Safety of Foods was held at the University of Reading, England, in September, 1972. This was the 8th in a series of Symposia on Food Microbiology sponsored by the Committee on Food Microbiology and Hygiene of the International Association of Microbiological Societies.

The Committee is grateful to the University of Reading for the facilities placed at its disposal and to Professor F. Aylward and Dr R. K. Robinson for their help in organizing the meeting and for their editorial assistance with the papers on Education.

In the belief that discussions play an important role in Symposia such as this, full accounts of the questions, answers and comments which followed each paper are printed in these Proceedings. The Editors are indebted to Dr B. Jarvis and Miss R. Blood of the British Food Manufacturing Industries Research Association and Dr M. O. Moss of the University of Surrey for their diligence in recording and collating the discussion.

We are most grateful to Miss Nancy Cockman and her colleagues of the Food Hygiene Laboratory, Colindale, for all the typing and other secretarial labours, particularly for the verification of references, essential in the production of a collection of technical papers, Mrs. H. Christian for much help in proof reading and to Academic Press for advice and assistance.

The Symposium was supported by generous contributions from I.A.M.S., the British Council and Air Corporation Joint Medical Service. The Committee is thankful for these donations.

BETTY C. HOBBS
Food Hygiene Laboratory,
Central Public Health Laboratory,
Colindale Avenue,
London, NW9 5HT, England

J. H. B. CHRISTIAN
Commonwealth Scientific and
Industrial Research Organization,
Division of Food Research, North Ryde,
New South Wales, Australia

August 1973

Introductory Address

SIR GRAHAM WILSON

MR PRESIDENT, Ladies and Gentlemen,

It is my pleasant duty, like that of Professor Aylward, to welcome you from the 4 corners of the earth, and to express the hope that this symposium will prove both interesting and fruitful. Knowing how much labour is expended on the organization of a meeting such as this, I am sure you will agree that we owe a debt of gratitude to those who have been responsible for ordering the necessary arrangements; and agree too that the best way in which we can express our gratitude is to do our utmost to make the meeting a success.

The microbial safety of food, or more specifically the prevention of food poisoning, is the main concern of our symposium. Little attention was paid to this subject before the war, and little evidence existed to show that it was a serious problem. But the war soon changed our attitude. Partly because of the introduction of communal feeding on an extensive scale; partly because of the increased consumption of pre-cooked foods; partly because of the bulk handling of foodstuffs; partly because of the importation of new sorts of food; and partly because of improved laboratory services, the number of recorded outbreaks of food poisoning leapt up in a way more suggestive of an epidemic disease than of a series of sporadic occurrences. Some countries with similar feeding habits experienced much the same increase in food poisoning, whereas others whose methods of handling and consuming food differed substantially from ours suffered much less.

One of our chief objects during the 3 or 4 days we are to be together is to compare the prevalence and type of food poisoning in different countries, to analyse the differences, and to study the ways in which the disease may be prevented.

Another object is to pool the knowledge and experience of workers—medical, veterinary, and industrial—in their various fields. Perhaps this is not the place to say so, but for years I have been troubled at the failure of medical and veterinary workers to co-operate in the way that common sense dictates. Almost wherever one goes one sees public health and veterinary hygiene laboratories established in separate buildings with separate staffs using much the same methods and the same equipment. Joint laboratories would not only ensure better buildings and better equipment at lower cost, but would greatly improve epidemiological co-ordination and stimulate, as discussion between workers with different backgrounds usually does, the generation of new ideas and point the way to further research.

Our third main object is to consider what may be done—technically, educationally, or by legislation—to lower the incidence of food-borne disease.

Here we may find ourselves on contentious ground, some members pressing for more and more bacteriological control and compulsory standards, and others relying rather on education, persuasion, example, and the use of advisory standards. This is where we need to keep an eye on the main chance, remembering that what we are all really concerned about it the prevention of food poisoning. We must keep asking ourselves whether any measure we should like to introduce would in fact appreciably diminish the incidence of the disease. When I look back on some of the great food-poisoning outbreaks with which I have been associated I sometimes wonder whether any routine bacteriological tests that might have been made would have prevented them; and I continue to wonder.

One last point, though it is rather for the Chairman than for me to make it. We belong to different nationalities and speak different languages and often different dialects. If we are to understand each other and profit fully from our discussions I suggest that we speak comparatively slowly, very distinctly, and sufficiently loud to be heard in all parts of the room. Personally I find it just as difficult to understand a foreigner speaking rapidly in English as I expect he would if I was to speak rapidly in his language—that is if I were able to do so. Remember too that all remarks should be addressed to the Chair, and not carried on in muffled tones between 2 members sitting next to each other.

I leave you now to get ahead with the business of the symposium and wish you well in your deliberations.

Contents

SESSION 1 Bacteriology of Various Commodities in Relation to Food
Poisoning

Part 1

Chairman Professor M. Ingram

Part 2

Chairman Professor E. Hess

Session 1

Bacteriology of Various Commodities in Relation to Food Poisoning

Part 1

Chairman: Professor M. Ingram

Hygiene During Meat Production

E. HESS

*Veterinar-Bakteriologisches Institut
der Universitat Zurich,
Zurich 11, Switzerland*

The hygiene of meat production has to begin with the prevention of zoonoses in the breeding and fattening herds. In the slaughtering process itself, there must be a clear division between the contaminated and clean sections both in the construction of the building and in all operational matters. The contaminated procedure ends with the removal of the hide and the bowel. *Conditio sine qua non* for a hygienic slaughter procedure is the absolute separation of the abdominal viscera cleaning department from all other sections. Further prophylactic measures during meat production consist of mechanizing the slaughtering procedure as far as possible from the removal of hooves and hides, through splitting of the carcasses and finally to their automatic conveyance in a hanging position.

Today's knowledge indicates that meat poisonings could most effectively be prevented by the enforcement of the following hygienic measures: (1) regular bacteriological meat inspection in all cases of anamnestic, clinical or pathologic-anatomic based suspicions of a zooanthroponosis; (2) uninterrupted chain of hygiene and cooling. Hygiene of meat production and constant cooling will also result in a product with significantly longer shelf life.

FOODS OF animal origin are potential vehicles of, and may often even be optimal media for, the growth of micro-organisms pathogenic to consumers. In many countries, therefore, milk has to be pasteurized before distribution either because of prevailing mastitis, the hazard of brucellosis or since the producers failed to meet all the necessary hygienic requirements during milking and storage. The same may be said for egg products. With meat, on the other hand, except in the case of canned and certain other special products, there is no heating process before distribution. Hygienic measures during its processing therefore assume prime importance.

The hygiene of meat production has to begin with the prevention of zoonoses in the breeding and fattening herds. Prevention of salmonella infections must have already begun by the enforcement of sufficient controls on feedstuffs.

The next link in the chain of hygiene is the care taken of the animals during transportation to the slaughter house. All unnecessary stress is to be avoided and the animals' subsequent stabling should be in regularly cleaned and disinfected holding pens. Nowadays, the transportation of meat over longer distances has, with ever increasing frequency, taken precedence over the transportation of living animals. If the cooling chain is not interrupted, this procedure is welcome from a hygienic standpoint.

3

In the slaughtering process itself, there must be a clear division between the contaminated and clean sections both in the construction of the building and in all operational matters. The contaminated procedure ends with the removal of the hide and the bowel. *Conditio sine qua non* for a hygienic slaughter procedure is the absolute separation of the abdominal viscera cleaning department from all other sections. This problem may be optimally resolved if hides and viscera can be conveyed from the carcass directly down to the ground floor. In older slaughter houses, without this type of arrangement, the contamination of the meat with intestinal germs during the horizontal transportation of offal could hardly be prevented.

In Switzerland the mesenteries of swine and cattle may be used only for industrial purposes, that is to say, for the production of tallow. This is an important prophylactic measure as was shown by Lott & Britschgi (1967) at the Zürich slaughter house in 1967. In that year, 9·2% of 500 mesenteric lymph nodes of healthy swine, taken at random, were found to be infected with salmonella. In consideration of these figures, one should at the very least avoid the incision of the mesenteric glands during a routine postmortem inspection. Further prophylactic measures during meat production consist of mechanizing the slaughtering procedure as far as possible, from the removal of hooves and hides, followed by the splitting of the carcasses and finally their automatic conveyance in a hanging position. This minimizes the contact of fresh meat with hands, tools and workers' clothing. Hands and tools may transmit tremendous quantities of micro-organisms from hooves, hides and faeces to freshly slaughtered carcasses which were originally germ-free (Hess & Lott, 1970) (see Table 1).

Table 1

Germ counts from surfaces of hooves, from faeces and from hides in groups of 100 animals* (germs/g)

Material	Species	Entero-bacteriaceae	Proteolytic bacteria	Spores	Psychrophilic bacteria
Filth from hooves	Cattle	3,069,000	126,300,000	3,843,000	43,265,000
Faeces	Cattle	1,821,000	623,000	846,000	3,000
	Calves	151,727,000	1,283,000	227,000	300
	Pigs	5,529,000	488,000	89,000	400
Hides	Cattle	2,000	12,280,000	1,261,000	304,000
	Calves	42,000	100,600,000	253,000	4,173,000

* The averages were obtained from 5 pools of 20 animals each, whereby the animals in the first pool were examined individually.

The frequent cleansing and disinfection of hands and tools is made practicable by the installation of a combination unit for washing and disinfection as close as possible to the working areas. A combination consisting of a hot water jet (for defatting and cleansing tools), a movable spray (for the cleansing of aprons, boots and hands) and an automatic disinfectant injector has thus far stood the test of time.

With regard to the processing of meat, one must recall that every cut made increases the risk not only of contamination but also of an intensive multiplication of micro-organisms. Our own investigations have shown that bacterial multiplication is directly proportional to the degree of mincing (Fig. 1). Therefore, in many countries, the storage of unfrozen minced meat is prohibited (Hess & Marthaler, 1962; Hess & Lott, 1965; Lott, 1966).

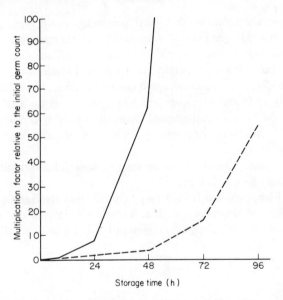

Fig. 1. Multiplication of the normal bacterial flora in fresh meat stored at a temperature of 0 ± 1°. ———, Minced meat; – – – 100 g pieces.

An essential step in the prevention of meat poisoning is the regular inspection of meat production plants. An objective measure of plant hygiene is provided by the qualitative and quantitative bacteriological examination of such semifinished products as raw chopped or sausage meat. In Switzerland we have introduced a regular bacteriological examination of raw products like minced meat and sausage meat on a voluntary basis. The examination of such products acts as an indicator of hygiene on a plant-wide scale. Meat plants with repeatedly good bacteriological results are awarded a "Certificate of Hygiene" and are permitted to make use of this certification in subsequent advertising.

We are now in the process of planning a screening test which would measure the bacterial enzymatic activity in raw products. This should enable us to monitor the procedures in many more meat packing operations with a minimum of additional effort. Our aims for the improvement of plant hygiene are being supported by a large insurance company. The company accepts the final risk of a plant being closed due to a proven contamination with meat poisoning agents. The exclusion of firms that do not fulfill the hygienic standards and the reduction of insurance premiums by up to 30% for plants with a higher standard of hygiene are 2 measures which have without doubt had a positive educational effect.

Summary

Today's knowledge indicates that meat poisonings could most effectively be prevented by the enforcement of the following hygienic measures:

(1) Bacteriological meat inspection in all cases of anamnestic, clinical or pathological-anatomical based suspicions of a zooanthroponosis.

(2) Maintenance of an uninterrupted chain of hygiene, from slaughtering and manufacturing up to and including ultimate preparation in the kitchen.

(3) Provision for an uninterrupted chain of cooling with temperatures not exceeding $+5°$. The very dangerous multiplication of meat poisoning bacteria can thus be prevented.

Hygiene of meat production and constant cooling will also result in a product with significantly longer shelf life.

Barnes & Kampelmacher (1964) have estimated that spoilage represents a loss of up to 20% of total production. This is intolerable in view of the fact that a great part of the world's population constantly suffers from protein deficiency.

References

Barnes, E. M. & Kampelmacher, E. H. (1964). The possibility of and risks in the use of antibiotics (particularly the tetracyclines) for meat preservation under tropical conditions. In *Ier Congr. Internat. Industr. Aliment. Agric., Abidjan* 1964, p. 1093.

Hess, E. & Marthaler, A. (1962). Keimzählungen an abgepacktem Frischfleisch. *Fleischwirtschaft* 14, 497.

Hess, E. & Lott, G. (1965). Die Zerkleinerung des Fleisches, ein haltbarkeitsvermindernder Faktor. *Archiv. Lebensmittelhyg.* 16, 265.

Hess, E. & Lott, G. (1970). Kontamination des Fleisches während und nach der Schlachtung. *Fleischwirtschaft* 50, 47.

Lott, G. (1966). Untersuchungen über die Keimvermehrung in Hackfleisch. Thesis, University of Zürich, 1966.

Lott, G. & Britschgi, T. (1967). Salmonellenprobleme im Schlachthof. *Schweizer. Arch. Tierheilk.* 109, 363.

Discussion

Wilson

I have often wondered to what extent it would be possible to use local hospital and other laboratory facilities to assist in the hygienic control of food processing plants.

Linderholm

I should like to underline what Prof. Hess said about the risk of contamination if areas of food processing plants are not absolutely separated. In many countries the separation between, for example, the gut room and the slaughtering room in slaughterhouses is not sufficient. There are many dangerous communications between areas and rooms of different hygienic standard in other factories also. Further, the rooms for personnel must be separated if persons are working with materials of different hygienic standard. A main objective for food hygienists in the future must be to require sufficient separation whatever the architects may say that it is more practicable to have all operations in one big room or to have room connections with open doors. It is not enough to ensure that each room has a hygienic standard for its own purpose. The main task in food hygiene is to prevent both food poisoning and food spoilage and therefore the lay-out of the premises is most important.

For preventive food hygiene, it is essential that all food processing plants (including slaughterhouses) should have well-equipped laboratory facilities. In Sweden we found that the veterinary inspectors who have their own laboratories in the process plants are best able to work in a preventive way. It must be mentioned that the veterinary inspectors are not doing only the meat inspection but also supervising the hygiene of the food processing; this applies also in many processing plants that do not do their own slaughter. I am sorry to say that I cannot agree with Sir Graham Wilson's suggestion that human bacteriological and veterinary bacteriological laboratories should go together if that means that laboratories cannot be established under veterinary control in factories. However, in central laboratories working with principal public health investigations a co-operation can be successful.

Hess

Veterinarians being responsible for meat hygiene should concentrate more and more on laboratory diagnosis, which will in future be the most important tool in the protection of the consumer and in the prevention of food spoilage.

Barrow

What kind of laboratories will undertake the frequent bacteriological checks on food processing and catering employees envisaged under the new Swedish law?

Linderholm

All official laboratories authorized by the National Food Administration for food control may be used in faeces control of salmonellae-shigellae examination for personnel in food processing. Some of the laboratories in hospitals can be used although they are not authorized to analyse food samples. Private laboratories may not be used but the laboratories in all bigger slaughterhouses and food processing plants are regarded as official because they are headed by the chief veterinary inspector (appointed by the National Food Administration) of the respective plant.

Ingram

Laboratory comparison of rates of growth of organisms in wet minced meat and on sliced meat at 100% relative humidity, indicated *slower* growth in mince due to lack of oxygen. In Dr Hess's experiments the slower growth on larger pieces of meat may possibly have been due to surface drying.

Seaman

The difference between your situation and that of Dr Hess is that your samples showed the same start of growth but diverged in rate whereas his showed a difference of 3 days in the lag period. Can the second state be explained by relative humidity?

Hess

We examined the multiplication factor of the normal bacterial flora in minced meat and in 100 g pieces of fresh meat during storage at $0 \pm 1°$. There was no drying on the surface because both types of samples were packaged and sealed. The lag phase in minced meat was *c.* 48 h shorter than in meat pieces. Our results were confirmed statistically. Our findings would explain, to cite an example, the well-known danger of the rapid multiplication of salmonellae in minced meat.

Salmonella Contaminated Animal Feed in Relation to Infection in Animals and Man

R. W. S. HARVEY

Public Health Laboratory,
University Hospital of Wales,
Heath Park, Cardiff, Wales

Major contributions on this subject have been made by several countries, but there is no unanimity on the epidemiological significance of salmonellae in animal feed. The incidence of contamination is an important measurement, but its numerical value is influenced by several factors. Some of these are discussed: type of raw material, size of sample examined, enrichment medium, enrichment technique and type of rendering plant producing the feed ingredients. The problem of cross contamination is discussed with reference to dust, water and rodent borne contamination and plant contamination. Correlation of salmonella serotypes found in animal feed and in animals is considered in the light of some recent collaborative work. An experiment with pigs fed on infected food prepared from naturally contaminated raw ingredients is described. Duration of excretion in the animals depended on salmonella dosage. Studies in man in Cardiff in 1971 showed that 47% of salmonella infections were caused by serotypes found in pigs in the same area and in the same year. This pattern suggests that an animal, to which the salmonella content of animal feed is relevant, is important to salmonella epidemiology in man. The control of salmonella infected animal feed is briefly discussed.

Introduction

IT MIGHT SEEM natural to preface this paper using the historical approach but the time at my disposal is strictly limited. The salmonella content of animal, marine and vegetable products has aroused great interest in various countries and major contributions in this field have been made in Norway, Sweden, Denmark, the Netherlands, the Federal Republic of Germany, the United Kingdom and the United States of America.

There is, however, a lack of unanimity on the relative importance of salmonella contaminated animal feed in salmonella epidemiology despite the interest which the subject has aroused. For instance, Taylor & McCoy (1969) concluded that, even if all animal feed were to be heat treated, the incidence of salmonella infections in animals due to host specific and host associated serotypes would be unchanged.

In view of the controversial nature of the subject, the facts which do not seem to be in dispute will be considered.

* Paper presented by Dr A. H. Linton, Department of Bacteriology, The Medical School, University Walk, Bristol, England.

First, how do we measure the salmonella content of animal feed? What factors influence this measurement?

The materials

Some ingredients are lightly contaminated, others contain relatively large numbers of salmonellas. The incidence of infected samples varies with the material. Between November 1968 and January 1970, samples coming from a single factory were examined. The samples were ingredients which could be used in the manufacture of pig feed.

Table 1

Incidence of salmonellae in different ingredients of animal feed

Raw material	No. of samples	Positive %
Feather meal	99	27
Meat and bone meal	704	23
Fish meal	31	23
Fish pellets	264	20
Herring meal	60	5

Figures taken from P.H.L.S. Working Group, Skovgaard & Nielsen (1972).

Sample size

The term incidence of salmonellas in animal feed is highly artificial and depends largely on the size of sample examined. This is well illustrated in a paper by Jacobs et al. (1963). They examined 5 x 10 g samples from each of 7 bags of fishmeal—the normal level of sampling—and failed to demonstrate the presence of salmonellae. However, when all the contents of the 7 bags were sampled in 50 g quantities (i.e. 1000 x 50 g), 6 of the 7 bags were found to be contaminated. By examining statistically significant samples they estimated that 26% of all bags imported into the Netherlands in 1962 were contaminated with salmonellae. It was concluded that only the examination of a statistically useful random sample could give reliable information on the probable incidence of contamination. They also reported that 10% of bags were contaminated with *Salmonella typhimurium*—a serotype often said to be uncommon in animal feed ingredients.

In Cardiff also, the effect of the size of sample tested was investigated. In a series of 204 specimens the salmonella contamination rate doubled when the quantity cultured was quadrupled from 25 to 100 g (Harvey & Price, 1967).

Enrichment medium used for salmonella isolation

While the total incidence of salmonellas in ingredients of animal feed is influenced by sample size and isolation technique (temperature of incubation of enrichment cultures; use of secondary enrichment methods; Harvey & Price, 1967, 1968), the incidence of single serotypes of epidemiological importance must also be considered. Prost & Riemann (1967) suggested that the present techniques of culture may favour the isolation of particular serotypes. In 1957, we began to investigate the significance of salmonellas in sewage (Harvey, 1957). Current techniques of salmonella isolation are *not* unbiased and epidemiological thinking founded on biased data may be criticized and rightly so. The apparent absence of certain serotypes from animal feed suggests to some workers that animal food plays little part in their distribution. This may be so, but *S. typhimurium* and *S. dublin* are not the easiest serotypes to isolate from materials containing other salmonella species. In a study of salmonellae present in a polluted river in our area, the performance of selenite F broth was compared with that of Kauffmann Muller tetrathionate broth. Both media were incubated at 43°. Water samples were paired. From selenite F, 46 isolations of *S. typhimurium* were made; from Kauffmann Muller tetrathionate only 24 strains of the serotype were cultured. Certain serotypes were better isolated from Kauffmann Muller, but *S. dublin* was more efficiently recovered using selenite F broth.

The use of certain isolation techniques can actually bias the result in favour of recovery of a particular serotype and this bias is to some extent under the control of the bacteriologist. Imported bone meal was examined in Cardiff by two different methods, one using agglutinating sera to suppress unwanted serotypes. With this serological technique we isolated 60 strains of *S. typhimurium* from 268 samples of bone meal. With the orthodox method, not employing sera, we cultured only 12 strains of *S. typhimurium* from the same samples.

Type of rendering plant producing feed ingredients

Certain sources of supply to major food compounders can provide products relatively free from salmonellae. Others seem to have difficulty in doing so. This is illustrated in a paper by Timoney (1968) contrasting the results of hot solvent extraction in a single plant with four plants using "dry rendering". No isolations of salmonellas were made from the final product in the hot solvent extraction plant, but many isolations were made in the percolation areas of dry rendering plants. Rendering plant construction has a bearing on salmonella incidence and sound design should be encouraged as one alternative to heat sterilization of animal feed ingredients. This changing incidence is shown in Table 2.

Table 2

*Annual incidence of salmonella contamination
meat and bone meal 1967–1971*

Year	Total samples	Positive samples
1967	298	24 (8)
1968	396	105 (27)
1969	456	143 (31)
1970	464	83 (18)
1971	367	68 (19)

Figures in parenthesis are percentages as integers.

Cross Contamination of the Final Product

In any process involving finely ground salmonella-containing material, the problem of cross contamination of a finished product by its infected raw ingredients is a very real one.

Dust infected with salmonellae can spread through the air unless precautions are taken. As illustration, one may cite an interesting outbreak of human food poisoning caused by baker's confectionery (Taylor, 1960). Many of the victims excreted multiple serotypes. It was believed that the foods were sterile on removal from the ovens, but were contaminated by dust itself contaminated with powdered American egg albumen. Bacteriological plates were placed in parts of the factory where food was prepared and salmonellae were isolated. This incident did not involve animal feed, yet the dust situation in the processing of animal food is similar. Another human food prepared from an infected raw material was investigated. Swabs taken from dusty surfaces in the room where the final product was blended often produced salmonella-positive samples. The same serotype, *S. cubana,* was repeatedly isolated over several months.

The possibility that some of the isolations of salmonellae made from samples infrequently found positive were examples of dust-borne infection from more heavily contaminated products such as meat and bone meal cannot be ignored. When the salmonella species was identical in both products and was an uncommon type, this hypothesis became more probable.

Fish, as taken from the sea, rarely contain salmonellae. Catches of fish, however, sometimes come into contact with sewage polluted water or surfaces of ships washed down with polluted water. Transfer of salmonellae occurs and the environment of the raw areas of fish meal plants may become contaminated. This danger and its control has been discussed by Morris *et al.* (1970).

Rats are an obvious danger in the problem of cross contamination. This hazard was recorded many years ago in a butchers' by-products factory by Ludlam (1954). In the last 3 months of 1953, the incidence of salmonellae in rats caught inside the factory rose to 40%. This contrasted with the incidence found in rats caught outside the factory. The high incidence in factory caught rats was related to over loading of the premises with offal from the autumn killings.

Correlation of Serotypes Present in Animal Feed and in Animals

Between November 1968 and January 1970, certain P.H.L.S. laboratories in England and Wales studied the salmonella incidence in pigs and in certain ingredients which can be included in pig feed (P.H.L.S., Skovgaard & Nielsen, 1972). Twenty-five species were common to pigs and feed and these included several types prevalent in man. Of the 10 serotypes most frequently found in man in 1969, 9 were common to pigs and to feeding stuffs.

It is difficult to observe directly the pathway: Animal Feed →Pigs →Man and evidence for existence of such a pathway usually comes from comparative studies. One recent paper by Linton et al. (1970), however, has emphasized the relationship between infected animal feed and the excretion of salmonellae by pigs. In this study, liquid feed prepared from pooled, naturally contaminated ingredients was shown to contain c. 1 salmonella/ml. During holding at 20° and 28° for 48 h, multiplication of salmonellae occurred and counts increased to the order of 200,000 organisms/ml. Liquid feed made from the same batch of ingredients was fed to one group of pigs immediately after preparation and to another group after holding for 24 h. The pigs in the investigation were shown to be salmonella-free prior to feeding. In the first group, pigs excreted salmonellae only on the day following feeding. In the second, excretion continued for 34 days and thereafter intermittently until the experiment ended. The second group of pigs showed slight looseness of stool but no other symptoms of salmonella infection. The danger of holding liquid feed for any length of time and the risk of cross contamination of fresh mixtures with the remains of a previous batch are emphasized by this investigation.

The association between salmonella serotypes found in pigs and in spraydried egg fed to pigs was shown many years ago (Report, 1947), and between salmonella species harboured by rats caught in a butcher's by-products factory and those present in the offal processed by the factory (Ludlam, 1954).

Evidence Derived from Studies in Man

It is often maintained that infections caused by S. typhimurium are more likely to be spread from animal foci than from animal feed. These infections are of great importance, but in 1968, for the first time, the number of infections caused by serotypes other than S. typhimurium was greater than those due to this serotype. The "other serotype" infections have, therefore, assumed considerable importance. If some of these incidents are initiated from an animal feed source one would expect annual serotype patterns in man to change with the entry into the United Kingdom of different serotypes imported from abroad. This changing pattern is recorded in a recent communication (P.H.L.S., Skovgaard & Nielsen, 1972).

We have stated that a similarity can be demonstrated between animal feed and pig isolations. Is there any similar correspondence between pig and human salmonellosis? In 1971, we examined the incidence of salmonellosis in pigs slaughtered in Cardiff. We found that 47% of United Kingdom infected cases of salmonellosis in man in our area in 1971 were caused by serotypes isolated from pigs slaughtered in Cardiff in the same year. The date of isolation of a salmonella type from a pig was not infrequently separated by only a brief interval from its isolation from man. This similarity of pattern suggested a possible causal association.

Selected sewage examination has been used as a guide to total salmonella infection in a population of 4000 persons (Harvey et al., 1969). This study showed that man regularly encountered a wide range of salmonellae and that these organisms were regularly excreted by him. From this investigation, evidence was obtained of infection by subgenus II salmonella serotypes. It is possible that these strains may have originated in South Africa and entered the United Kingdom in imported ingredients of animal feed. Subgenus II salmonellae are relatively common in South Africa and trade in fish meal between that country and the United Kingdom appears to have increased in recent years (Harvey, 1971).

Control

What control measures are possible to prevent contaminated animal feed spreading salmonellosis?

Selection of reputable sources of raw ingredients is an obvious precaution. This would seem to be practised and material such as crushed bone from India and Pakistan would rarely enter into animal feed nowadays because of the danger of anthrax (Davies & Harvey, 1953, 1955). It is possible, however, that this commodity, which is heavily contaminated with salmonellae, could cross infect safer products if processed in the same factory. Large compounders are aware of the salmonella problem and have become careful in selecting ingredients from rendering plants providing relatively samonella free material (Williams, 1971). The recent reduction in incidence of contamination is encouraging.

Improvements in rendering plant design could be expected to have an effect on the quality of the final product (Albertsen, 1957). Unfortunately several plants were built before the dangers of cross contamination were realized (Timoney, 1968).

Cross contamination is one of the hazards of this trade and it can occur at several stages in the process. Some rendering plants receive raw materials in the same area from which finished products are dispatched. This is obviously unsound. Cross contamination can also occur in the laboratory if a dusty heavily contaminated product is handled in close proximity to a salmonella-free sample.

Heat treatment, if cost is not prohibitive, is known to be effective. Danish experience in this connection is useful (P.H.L.S. Working Group, Skovgaard & Nielsen, 1972). British investigations on pelleting combined with heat treatment also show the advantages of these methods (Riley, 1969). It would appear, however, to be difficult for smaller rendering plants in the United Kingdom to heat treat their final products. Supervision would pose problems.

Can the chemical treatment of animal feed be useful? Some work has been done in the United States of America using nontoxic fatty acids (Khan & Katamay, 1969). The method would seem to decrease the possibility of cross contamination of the treated product, but the process is probably more costly than heat treatment.

I have tried in this short paper to avoid speculation. My own view is that infected feed plays a part in salmonella epidemiology. The importance of this probably varies from time to time and the element of chance is likely to be involved. In the pig industry infected animal feed is possibly of little veterinary importance. I am much less sure that this is true with the poultry industry. It is certain that infected pigs and poultry are a hazard to man and that any measure designed to lessen this hazard is worthy of serious consideration.

References

Albertsen, V. E. (1957). Disposal and reclamation of by-products. In *Monograph. Ser. Wld. Hlth. Org.* No. 33, p. 263. Geneva: W.H.O.

Davies, D. G. & Harvey, R. W. S. (1953). Dried bones as a source of anthrax. *Lancet ii,* 880.

Davies, D. G. & Harvey, R. W. S. (1955). The isolation of *Bacillus anthracis* from bones. *Lancet ii,* 86.

Harvey, R. W. S. (1957). The epidemiological significance of sewage bacteriology. *Br. J. clin. Pract.* 11, 751.

Harvey, R. W. S. (1971). Salmonellosis. In *University of Nottingham Nutrition Conference for Feed Manufacturers.* Eds. H. Swan and D. Lewis. Edinburgh and London: Churchill Livingstone.

Harvey, R. W. S. & Price, T. H. (1967). The isolation of salmonellas from animal feeding stuffs. *J. Hyg., Camb.* 65, 237.

Harvey, R. W. S. & Price, T. H. (1968). Elevated temperature incubation of enrichment media for the isolation of salmonellas from heavily contaminated materials. *J. Hyg., Camb.* 66, 377.

Harvey, R. W. S., Price, T. H., Foster, D. W. & Griffiths, W. C. (1969). Salmonellas in sewage: a study in latent human infection. *J. Hyg., Camb.* 67, 517.

Jacobs, J., Guinee, P. A. M., Kampelmacher, E. H. & van Keulen, A. (1963). Studies on the incidence of salmonella in imported fish meal. *Zentbl. VetMed.* 10, 542.

Khan, M. & Katamay, M. (1969). Antagonistic effect of fatty acids against salmonella in meat and bone meal. *Appl. Microbiol.* 17, 402.

Linton, A. H., Jennett, Nada E. & Heard, T. W. (1970). Multiplication of Salmonella in liquid feed and its influence on the duration of excretion in pigs. *Res. vet. Sci.* **11**, 452.

Ludlam, G. B. (1954). Salmonella in rats, with special reference to findings in a butcher's by-products factory. *Mon. Bull. Minist. Hlth.* **13**, 196.

Morris, G. K., Martin, W. T., Shelton, W. H., Wells, Joy, G. & Brachman, P. S. (1970). Salmonellae in fish meal plants: relative amounts of contamination at various stages of processing and a method of control. *Appl. Microbiol.* **19**, 401.

P.H.L.S. Working Group, Skovgaard, N. & Nielsen, B. B. (1972). Salmonellas in pigs and animal feeding stuffs in England and Wales and in Denmark. *J. Hyg., Camb.* **70**, 127.

Prost, E. & Riemann, H. (1967). Food-borne Salmonellosis. *Ann. Rev. Microbiol.* **21**, 495.

Report (1947). The bacteriology of spray-dried egg, with particular reference to food poisoning. *Spec. Rep. Ser. med. Res. Coun.*, No. 260. London: H.M.S.O.

Riley, P. B. (1969). Salmonella infection: the position of animal food and its manufacturing process. In *Bacterial Food Poisoning*, London: Royal Society of Health.

Taylor, Joan (1960). The diarrhoeal diseases in England and Wales. *Bull. Wld Hlth Org.* **23**, 763.

Taylor, Joan & McCoy, J. H. (1969). Salmonella and Arizona infections. In *Food-borne Infections and Intoxications.* Ed. H. Riemann. New York and London: Academic Press.

Timoney, J. (1968). The sources and extent of salmonella contamination in rendering plants. *Vet. Rec.* **83**, 541.

Williams, D. R. (1971). Salmonella and animal feeds. In *University of Nottingham Nutrition Conference for Feed Manufacturers.* Eds. H. Swan and D. Lewis. Edinburgh and London: Churchill Livingstone.

Discussion

Pye

In respect to certain comments made in Dr Harvey's paper, isolates of salmonellae can obviously be made from curious sources. Recently I read a short paper giving data on isolation of salmonellae from spiders' webs. The Somerset area provides considerable trouble with sickness in pigs and the farmer seeks compensation from the manufacturer of the feedingstuff. Scouring organisms and salmonellae were not found in the feed but were abundant around the farm (the results obtained were supported by the public health laboratory). The organisms seem to be endemic on the farm but fortunately not in the feed.

In his paper Dr Harvey reported on the importance of dust-borne contamination by salmonellae. In any environment in which salmonella contaminated materials are being agitated, as in mixing the feed, contamination of any surface is a real possibility. It is always difficult to be absolutely sure that any single supply of food initiated an infection with salmonellae. Most of our evidence is circumstantial only. In my experience with a disease-free herd of pigs salmonellae when introduced (possibly in the feed) persisted for years and were only eradicated by slaughter of the infected pigs, segregation of clean stock and repopulation of the farm with other clean stock. These facts, however, do not invalidate the important observation that batches of food can initiate infection on farms.

Rehm

Can Dr Linton tell us the nature of the non-toxic fatty acid used in the work described in Dr Harvey's paper?

Linton

I regret that I am not familiar with the work which Dr Harvey reports having been carried out in the U.S.A. Since Dr Harvey will be attending later in the conference an opportunity may be available to put the question to him.

Ingram

Prof Lerche (?) of Berlin developed a procedure using fumigation with ammonia gas to kill salmonellae in dry foods. Has it been tried for animal feedstuffs?

Linton

I have no information on this point.

Recent Trends of *Vibrio parahaemolyticus* as a Causative Agent of Food Poisoning

R. SAKAZAKI

*National Institute of Health,
Tokyo, Japan*

Vibrio parahaemolyticus, a polar-flagellated monotrichous rod conforming to the definition of the genus *Vibrio*, is characterized by the following properties: acid production from mannitol, but not from 1% sucrose; negative Voges-Proskauer and hydrogen sulphide tests; positive lysine and ornithine decarboxylase but negative arginine dihydrolase tests; no growth in peptone water without salt or with 10% NaCl, but good growth in peptone water containing 8% NaCl; and growth at 43°. An antigenic schema consisting of 11 O antigen groups and 52 K antigens has been established for *V. parahaemolyticus*. The Kanagawa phenomenon, a haemolytic reaction on a special blood agar devised by Wagatsuma, is associated with enteropathogenicity of the vibrio, but the Kanagawa haemolysin itself is probably not directly concerned with gastro-enteritis. Enterotoxin production by the vibrio has not yet been demonstrated. Food poisoning due to *V. parahaemolyticus* is an infection and gastro-enteritis develops after an incubation period of 12–15 h. In Japan, infection occurs only in the summer season, and is always food-borne, usually from seafoods. Recently, infection with *V. parahaemolyticus* has been recognized not only in Japan but also in other countries in South-east Asia, Australia and the United States. In view of this, further reports of infection with this organism can be expected in other parts of the world.

Introduction

HUMAN INFECTION with *Vibrio parahaemolyticus* has been reported in many countries in recent years. This report concerns illness caused by this vibrio in Japan. It is of interest not only to medical bacteriologists but also to food microbiologists. The organism was first named *Pasteurella parahaemolytica* by Fujino *et al.* (1951) who isolated it from autopsy specimens in an outbreak of food poisoning associated with the eating of semi-dried young sardines. A similar organism isolated from numerous subsequent outbreaks was given other names such as *Pseudomonas enteritis* (Takikawa, 1958) and *Oceanomonas parahaemolytica* (Miyamoto *et al.*, 1961), but after extensive taxonomic studies Sakazaki *et al.* (1963) suggested the name *V. parahaemolyticus* for this organism. This name has been widely accepted throughout the world.

Bacteriology

V. parahaemolyticus is a Gram-negative, facultatively anaerobic rod conforming to the definition of the genus *Vibrio* (Feeley, 1966). It grows profusely on/in

19

ordinary nutrient media with 2–4% NaCl, but grows very poorly or not at all on/in media containing no salt. Although growth is most abundant in media with 2–4% NaCl, it also grows in peptone water containing 8% NaCl. Little or no growth, however, occurs in peptone water with 10% NaCl. It prefers alkaline conditions (pH 7·6–8·6), and grows between 15° and 43°, but best at 37°.

The biochemical characteristics of *V. parahaemolyticus* are summarized in Table 1. Although aberrance may occur in some cultures, especially in those isolated from sea sources, the most important features differentiating it from related halophilic organisms are: the negative V. P. reaction, negative arginine dihydrolase but positive lysine decarboxylase tests, and negative fermentation of sucrose (1%), inositol, rhamnose, and dulcitol.

Table 1

Physiological and biochemical characteristics of
Vibrio parahaemolyticus

			+%
Growth in peptone water with	0% NaCl	–	0
	3% NaCl	+	100
	8% NaCl	+	100
	10% NaCl	–	0·6
Growth at 5°C		–	0
37°C		+	100
43°C		+	100
Catalase		+	100
Oxidase		+	100
Indole		+	98·4
Methyl red		+	84·3
Voges-Proskauer		–	0
Citrate utilization (Simmons)		+	96·6
Citrate utilization (Christensen)		+	100
Malonate utilization (Ewing)		–	0
Hydrogen sulfide (TSI)		–	0
Nitrate reduction to nitrite		+	100
Urease (Christensen)		–	0
Phenylalanine deaminase		–	0
Lysine decarboxylase		+	100
Arginine dihydrolase		–	0
Ornithine decarboxylase		+	97·3
Gelatinase		+	99·6
Milk digestion		+	100
DNase		+	100
Phosphatase		+	100
Lipase		+	100

Table 1—*continued*

		+%
O-F medium glucose, open	+	100
O-F medium glucose, sealed	+	100
O-F medium glucose, gas	−	0
Acid from: Glucose	+	100
Mannitol	+	99·6
Lactose 1%	−	4·3
Lactose 10%	+	100
Sucrose 1%	.−	5·3
Sucrose 10%	+	100
Arabinose	d	65·9
Cellobiose	d	28·6
Galactose	+	100
Levulose	+	100
Maltose	+	100
Mannose	+	100
Melezitose	−	3·6
Melibiose	−	15·3
Raffinose	−	0
Rhamnose	−	0
Sorbose	−	0
Trehalose	+	100
Xylose	−	0
Adonitol	−	0
Dulcitol	−	0·3
Inositol	−	0
Sorbitol	−	0·5
Salicin	−	0·1
Esculin hydrolysis	d	45·9
Starch hydrolysis	+	100
β-Galactosidase	+	100
Luminescence	−	0

Note: +, positive in 80% or more strains; −, negative in 80% or more strains; d, different reaction in various strains.

V. parahaemolyticus possesses 3 antigenic components, the O, K, and H antigens. An antigenic schema for *V. parahaemolyticus* in which 11 O groups and 41 K antigens were recognized was established by Sakazaki *et al.* (1968*a*). The H antigen was not included in the schema since it was found that all the H antigens of *V. parahaemolyticus* were serologically identical (Sakazaki *et al.*, 1968*a*; Terada, 1968). The antigenic schema was extended later by the Committee for Serotyping of *V. parahaemolyticus* (1970), and at the present time it consists of 11 O groups and 52 K antigens (Table 2). K antigens 2, 14, 16, 27 and 35 were excluded because they were found to be identical with others already established.

Table 2

Antigenic schema of Vibrio parahaemolyticus *(1972)*

O group	K Antigen	O group	K Antigen
1	1	5	15
	25		17
	26		30
	32		47
	38		
	41	6	18
	56		46
2	3	7	19
	28		
		8	20
3	4		21
	5		22
	6		39
	7		
	29	9	23
	30		44
	31		
	33	10	24
	37		52
	43		
	45	11	36
	48		40
	54		50
	57		51
4	4		
	8		
	9		
	10		
	11		
	12		
	13		
	34		
	42		
	49		
	53		
	55		

Enteropathogenicity

Early in the study, it was thought that all cultures of *V. parahaemolyticus*, regardless of their source, might be enteropathogenic for man. Kato *et al.* (1965) found, however, that vibrio strains isolated from diarrhoeal stools gave a haemolytic reaction on unautoclaved brain heart infusion agar containing 5%

human blood, 3% NaCl, and 0·001% crystal violet, whereas strains isolated from marine sources were nonhaemolytic. This medium was modified by Wagatsuma (1968) to give more clear-cut haemolysis by *V. parahaemolyticus,* and the test was named the "Kanagawa reaction". Sakazaki *et al.* (1968*b*) carried out a study of the Kanagawa reaction with 3370 cultures of the vibrio, comprising 2720 strains isolated from patients with gastroenteritis and 650 strains from sea sources. Using Wagatsuma agar, they found that 96·5% of the 2720 human cultures gave a positive Kanagawa reaction, whereas only 1% of the 650 cultures from marine sources were haemolytic. The results are shown in Table 3. These findings have since been confirmed by many Japanese workers, and they apply also to strains isolated in countries other than Japan. Thus, there seems to be a relationship between the Kanagawa reaction and enteropathogenicity, although no correlation was found between the Kanagawa reaction and the serotype of the vibrio.

Table 3

Kanagawa reactivity of 3370 cultures of Vibrio parahaemolyticus

Source of strains	Kanagawa reaction		Total
	+	−	
Human patient	2655 (96·5%)	75 (3·5%)	2720
Sea source	7 (1·0%)	643 (99·0%)	650

Feeding tests with *V. parahaemolyticus* were carried out by Takikawa (1958) and by Aiiso & Fujiwara (1963). They reported that some cultures of the vibrio produced gastroenteritis in volunteers but others did not. When the present author analysed their results several years later, it was realized that the cultures which had produced illness were human in origin and were Kanagawa positive, and that volunteers given Kanagawa negative cultures from sea fish were not ill. In addition, a case of laboratory infection with the vibrio occurred in our laboratory in the winter of 1966. The patient, a young male research technologist, accidentally ingested about 0·3 ml of a suspension which contained 10^6/ml viable cells of a Kanagawa positive cultures. He developed acute gastroenteritis after an incubation period of 6 h. Sakazaki & Tamura (1971, unpublished data) also confirmed the enteropathogenicity of a Kanagawa positive strain with a volunteer who received 10^6/ml viable cells and developed gastroenteritis after 5 h. In contrast, feeding tests with Kanagawa negative strains of *V. parahaemolyticus* carried out with 15 adult volunteers by Sakazaki *et al.* (1968*b*) failed to induce any clinical signs, although 10^9/ml or more viable organisms were administered.

Haemolytic substances produced by *V. parahaemolyticus* have been studied by several Japanese workers (Fujino *et al.*, 1969; Sakazaki & Tamura, 1969; Yanagase *et al.*, 1970; Zen-Yoji *et al.*, 1971; Obara, 1972). At lest 4 categories of haemolytic substances have been recognized. One of these is present in the supernatant fluid of broth cultures of Kanagawa positive strains but not in those of Kanagawa negative strains.

The haemolytic fraction concerned in the Kanagawa reaction was studied by Zen-Yoji *et al.* (1971) and Obara (1972). This Kanagawa haemolysin was precipitated as a sugar and nucleic acid-free protein fraction by 60% saturation with ammonium sulphate at pH 4·5 and purified by zone chromatography with Sephadex G200. The haemolytic activity remained after heating at 100° for 30 min, but was inactivated by trypsin digestion. The haemolysin was strongly active against dog, rat and human red blood cells, intermediate and weak with those of rabbit and sheep, respectively, but negative with horse blood (Table 4).

Table 4

Properties of the Kanagawa haemolysin

Lowry reaction	+
Molisch reaction	−
Dimethylamine reaction	−
Haemolytic activity after treatment by	
Trypsin	−
Heating at 100° for 5 min	+
Human red cells	++
Dog red cells	+++
Rabbit red cells	++
Rat red cells	+++
Sheep red cells	+
Horse red cells	−
Ca^{2+}	activate
Lecithinase A activity	−
Lecithinase B activity	−
Cytotoxicity (HeLa cells)	+
LD_{50} (mice)	0·6–0·8 mg/N

Kato *et al.* (1970) demonstrated antibody to the Kanagawa haemolysin in hospital patients recovering from *V. parahaemolyticus* gastroenteritis. The antibody titre rose in 50% of the patients after the first week and in 70% after 2 weeks. In contrast, antibody was not demonstrated in 200 normal sera and in 50 sera from cases of diarrhoea due to other causes.

From these results, it is clear that the Kanagawa reaction is associated with enteropathogenicity of *V. parahaemolyticus*. At the present time, however, it is not known whether Kanagawa negative strains of *V. parahaemolyticus* found in seafoods are potentially pathogenic or can be converted into enteropathogenic types. Twedt (1972) investigated the enteropathogenicity of the vibrio with the ligated rabbit gut test of De & Chatterje (1953). He showed that ileal loop reactions are associated with positive Kanagawa tests. In the study of Sakazaki *et al.* (1963), most strains of *V. parahaemolyticus* produced positive ileal loop reactions, irrespective of their sources. From this data, Sakazaki *et al.* (1971) suggested that the ileal loop reaction had limited value in testing for pathogenicity of the vibrio. However, in further work using the revised ligated loop method of Gohda *et al.* (1971), Sakazaki *et al.* (1972, unpublished data) found some correlation between ileal loop reactivity and the Kanagawa reaction. In this study, 10 of 25 Kanagawa positive cultures isolated from human patients produced a severe reaction in ligated loops, whereas 24 of 25 Kanagawa negative cultures failed to show any response (Table 5).

Table 5

Correlation of Kanagawa reaction and ileal loop test for
Vibrio parahaemolyticus

Investigator	Number of strains tested	Kanagawa reaction	Number of strains gave positive loop test
Twedt	19	+	17
	10	−	1
Sakazaki *et al.*	20	+	10
	25	−	1

Teramoto *et al.* (1969) reported an outbreak of *V. parahaemolyticus* food poisoning in which only Kanagawa negative strains were isolated, and Zen-Yoji *et al.* (1970) also recognized similar cases. When examined by the ligated loop test, only one of the 10 of these Kanagawa negative cultures gave a positive reaction.

The Kanagawa haemolysin is toxic to mice. Obara (1972) reported that the LD_{50} of the purified haemolysin was 4·9 µg/N. However, the author found that the purified Kanagawa haemolysin gave a negative reaction in the ligated loop test, thus suggesting that it does not play an important role in producing gastroenteritis. Also, it could not be affirmed that all Kanagawa negative cultures are nonpathogenic, although Sakazaki *et al.* (1968*b*) pointed out that 15 Kanagawa negative strains, isolated from fish, failed to produce illness in human volunteers. In studies by Twedt (1972) cell-free culture filtrates of *V.*

parahaemolyticus caused no effect in the ligated loop test, thus suggesting absence of enterotoxin. The mechanism of enteropathogenicity in *V. para-haemolyticus* has therefore yet to be explained.

The actual source of Kanagawa positive strains of *V. parahaemolyticus* is still not known, as the majority of cultures from marine sources are Kanagawa negative. Even in outbreaks of food poisoning, the isolation of Kanagawa positive strains from the causative foods is very rare, although all isolates from faecal specimens from patients are Kanagawa positive.

Transformation of Kanagawa reactivity by genetic recombination and transduction has been attempted by several Japanese workers, but without success. Attempts were also made by Sasaki (1969) to induce changes of Kanagawa reactivity *in vivo* using germ-free mice. After administration, *V. parahaemolyticus* persisted in the intestines for 6 months or more, and it was found that Kanagawa negative mutants dissociated from Kanagawa positive cultures, but dissociation of Kanagawa positive mutants from Kanagawa negative strains was never observed. Similar results were obtained with rabbits, dogs, and human volunteers by Sakazaki *et al.* (unpublished data).

Epidemiology

Food poisoning due to *V. parahaemolyticus* is an infection and gastroenteritis develops in most cases. The symptoms usually appear within 10–15 h after eating infected food, although the incubation period may be as short as 2 h or as long as 48 h. The main symptoms are nausea, vomiting, abdominal pain and diarrhoea. Mild fever, headache, and chills may occur. These symptoms are very similar to those of *Salmonella* infection.

Stools are usually watery, but mucus and blood may be observed and such cases could be misdiagnosed as bacillary dysentery, though tenesmus is usually absent. Symptoms usually subside in 1 or 2 days. The fatality rate is low, most deaths from this type of food poisoning occurring in old and debilitated persons.

Outbreaks of this vibrio infection occur only in the summer season in Japan. Infection is always food borne, man to man cases have not been recognized. It is one of the most important causative agents of food poisoning in Japan and was considered a local problem until recently, but it has now been recognized in many countries, especially in South-east Asia (Chatterjee *et al.*, 1970; Sakazaki *et al.*, 1971; M'cMinn, 1972; Inaba, 1972). These workers reported the isolation of *V. parahaemolyticus* from up to 15% of patients with diarrhoea in Calcutta, Thailand and the Philippines, an isolation rate greater than that of *Salmonella* and *Shigella* in these countries. Battey *et al.* (1970) reported an outbreak of gastroenteritis in Australia caused by *V. parahaemolyticus*. More recently, 2 outbreaks of infection with *V. parahaemolyticus* were reported in the United States (Sumner *et al.*, 1971). This infection may also occur in travellers returning to Western countries from the Orient. Such incidents have been reported by Freedman & Francis (1969) and by Barrow (1972). The vibrios have also been isolated from infections of the hands and feet, eyes and ears of persons who have been in contact with marine shore areas (Weaver, 1970). All the cultures isolated from patients in the countries mentioned have been confirmed as *V. para-haemolyticus* in the author's laboratory. Of these cultures, isolates from diarrhoeal stools were all Kanagawa positive, but those from wound infections were Kanagawa negative. Seligmann & Reitler (1967) described a case of food poisoning in Israel which Velimirovic (1972) thought was due to *V. para-haemolyticus* infection. However, when examined by the author, the Israel strains were found to be *Vibrio costicolus*-like organisms which occur usually in curing brine.

Vibrio parahaemolyticus seems to live in the marine environment. It has been found in coastal and estuarine waters and sediments, and on marine fishes, crustaceans, and shellfish in many areas of the world. The vibrio is therefore associated almost exclusively with seafoods, and has been found in practically all sea fish. Food poisoning is associated directly or indirectly with sea fish or sea water, and outbreaks in Japan usually follow the eating of seafoods, the high incidence undoubtedly being due to the national custom of eating raw fish products. Sometimes cured vegetables are the source of infection: they become contaminated with the vibrio from a chopping board or a kitchen knife also used for sea fish. In other countries, infection from raw seafoods, except shellfish, is unlikely but cooked food may be contaminated with the vibrio from raw sea materials or with sea water after cooking. The incidence of *V. parahaemolyticus* food poisoning is directly associated with atmospheric temperature. In Japan, it is confined to the warmer months of the year, May to October, and is not found during the cold season. The number of *V. parahaemolyticus* in coastal sea water decreases during the winter season.

In the light of this knowledge, more reports of *V. parahaemolyticus* infection may be expected in other parts of the world, including Europe, South America and Africa, in the future.

References

Aiiso, K. & Fujiwara, K. (1963). Feeding test of the pathogenic halophilic bacteria. *Ann. Res. Inst. Fd Microbiol., Chiba Univ.* **15**, 34.

Barrow, G. I. (1972). Personal communication.

Battey, Y. M., Wallace, R. B., Allan, B. C. & Keeffe, B. M. (1970). Gastroenteritis in Australia caused by a marine vibrio. *Med. J. Aust.* **1**, 430.

Chatterjee, B. D., Gorbach, S. L. & Neogy, K. N. (1970). *Vibrio parahaemolyticus* and diarrhoea associated with non-cholera vibrios. *Bull. Wld Hlth Org.* **42**, 460.

Committee on the Serological Typing of *Vibrio parahaemolyticus*. (1970). New serotypes of *Vibrio parahaemolyticus*. *Jap. J. Microbiol.* **14**, 249.

De, S. N. & Chatterje, D. N. (1953). An experimental study of the mechanisms of action of *Vibrio cholerae* on the intestinal mucous membrane. *J. Path. Bact.* **66**, 559.

Feeley, J. C. (1966). Minutes of IAMS Subcommittee on Taxonomy of Vibrios. *Int. J. Syst. Bact.* **16**, 135.

Freedman, D. K. & Francis, M. (1969). *Morbidity & Mortality Weekly Rept.* **18**, 150. (Cited by Velimiroviv, B., 1972.)

Fujino, T., Okuno, Y., Nakada, D., Aoyama, A., Fukai, K., Murai, K. & Ueho, T. (1951). On the bacteriological examination of shirasu food poisoning. *J. Japan. Ass. Infect. Dis.* **25**, 11. (Text in Japanese.)

Fujino, T., Miwatani, T., Takeda, Y. & Tomaru, A. (1969). A thermolabile direct hemolysin of *Vibrio parahaemolyticus. Biken J.* **12**, 145.

Gohda, A., Kazuno, Y. & Sasaki, S. (1971). Methodology of the rabbit ileal loop test. *J. Japan. Ass. Infect. Dis.* **45**, 196. (Text in Japanese.)

Inaba, H. (1972). Personal communication.

Kato, T., Obara, Y., Ichinohe, H., Nagashima, K., Akiyama, S., Takizawa, K., Matsushita, A., Yamai, S. & Miyamoto, Y. (1965). Subdivision of *Vibrio parahaemolyticus* due to hemolytic reaction. *Shokuhin Eisei Kenkyu* **15**, 839.

Kato, T., Kuwahara, N., Takahashi, R., Tanaka, E., Yamaguchi, T. & Sato, M. (1970). Studies on antihemolysin in diarrheal patients with *Vibrio parahaemolyticus* infection. *Media Circle* **15**, 109. (Text in Japanese.)

M'cMinn, M. T. (1972). Personal communication.

Miyamoto, Y., Nakamura, K. & Takizawa, K. (1961). Pathogenic halophiles. Proposal of a new genus *"Oceanomonas"* and of the amended species names. *Japan. J. Microbiol.* **5**, 477.

Obara, Y. (1972). Studies on hemolytic substance in *Vibrio parahaemolyticus.* II. Partial purification of hemolytic substance. *J. Japan. Ass. Infect. Dis.* **45**, 392. (Text in Japanese.)

Sakazaki, R., Iwanami, S. & Fukumi, H. (1963). Studies on the enteropathogenic, facultatively halophilic bacteria, *Vibrio parahaemolyticus.* I. Morphological, cultural and biochemical properties and its taxonomical position. *Japan. J. med. Sci. Biol.* **16**, 161.

Sakazaki, R., Iwanami, S. & Tamura, K. (1968a). Studies on the enteropathogenic, facultatively halophilic bacteria, *Vibrio parahaemolyticus.* II. Serological characteristics. *Japan. J. med. Sci. Biol.* **21**, 313.

Sakazaki, R., Tamura, K., Kato, T., Obara, Y., Yamai, S. & Hobo, K. (1968b). Studies on the enteropathogenic, facultatively halophilic bacteria, *Vibrio parahaemolyticus.* III. Enteropathogenicity. *Japan. J. med. Sci. Biol.* **21**, 325.

Sakazaki, R. & Tamura, K. (1969). A thermostable hemolytic substance in *Vibrio parahaemolyticus. Modern Media* **15**, 220. (Text in Japanese.)

Sakazaki, R., Tamura, K., Prescott, L. M., Bencic, Z., Sanyal, S. C. & Sinha, R. (1971). Bacteriological examination of diarrheal stool in Calcutta. *Ind. J. Med. Res.* **59**, 1025.

Sasaki, S. (1969). Personal communication.

Seligmann, R. & Reitler, R. (1967). *Israel J. Med. Sci.* **3**, 456 (Cited by Velimirovic, B., 1972).

Sumner, W. A., Moore, S. J., Bush, M. A., Nelson, R., Molenda, J. R., Johnson, W., Garber, H. J. & Wentz, B. (1971). *Vibrio parahaemolyticus* gastroenteritis-Maryland. *Morbidity & Mortality Weekly Rpt.* **20**, 356.

Takikawa, I. (1958). Studies on pathogenic halophilic bacteria. *Yokohama Med. Bull.* **2**, 313.

Terada, T. (1968). Serological study on antigens of *Vibrio parahaemolyticus.* II. Flagellar antigens. *Japan. J. Bact.* **23**, 767. (Text in Japanese.)

Teramoto, T., Nakanishi, H. & Maejima, K. (1969). Kanagawa reaction of *Vibrio parahaemolyticus* isolated from food poisoning. *Modern Media* **15**, 215. (Text in Japanese.)

Twedt, R. M. (1972). Personal communication.

Velimirovic, B. (1972). The geographical distribution of the human disease due to *Vibrio parahaemolyticus* in South-east Asia and the Pacific. *Zentbl. Bakt. Abt.* **227**, 91.

Wagatsuma, S. (1968). A medium for the test of hemolytic reaction of *Vibrio parahaemolyticus*. *Media Circle* **13**, 159. (Text in Japanese.)

Weaver, R. E. (1970). Personal communication.

Yanagase, Y., Inoue, K., Ozaki, M., Ochi, T., Amano, T. & Chadono, M. (1970). Hemolysins and related enzymes of *Vibrio parahaemolyticus*. I. Identification and partial purification of enzymes. *Biken J.* **13**, 77.

Zen-Yoji, H., Sakai, S., Kudo, Y., Itoh, T. & Maruyama, T. (1970). Food poisoning due to Kanagawa negative *Vibrio parahaemolyticus*. *Media Circle* **15**, 82.

Zen-Yoji, H., Hitokoto, H., Morozumi, S. & LeClair, R. A. (1971). Purification and characterization of a hemolysin produced by *Vibrio parahaemolyticus*. *J. Infect. Dis.* **123**, 665.

Discussion

Malkki

I would like to ask Dr Sakazaki whether *Vibrio parahaemolyticus* can survive in heavy concentrations of sodium chloride (above 12%) containing nitrates?

Sakazaki

I am not certain but I think it could survive in such conditions.

Ingram

Do all strains of *V. parahaemolyticus* react similarly to different salt concentrations?

Sakazaki

Yes, so far as I know.

Woodbine

In relation to the susceptibility to salt of *V. parahaemolyticus,* does the virulence alter under different salt concentrations (e.g. 2–12%) compared with *E. rhiziopathiae* which is susceptible to 1–2% and *Listeria monocytogenes* which also loses virulence in salt?

Sakazaki

It is not possible to answer this question as all our experiments on pathogenicity have been done with cultures grown in media containing 3% NaCl.

Olson

In a previous conversation you indicated to me that an unusually large number of "cholera" cases in India were misdiagnosed when in fact they were parahaemolyticus food poisoning or infections. In your paper you emphasized the difference in symptoms between the 2 diseases. Would you comment on why there was such a high percentage of misdiagnosis?

Sakazaki

In my experience most cases of diarrhoea in Calcutta are recognized clinically as "cholera" during the cholera season.

Ashton

Was the data presented in the final Table of your paper gathered in Calcutta?

Sakazaki

Yes.

Hobbs

Are Kanagawa positive and negative strains present together in cases (diarrhoea samples) and in foods? Do the positive strains change to negative outside the body?

Sakazaki

In our experience, weakly haemolytic colonies may dissociate from Kanagawa positive strains, but Kanagawa positive mutants have never been obtained either in cultures of Kanagawa negative organisms nor from man to whom Kanagawa negative strains were administered.

Monty

Can Dr Sakazaki tell me whether *V. parahaemolyticus* can exist under deep frozen conditions for considerable periods?

Sakazaki

It can survive for at least 1 year at $-20°$.

Riemann

Practically all strains isolated from marine sources are non-haemolytic, but only haemolytic strains cause food poisoning. Are circumstances known that will preferentially promote growth of haemolytic strains?

Sakazaki

No.

Riemann

What is the infective dose for humans?

Sakazaki

Probably 10^6 or more viable cells.

Olson

Regarding the difference in the distribution of Kanagawa positive strains in the marine environment as compared with the human stool, could this be explained as being the result of the selection for Kanagawa positive strains in the human gut, since such strains are so highly correlated with pathogenicity?

Sakazaki

Yes. Selection may occur in the human intestine. However, seafoods contaminated with the vibrio do not always produce illness. Possibly only foods contaminated with Kanagawa positive organisms cause gastroenteritis.

Skovgaard

V. parahaemolyticus is common in Japan and the U.S.A. but isolations have only seldom been reported from Scandinavian waters and there are no reports of isolations from the Mediterranean area. From an ecological standpoint I would like to ask you whether you have an explanation for this. We can of course explain the absence in Scandinavia by the rather low temperature, *V. parahaemolyticus* being a thermophilic bacterium.

Sakazaki

V. parahaemolyticus is usually easily isolated from sea sources only in the summer season. I believe that the temperature of the sea water in Scandinavian areas is probably too low to permit persistence of *V. parahaemolyticus* in significant numbers.

Bacteriophages and Toxigenicity of *Clostridium botulinum**

M. W. EKLUND AND F. T. POYSKY

National Marine Fisheries Service,
Pacific Fishery Products Technology Center,
Seattle, Washington, U.S.A.

Bacterial cultures sensitive to bacteriophage have been isolated from toxigenic strains of *Clostridium botulinum* after culture in medium containing acridine orange or by ultraviolet light treatment. Changes from nontoxigenicity to toxigenicity (production of dominant toxic components C_1 and D) in types C and D require the active and continued participation of specific bacteriophages designated as CEβ in type C and DEβ in type D. Some of the type C and D cultures that are "cured" of their prophages and cease to produce the C_1 and D toxins, however, continue to produce low levels of another toxin that is antigenically monospecific. This toxin designated as C_2 is produced as a protoxin and the culture filtrates are toxic only after trypsin treatment. These "cured" strains of types C and D therefore become indistinguishable with respect to the toxin produced. Non-proteolytic cultures of types B and F have each been "cured" of one of their prophages but they continue to carry a second bacteriophage and produce toxin. The Langeland strain of proteolytic type F has been "cured" of 2 of its prophages but it continues to produce toxin. It is possible either that these cultures carry other bacteriophages or that not all *C. botulinum* toxins are induced by bacteriophages.

Introduction

NONTOXIGENIC CLOSTRIDIA resembling toxigenic *Clostridium botulinum* have frequently been isolated from the aquatic and terrestrial environments and, occasionally, pure cultures of toxigenic strains of *C. botulinum* have become nontoxigenic when transferred in laboratory media. It has often been suggested that the production of toxin by *C. botulinum* might be governed by bacteriophages in a system analogous to the production of diphtheria toxin by *Corynebacterium diphtheriae* (Freeman, 1951; Groman, 1955).

The data of Inoue & Iida (1971) strongly suggested that bacteriophages are involved in the toxigenicity of *C. botulinum* types C and D. They were able to recover toxigenic isolates from nontoxigenic cultures incubated in broth containing filtrates of the toxigenic parent strain (Stockholm strain of type C and strain 1873 of type D). Lysis was observed in the cultures but plaques were not demonstrated on solid medium.

This report provides evidence for the involvement of specific bacteriophages in the toxigenicity of *C. botulinum* types C and D. The "curing" of other types of *C. botulinum* of their prophages is also discussed.

* Supported by AEC Contract No. AT(949-7)-2442.

Materials and Methods

Isolation of nontoxigenic cultures from toxigenic strains

Nontoxigenic bacterial cultures "cured" of their prophages were isolated from the toxigenic parent cultures (type C strain 468C and type D strain South African) after culturing in medium containing acridine orange (AO), after treatment of toxigenic strains with ultraviolet irradiation and also from sporulated cultures.

For irradiation, about 10^6 cells of the toxigenic strain in the logarithmic phase of growth were spread on to the surfaces of trypticase, yeast-extract, glucose (TYG) blood agar plates and irradiated for 20 or 60 sec at a distance of 20 cm (General Electric Germicidal lamp, 15 W). The agar plates were incubated anaerobically in Brewer jars at 37°. Colonies from cells surviving the treatment were cultured in Segner's fortified egg meat medium (SFEM) and tested for toxigenicity by the mouse assay (Eklund et al., 1967) and for sensitivity to the bacteriophage of the toxigenic parent culture by the agar-layer procedure (Adams, 1959). With type C cultures the base agar used in the agar-layer procedure was TYG agar and the overlay was soft agar (0·7%) prepared from the filtered broth of SFEM medium. The addition of 0·2 ml blood to the overlays was essential for confluent growth.

With type D cultures the base agar was trypticase, peptone, yeast-extract, glucose (TPGY) agar and the overlay was TPGY soft agar (0·7% agar) containing 2·5% sodium chloride and 0·5% lactalysate. The addition of catalase (Sigma, 3600 U/mg) to 6 ml overlays improved confluent growth. Cysteine hydrochloride (0·1%) was used as the reducing agent in all media.

For curing with acridine orange, c. 10^5 cells of the toxigenic parent culture in the logarithmic phase of growth were added to TYG broth containing 5–30 μg of the dye/ml and incubated at 37° for 18 h. This culture was then diluted and plated with TYG blood agar and isolated colonies were tested for toxigenicity and sensitivity to bacteriophages using the same procedures that were described for colonies treated with ultraviolet light.

Nontoxigenic, bacteriophage-sensitive, bacterial cultures were also isolated from the sporulated toxigenic cultures. The toxigenic strains were cultured in SFEM medium at 33° until sporulation. The spores were heated at 70° for 20 min to inactivate the free background phage. The spores were then diluted and plated with TYG blood agar. Isolated colonies were tested for toxigenicity and sensitivity to bacteriophages of the toxigenic parent culture using procedures described above.

Bacteriophage

Bacteriophages were isolated from the toxigenic parent. The broth culture was centrifuged at 6000 g for 10 min and sterilized by filtration. The bacteriophages were purified by 6 successive single-plaque isolations on one of the "cured" nontoxigenic isolates.

Bacteriophage stocks used in the studies of the conversion of the non-toxigenic cultures to toxigenicity were produced by propagating the purified bacteriophages with the "cured" nontoxigenic culture in TYG broth. Bacterio-phage stocks were treated with 40 μg of crystalline desoxyribonuclease II (Sigma)/ml for 3 h at 37° and then sterilized by filtration. Desoxyribonuclease was used to rule out the possibility of a transformation principle of the desoxyribonucleic acid type. All filtrates were checked for sterility by inoculating several millilitres of the filtrate into TYG broth and incubating for several weeks at 33°.

Relation of bacteriophages to toxigenicity

Two methods were used to study the relation of the purified bacteriophages to the toxigenicity of the nontoxigenic cultures. With the first procedure, the bacteriophages were diluted and plated with the recipient nontoxigenic cultures using the agar-overlay procedure. Material from the plaques was transferred into TYG broth and incubated at 33° for 3 days and tested for toxigenicity and bacteriophage production. In the second procedure, the bacteriophage was added to actively growing nontoxigenic cultures. After a 40- or 240-min exposure to the bacteriophage, the cultures were plated on to TYG blood egg yolk agar plates and isolated colonies were transferred into TYG broth. After incubation at 33° the culture was assayed for toxin production and tested for bacteriophage production with one of the "cured" nontoxigenic cultures as the indicator strain.

Results and Discussion

C. botulinum types C and D produce at least 3 toxins which have been designated as C_1, C_2, and D (Jansen, 1971; Mason & Robinson, 1935). The production of the C_1 and D toxins are governed by specific bacteriophages. When type C strains and some type D strains are "cured" of their prophages they cease production of the C_1 and D toxins, but they continue to produce the C_2 toxin. The production of the C_2 toxin therefore does not appear to be governed by bacteriophages. The C_2 toxin is produced as a protoxin and the culture filtrates are toxic only after trypsin treatment.

Relation of bacteriophages to toxigenicity of type C

Different strains of type C have been used in our studies. Type C strain 468C will be used as an example in this discussion. Strain 468C was "cured" of its prophages and simultaneously became nontoxigenic after treatment with

Table 1

Curing type C (468C) of prophages and loss of toxigenicity

| | | No. of colonies | |
	Tested	"Nontoxic" and "cured" of phage (CEβ)	"Nontoxic"* and "cured" of phages (CEβ and CEγ)
U.V. 20 sec	117	7	0
U.V. 60 sec	106	15	0
A.O. 20 μg/ml	68	2	1 (AO28)

* "Nontoxic" means that cultures no longer produce dominant C_1 and D toxins but continue to produce C_2 toxin.

ultraviolet light or by culturing in TYG medium containing acridine orange. Table 1 summarizes the results of these experiments. After a 20-sec exposure to ultraviolet light, 7 of the isolates were "nontoxic" and "cured" of bacteriophage designated as CEβ. After a 60-sec exposure, 15 cultures were "nontoxic" and sensitive to the CEβ phage. When strain 468C was cultured in medium containing acridine orange, two of the cultures were simultaneously "cured" of prophage CEβ and toxigenicity. Each of the nontoxigenic isolates was tested for its sensitivity to the lysates of the other "cured" cultures. Only culture AO28 was "cured" of both the CEβ and CEγ phage. These "cured" cultures have been subcultured over 40 times during a 2-year period and they remain nontoxigenic and sensitive to the phages of the toxigenic parent culture. "Nontoxic" in Table 1 means that the "cured" cultures cease to produce the C_1 and D toxins but continue to produce the C_2 toxins.

Each of the nontoxigenic isolates was studied. Results from culture AO28 will be used as an example.

Electronmicrographs were prepared of the 2 bacteriophages according to the procedures of Eklund *et al.* (1969). Bacteriophage CEβ exhibited a hexagonal head 100 nm in diameter and a tail 400 nm long and 11 nm in diameter. The tail was surrounded by a contracted sheath 30 nm in diameter. Bacteriophage CEγ was smaller than CEβ and exhibited a head 60 nm in diameter and a tail 185 nm long and 6 nm in diameter surrounded by a contracted sheath 20 nm in diameter. Bacteriophage CEβ produces colony-centered plaques on indicator strain AO28, whereas CEγ produces turbid plaques. Both bacteriophage CEβ and CEγ were studied for their relation to the toxigenicity of type C strain 468C.

When bacteriophage CEβ was plated on AO28 and material from plaques was transferred into TYG broth, each of the cultures arising from the plaques was toxigenic (produced C_1 and D toxins in addition to C_2 toxin) and produced bacteriophage CEβ (Table 2).

Table 2

*Reinfecting type C culture AO28 (468C) with phage CEβ
and effect on toxigenicity*

Method	Tested	No. of cultures producing phage CEβ and C_1, C_2 and D toxins
Plaques*	80	80
30 min†	40	4
240 min†	40	40

* Material from plaques produced by bacteriophage CEβ was cultured, checked for toxigenicity and bacteriophage production.

† Culture AO28 was exposed to bacteriophage and isolated colonies were checked for toxigenicity and bacteriophage production.

Exposing culture AO28 to bacteriophage CEβ (at a multiplicity of infection of 1·8) for 30 min yielded 4 toxigenic, bacteriophage-producing cultures. After a 240-min exposure, each of the 40 cultures tested produced bacteriophage CEβ and were toxigenic. The cultures remaining non-toxigenic after the 30-min exposure continued to be sensitive to the bacteriophage CEβ and when infected with CEβ they also became toxigenic.

Each of the above experiments was repeated with bacteriophage CEγ and indicator strain AO28. Cultures arising from plaques produced by CEγ remained "nontoxigenic". After a 240-min exposure with bacteriophage CEγ, 38 of the 40 cultures tested produced CEγ, but all of them remained "non-toxigenic".

Table 3 summarizes the relation of type C (468C) bacteriophages to indicator strain AO28. Culture AO28 is nontoxigenic and sensitive to CEβ and CEγ. When AO28 carries CEβ it becomes toxigenic, it is sensitive to CEγ and immune to CEβ. Culture AO28 carrying both CEγ and CEβ is toxigenic and is immune to both phages.

Table 3

*Relation of type C (468C) phages to indicator
strain AO28*

Bacterial strain	Toxigenicity	Sensitivity to phage	
		CEβ	CEγ
AO28	−	+	+
AO28 (CEβ)	+	−	+
AO28 (CEγ)	−	+	−
AO28 (CEγ, CEβ)	+	−	−

Strain 468C was subcultured in TYG broth containing 50% antiserum against CEβ to determine whether bacteriophage sensitive, nontoxigenic cultures could be isolated. After 7 daily transfers in antiserum, 10 of the 37 cultures ceased to produce the dominant C_1 and D toxins and also were cured of prophage CEβ (Table 4). Spores of 468C were heated at 70° for 20 min and plated on TYG blood agar plates. This study was made to determine whether the spores carried the bacteriophages. Table 4 shows that 10 of the 37 colonies lost the bacteriophage CEβ and toxigenicity through the spore state. When these nontoxigenic cultures were reinfected with CEβ they again became toxigenic and produced bacteriophage. The "curing" of 468C of the CEβ bacteriophage by subculturing in antiserum against CEβ indicated that this relation between bacteriophage CEβ and host is probably pseudolysogeny instead of true lysogeny.

Table 4

Loss of prophages by type C (468C) through spore state and by culturing in CEβ antiserum

Culture	No. tested	No. producing		
		Toxin	CEβ	CEγ
Spores	37	27	27	37
7th daily transfer in antiserum	39	12	12	39

Culture AO28 was also used as an indicator organism for bacteriophages of different toxic type C strains. Of the 30 Type C cultures tested, 10 produced bacteriophages that infected AO28 and each of these bacteriophages also converted AO28 to the toxigenic state. Bacteriophages which did not induce toxigenicity were also isolated from 4 of the type C strains.

In our studies we have also encountered a second group of type C strains. These cultures lose toxigenicity when treated with acridine orange or ultraviolet light and also lose toxigenicity while in the spore state in a manner similar to that observed for strain 468C. Filtrates of the toxigenic parent culture will not lyse broth cultures, will not produce plaques on the nontoxigenic isolates and will not induce toxigenicity. It is possible that the medium used is inadequate for demonstrating the bacteriophages of these strains.

Relation of bacteriophages to toxigenicity of type D

Many of the same experiments carried out with type C cultures were also used to study the relation of bacteriophages to toxigenicity of type D. The South African strain of type D was used in most of these studies. This South African strain was "cured" of its prophage and concomitantly became nontoxigenic by culturing in acridine orange; it does not produce the C_2 toxin. Only one bacteriophage has been found in this culture; it has been designated as DEβ and is very similar in size and morphology to the CEβ phage of type C strain 468C.

When the "cured", nontoxigenic cultures were reinfected with bacteriophage DEβ they again produced toxin and bacteriophage DEβ. The South African strain was incubated in SFEM Medium (containing 0·5% glucose and 0·5% ammonium sulphate) until sporulation and the spores were heated at 70° for 20 min to inactivate the free phage. The heated spores were plated on TYG blood agar and isolated colonies were tested for toxigenicity and bacteriophage production. Of the 39 isolates tested, 19 were nontoxic and sensitive to bacteriophage DEβ. When these nontoxigenic isolates were infected with DEβ they again produced toxin and bacteriophage.

Type D strain 1873 was also studied using similar experiments to that carried out used with the South African strain. Strain 1873 carries two bacteriophages, one induces toxicity and the second bacteriophage has no effect on the toxigenicity. When strain 1873 is "cured" of its prophages, it ceases to produce the D toxin, but it continues to produce the C_2 toxin. The C_2 toxin can be detected in the "cured" cultures only after trypsin treatment. Therefore when type C strains and type D strain 1873 are "cured" of their prophages, they cease to produce their dominant toxins. These "cured" isolates then become indistinguishable because they continue to produce only the C_2 toxin.

There is little doubt from our experiments that the conversion to toxigenicity of *C. botulinum* type C and D depends on the active and continued participation of specific bacteriophages.

The pseudolysogenic relationship of bacteriophage and host results in the loss of the bacteriophage in some of the cells while in the spore state. This then explains the reasons that type C and D cultures will lose toxigenicity when transferred in laboratory medium.

Relation of bacteriophages to toxigenicity of types B and F

Two cultures of nonproteolytic type B have each been "cured" of one of their prophages after ultraviolet light treatment. These "cured" isolates however continue to carry a second prophage and they continue to produce toxin. Similar results have been obtained with a nonproteolytic strain of type F in that the culture has been "cured" of one prophage and the culture continues to be toxigenic. Studies are currently in progress to "cure" these cultures of their second prophage. The host-bacteriophage relationship in these types appears to be true lysogeny in comparison to the pseudolysogenic state observed with types C and D.

The Langeland strain of proteolytic type F has been "cured" of 2 of its prophages; however the "cured" culture continues to produce toxin. It is possible either that it carries other bacteriophages or that the toxins of all types of *C. botulinum* are not induced by bacteriophages.

References

Adams, M. H. (1959). *Bacteriophages.* New York: Interscience.
Eklund, M. W., Poysky, F. T. & Wieler, D. I. (1967). Characteristics of *Clostridium botulinum* type F isolated from the Pacific Coast of the United States. *Appl. Microbiol.* 15, 1316.
Eklund, M. W., Poysky, F. T. & Boatman, E. S. (1969). Bacteriophages of *Clostridium botulinum* types A, B, E and F and nontoxigenic strains resembling type E. *J. Virol.* 3, 270.
Freeman, V. J. (1951). Studies on the virulence of bacteriophage. Infected strains of *Corynebacterium diphtheriae. J. Bact.* 61, 675.
Groman, N. B. (1955). Evidence for the active role of bacteriophage in the conversion of nontoxigenic *Corynebacterium diphtheriae* to toxin production. *J. Bact.* 69, 9.
Inoue, K. & Iida, H. (1971). Phage conversion of toxigenicity in *Clostridium botulinum* types C and D. *Japan J. med. Sci. Biol.* 24, 53.
Jansen, B. C. (1971). The toxic antigenic factors produced by *Clostridium botulinum* types C and D. *Onderstepoort J. vet. Res.* 38, 93.
Mason, J. H. & Robinson, E. M. (1935). The antigenic components of toxins of *Clostridium botulinum* types C and D. *Onderstepoort J. vet. Sci. Anim. Ind.* 5, 65.

Discussion

Hobbs

Will Dr Eklund please tell us whether phage induced toxicity is also observed with *Clostridium botulinum* types, A, B, E and F?

Eklund

We are now studying these types. Strains of non-proteolytic B and F have been "cured" of one of their prophages but they continue to carry other phages and they continue to be toxigenic. We must "cure" these strains of this other prophage before we can determine the relationship of the bacteriophage and the toxigenicity of these types. Proteolytic F has been cured of 2 of its prophages but it continues to produce toxin. It is possible that: (1) these prophages govern the production of some toxic components; (2) that the "cured" strains carry other prophages; or (3) that prophages do not govern the toxigenicity of all types of *Cl. botulinum.*

Rehm

Dr Eklund, what is your opinion about the inhibition of toxin formation by clostridia after phage infection? Is the DNA of the phage inserted into the DNA of the bacterium thereby causing synthesis of toxin?

Eklund

We do not know how the bacteriophages induce toxigenicity. We do know that the toxigenicity of types C and D depends on the continued presence of a specific bacteriophage. We have a collaborative study with Dr Boroff to determine whether the protein of the bacteriophage is the toxin.

Moss

Are "cured" strains physiologically distinguishable from uncured strains in any way other than toxin production?

Eklund

We are currently studying the characteristics of atoxigenic and toxigenic cultures. In the completed studies, the only difference that we have observed is the toxigenic characteristic.

Abattoir Practices and Their Effect on the Incidence of Salmonellae in Meat

E. F. WILLIAMS AND R. SPENCER

J. Sainsbury Ltd.,
Stamford House,
Stamford Street,
London S.E.1, England

Evidence is presented to show that the isolation of salmonellae from the lymph nodes of pigs can be reduced to a negligible level and that the incidence in pork and pork products can be reduced to a consistently low level of 1–2%. The practical methods used to achieve these low levels are described. Although considerable emphasis has been placed on the need to eliminate salmonellae from feeding-stuffs, it is believed that attention to the methods of transportation, a reduction in stress in the animals, improved systems of lairaging, and improved methods of slaughter, play an important part in the reduction of salmonellae in carcass meat and comminuted meat products. The design, construction, and hygiene of lairages, the spatial relationship between the lairage, the abattoir, and the meat products plant are also important in the reduction in the incidence of salmonellae. It is stressed that personnel in the lairage and the abattoir must not have access to any part of the production areas used for the preparation of pork products.

THIS PAPER does not, in my opinion, contribute anything fundamentally new from a scientific point of view, but it gives a brief account of a large number of investigations and observations over many years which have indicated that a consistently low incidence of salmonella can be maintained in pork meat and other pork products. The results presented are derived from the practical application of existing knowledge in the field of bacteriology to the animals on the farm, methods of transportation to the abattoir, condition of lairages and the time therein, methods of slaughter and evisceration and the cooling of carcass meat.

Interest in the overall problem of salmonellae in meat products commenced many years ago, at a time when published work indicated that pork, veal, and poultry—to a large extent—formed the reservoir of salmonellae which might ultimately lead to food poisoning in man. The early results on pork, however, were variable. Scott (1940) showed an incidence of 3·8% for salmonellae from the lymph nodes of 1000 pigs. A similar investigation 4 years later showed an incidence of 2·5% from *c.* 5000 pigs. The Public Health Laboratory Service Report in 1955 (Report, 1955) gave values of 0·4% from some 520 samples of lymph nodes; whilst Smith (1959) reported an isolation rate of about 12% from the lymph nodes of approximately 500 pigs. Other authors have reported higher figures for salmonellae in pigs. Galton *et al.* (1954) showed that, just prior to slaughter, results from rectal swabs indicated that about 80% of pigs were excretors although at farm level only 7% positives were obtained. There was also evidence that results were worse in the summer than in the winter.

In those early days there was little first-hand information concerning the incidence of salmonellae in pork products, and even less concerning the origin of salmonellae in pork meat, whether it stemmed from the farm, or spread as a result of cross-contamination in the lairage, or from gut contamination during slaughter, or from re-inoculation after slaughter and during subsequent processes.

When considerable modifications to an existing abattoir were proposed there was an opportunity to commence a long-term study of the incidence of salmonellae at all points from the farm to comminuted pork products.

The construction details of the abattoir, systems of drainage, methods of slaughter and handling of carcasses, and refrigeration systems, although somewhat poor by present standards, were nevertheless better than average. The methods for moving live animals and the standard of lairages were also typical of the period.

Preliminary studies showed results which were similar to those of other workers in that several serological types of salmonellae were isolated from faeces, abattoir drains and caeca. In the preliminary studies lymph nodes were not examined.

Method of examination consisted of the inoculation of faecal swabs into 10 ml of selenite or tetrathionate broth. After incubation at 37° for 48 h the enrichment cultures were subcultured on to sulphite and deoxycholate citrate agars for 48 h at 37°. Suspect colonies were subcultured and checked by slide agglutination using polyvalent salmonella sera.

25 g samples of lymph nodes were macerated with 150 ml of selenite broth, incubated and subcultured as for the faecal swabs.

There is reason to believe that the number of samples found positive by this method was too low and the method was changed to the incubation of the selenite broth suspensions for 24 h at 43° followed by subculture on to brilliant green and deoxycholate citrate agars, and incubation at 37° for 48 h.

The work was extended to the examination of the mesenteric, popliteal, suprarenal, superficial, external and internal iliac nodes. In the first 2000 examinations salmonellae were isolated from 8 to 14% and subsequent examination of comminuted pork meat gave positive results ranging from 2 to 6%; it was clear from these preliminary results that a more detailed study should be undertaken at various parts of the chain, starting at the farm.

Rectal swabs and faecal samples were taken on the farm when pigs were almost at killing weight; of *c.* 1500 samples only about 1% were positive, a surprisingly low figure. When these animals were transported under conditions which induced stress, and particularly in hot weather, over 50% were found to be excreting salmonellae on arrival at the abattoir. Furthermore, the pigs which arrived in a quiet condition with little or no evidence of salmonellae in the faeces, gave a high proportion of positive results when kept in lairages under poor conditions for periods in excess of 24 h.

Thus, the presumptive evidence suggested that, although pigs on the farm showed no clinical evidence of salmonellae and appeared to be negative with regard to rectal and faecal swabs, they became excretors during or after transportation, or after a lengthy period in lairages. There was sufficient evidence to suggest that there were probably a number of factors, from the live animal on the farm to the carcass in the abattoir, which might be responsible for the high incidence of salmonella excretors. There was considerable interest at the time in pre-slaughter stress and its effect on meat quality and it appeared that there might be a connection also between the degree of stress and the excretion of salmonellae.

It is clear from the behavioural pattern in pigs that stress commences during loading at the farm and that it is frequently increased during transportation and by methods of unloading and driving to slaughter. The situation is further aggravated by mixing producer lots which invariably results in excessive fighting.

Confirmation was obtained from films taken of the methods of droving and leading pigs with particular reference to the use of double-decker vehicles. Additionally, several films were taken of the behaviour of pigs during the actual period of transportation. It was interesting to note that whereas animals on the upper deck remained quiet during the entire journey, and even managed to lie down and sleep, those on the lower deck became excited within 15 min of loading and the restlessness of the animals culminated in considerable fighting, which continued throughout the journey.

Experiments showed that although the air temperature of the upper deck remained fairly constant, that of the lower deck rapidly increased. The differential between ambient temperature and the lower deck temperature became progressively greater as the ambient temperature increased. Furthermore, there was a small but definite increase in CO_2 in the atmosphere.

It is known that when pigs are anaesthetized with CO_2 they become agitated with the low levels of CO_2 given before the concentration required for complete anaesthetization is reached. The slightly increased levels of CO_2 in the vehicles may contribute to the overall restlessness of the animals on the lower deck. Correct ventilation of vehicles is therefore an important factor in reducing stress.

During the rebuilding of lairages, considerable attention was given to the details of construction. Floors consisted of well-finished concrete over which was laid a non-skid screed. Walls were smooth-finished concrete and the junctions between wall and floor were coved to facilitate washing. Special attention was paid to the elimination of projections and the pens were so arranged that the pigs could be moved easily and quietly without becoming jammed during droving. Movement of the animals was facilitated by the use of small electric goads and all forceful methods of driving were prohibited. Mechanical ventilation was designed to give an adequate number of changes of air/h. Fresh water was provided in each pen but no bedding. After each batch had been removed for slaughter, subsequent pigs were not allowed in the pen until it had been thoroughly washed with hot water and detergent, followed by chlorination with solutions containing 200 p/m chlorine.

It has been common practice for lairages, slaughter facilities and units manufacturing pork products to be housed in the same building and, furthermore, it was not uncommon for personnel working on the slaughter line to be transferred, at some period in the day, to the section preparing comminuted meat products. The authors believe that it was essential that the lairage and slaughter facilities should be totally divorced from the manufacturing part of the business, and essential also that staff working in the abattoir and lairages should not be allowed in the areas where pork products are prepared.

Slaughtering procedures, including gut washing, usually resulted in conditions of high humidity. It was essential therefore to ensure that ventilation systems were not only adequate but designed to prevent aerial contamination from lairage and gut washing passing into those areas holding clean carcasses.

The evisceration procedure should be carried out by methods which avoided contamination of the carcass by gut contents during dressing. The method of removing the intestines was therefore very important and, in particular, the method of detaching the rectal end of the alimentary tract, so that the interior of the carcass was not contaminated.

The modifications with regard to buildings, surface finishes, methods of slaughter and overall hygiene were followed at each stage with bacteriological analyses which provided a general indication of improvements. When stress was reduced during transportation, lairages were clean, properly designed and ventilated, and immediate pre-slaughter stress was reduced, a further survey of the incidence of salmonellae was made.

In this series the lymph nodes of about 2000 pigs were examined, together with a similar number of faecal specimens and about 500 surface samples from carcasses. There were 9 isolations of salmonellae from the faecal examination, 2 of S. heidelberg, 3 of S. typhimurium and 4 of S. dublin. No positives were found in the samples of lymph nodes or surface samples.

These results suggested that it was possible to obtain carcasses at the end of the slaughter-line which were apprently free from salmonellae, but clearly this did not necessarily prove that the incidence of salmonellae in pork meat was at the same level. An extensive survey of ground pork meat, however, showed that positive samples had been progressively reduced to between 1 and 2%.

These results were sufficiently encouraging to warrant a further look at the system and after minor modifications with regard to methods of evisceration, surface heat treatment of carcasses, and further improvement in hygiene, the study was repeated. Approximately 20,000 pig lymph nodes were examined and only 4 positive results were obtained. At the same time the incidence in ground pork remained at about 1–2% which suggested possible re-inoculation during the processing of carcass meat into meat products.

It should be noted in this context that more and more people take their summer holidays in the Mediterranean areas where it is not unusual for them to contract gastro-enteritis caused by salmonellae. It could, therefore, be possible for undiscovered symptomless carriers, returning from holiday, to resume work in the meat processing plants. This is a possible source of salmonellae in meat products and, although clearly it would not be reasonable or necessary to examine all meat handlers returning from holiday, there might well be a case for obtaining 3 negative stool specimens from those with a history of diarrhoea during the holiday period before allowing them to resume work.

It was established that after giving careful attention to all points in the chain from the farm to the finished meat product, it was possible to achieve a low incidence of salmonellae in pork products over a period of several years. In order to confirm this, independent observations were carried out by the Central Public Health Laboratory.

Samples of comminuted pork from 3 sources were examined each week over a period of about 2 years. From one abattoir 1·18% samples and from another 1·09% samples were positive. When samples of meat from the abattoir above were mixed with an outside source, 3·7% were shown to be positive. Overall a total of nearly 1000 samples were examined and 2·24% were shown to be positive with *S. typhimurium* phage-type 32, *S. bahati* and *S. panama*. The continued search for salmonella in comminuted pork meat indicated that between 1 and 2% of samples were positive.

It is not suggested that the problem is solved. The results indicated that, although no special precautions were taken with regard to the sterility of feeding-stuffs, levels of salmonellae in meat and meat products were approaching an acceptable standard. This can only be achieved providing considerable attention is paid to the entire operation from the farm to the product as exposed for sale. However, a number of authors have drawn attention to feeding-stuffs as a potential source of salmonellae in the live animal.

P.H.L.S. Working Group, Skovgaard & Nielsen (1972) compared the incidence of salmonella in Denmark, where sterilization of feed is compulsory, with feed in England. The incidence in Danish feed was 0·3% compared with 23% in meat meal used in England and bonemeal, and 20–27% of other ingredients of animal origin. Additionally, these authors found that 7% of caecal samples and 6% of lymph node samples from pigs in the south-west of England were positive compared with 3% of caecal samples and 4% of lymph nodes in Denmark. It is interesting to note that despite stock being fed on unsterilized feeding materials in the U.K. nevertheless the resulting level of salmonellae in pork products can be kept low.

Feed is clearly a source of infection on the farm. The examination of faecal samples and rectal swabs may give low or even negative results from the rested and undisturbed animals. However, the proportion of animals excreting salmonellae can increase rapidly when they undergo stress during transportation and marketing.

It would also appear that unstressed animals which are faecal negative become cross-contaminated and positive when held for periods in excess of 24 h under poor lairage conditions. For periods less than 24 h the cross-contamination appears to be negligible whether the animals are in good or poor lairage conditions. It should be noted however, that the situation appears to be worse with pigs severely stressed on arrival at the abattoir.

Consistent low levels of salmonellae in pork carcass meat and meat products can be obtained, provided strict attention is paid to transportation, reduction of stress, clean lairages, shorter resting time—i.e. less than 24 h—care in slaughtering procedures, particularly in evisceration, and a high standard of hygiene generally throughout processing.

References

Galton, M. M., Smith, W. V., McElrath, H. B. & Hardy, A. B. (1954). Salmonella in swine, cattle and the environment of abattoirs. *J. infect. Dis.* **95**, 236.
P.H.L.S. Working Group, Skovgaard, N. & Nielsen, B. B. (1972). Salmonellas in pigs and animal feeding stuffs in the United Kingdom and in Denmark. *J. Hyg., Camb.* **70**, 127.
Report (1955). Salmonella in carcass meat for human consumption. *Mon. Bull. Minist. Hlth* **14**, 132.
Scott, W. M. (1940). Salmonellas in healthy pigs at slaughter. *Proc. R. Soc. Med.* **33**, 366.
Smith, H. W. (1959). The isolation of salmonellae from the mesenteric lymph nodes and faeces of pigs, cattle, sheep, dogs and cats and from other organs of poultry. *J. Hyg., Camb.* **57**, 266.

Discussion

Elliott

Will Prof Williams please give further details on the 600° oven used for pork carcasses?

Williams

The oven consists of a rectangular tunnel, some 20 ft long by about 4 ft 6 in. wide and 8 ft high, through which passes a mechanized rail. At each end there are 2 pairs of swing doors designed to prevent heat loss. The air within the tunnel is heated electrically or with gas, via refractory bricks, to give an air temperature of 575–600° C. Carcasses pass through the tunnel at fixed intervals determined by the spacing on the rail and the duration within the oven is *c.* 12 sec. During that time the carcass expands slightly to eliminate all wrinkles and sagging skin. This oven was invented and is manufactured by a Swedish company.

Hygiene in Catering

H. Brodhage and B. Anderhub

Hygien.-mikrobiol. Institut,
Lucerne,
Switzerland

The catering for large communities has increased tremendously over the last 15–20 years. It not only presents difficulties to public health workers but to tourists also. Mistakes in hygiene can mean food poisoning to thousands of people. From the hygiene point of view, certain regulations should be designed for acceptable control of foodstuffs, particularly the raw materials. Special attention must be paid to meat, which is responsible for up to 80% of all poisonings. The delivery of the prepared dishes should take place within 4 h after cooking. Temperatures between 10° and 60° should be avoided. Salmonellae, staphylococci and *Clostridium welchii* are the most common agents of bacterial food poisoning. The risks of food poisoning can be reduced to a minimum by strict observance of the hygiene regulations.

IT IS WELL known that there has been an enormous increase in communal feeding over the last 20 years (Brodhage, 1969). One hundred and three million meals were served weekly in England in 1950. It is impossible to estimate the number of meals served communally at this time. In the London suburb of Camberwell there was an increase of 120% between 1955–1965 (Chalke, 1965). In Switzerland 2 million people, including alien workers, take a meal everyday outside their homes. In Western Germany the figure is estimated at 18 million (Eschmann, 1965). In Sweden 4 million meals are served daily outside the home, and thus there may be more meals eaten outside than in the home. In addition it should be mentioned that in 1968, 80 million tourists visited the various European countries. One per cent (that is 800,000) are estimated to have suffered from gastro-intestinal troubles (W.H.O., 1971). Because of this increase in the numbers of people eating out the processing, preparation and manufacture of food has become more highly centralized. Improper food-handling or lack of hygiene in food preparation used to affect the family circle only, today it may affect hundreds or perhaps thousands of people. In the U.S.A. alone, about 2 million people have salmonella infections yearly and these infections cause *c.* 500 deaths each year (Prost & Riemann, 1967). The indirect losses to the American economy were stated as 100 million dollars annually. It has been estimated that about 120 people including many young infants die yearly of food infections in England (Howie, 1969).

It seems that there is an increase of food poisoning, especially in the highly civilized countries, although such information is extremely difficult to derive

47

from official statistics. An illustration of this difficulty is the variation in government statistics and private estimates of salmonella infections in the U.K. in 1959. Officially there were only 5132 cases of salmonella infections reported, but Anderson (1962) estimated c. 50,000 cases of *Salmonella typhimurium* alone. Another example is provided by the 1970 estimate of salmonella infections in Switzerland. Our Institute alone, which is smaller than the average laboratory, reported 531 cases in that year. There are 11 far bigger institutes in Switzerland, yet the total national figure was 1086.

If we consider these facts, we cannot have more clear-cut evidence of the importance of hygiene in catering.

Food poisoning is caused predominantly by bacteria. Therefore it is important to know the main bacteria which are responsible. The most common food-poisoning agents are salmonellae (Taylor, 1969), enterotoxin-producing staphylococci and *Clostridium welchii* (Hobbs, 1969).

Table 1 lists the bacteria most likely to be incriminated in food poisoning.

Table 1

Bacteria most likely to be incriminated in food poisoning

(a) *Infections*		(b) *Intoxications*	
(1) *Salmonella*	+++	(1) Enterotoxin–producing	
(2) *Shigella*	+	*Staphylococcus aureus*	+++
(3) *Proteus*	+	(2) *Clostridium botulinum*	(+)
(4) *Pseudomonas*	+		
(5) *Escherichia coli*	+		

(c) *Intermediate position*
   (1) *Clostridium perfringens (welchii)*   ++
   (2) *Bacillus cereus*   +

Table 2 shows the reservoirs and modes of spread of the different types of food poisoning bacteria.

The statistics of Cockburn & Vernon (1969) show that, in the U.K. in the years 1949 to 1966, 75% of all food poisoning was connected with meat, 8% with sweetmeats, 5% with fish, 5% with egg or egg products and 3% with milk or milk products. As a consequence, new guide-lines and laws were worked out in various countries, especially England, and they have proved effective. In this connection the U.K. Government handbook, *Clean Catering* (1963) and *The Liquid Egg Regulations* (1963) may be mentioned.

Unfortunately such recommendations and laws do not exist in all countries. Therefore we would like to draw attention to the following points.

Table 2

Reservoirs and modes of spread of the different types of food poisoning

Bacteria	Main reservoir	Mode of spread
Enterotoxin-producing staphylococci	Man Staphylococcal mastitis of the cow	Meat, sausages, ham etc. Milk
Cl. perfringens (welchii)	Raw meat 10–20% faeces of man and animal 2–20%	Precooked meat, especially when meat is cooked the day before consumption
Salmonella	Food of animal origin	Meat, sausages, poultry, eggs, egg products etc.
	Human carrier	Contaminated food, direct infection seldom
Shigella	Man	Contaminated food, direct contact infections

Kitchen

The kitchen area should be at least half that of the dining room; in small cafés relatively greater space is required (Chalke, 1965). It should be possible to clean and disinfect all equipment easily. Metal parts should be of stainless steel. Once a day there should be a thorough cleaning of the kitchen and space-rooms, using efficient disinfectants (Seidel & Muschter, 1970). Mice, rats, flies and other vermin do not belong to the kitchen or space-rooms.

Toilets shall not be in direct connection with the kitchen. Sufficient lavatories with cold and warm water, soap, hand-disinfectants and paper napkins (no towels) should be provided. Good visible and readable hygiene directions should be displayed prominently.

Staff

Clean clothes should be worn by the staff during work periods and special attention must be paid to hair and fingernails. No one should be employed who has not passed a medical examination, but there are various points of view about the value of the bacteriological control of faeces. In our opinion these controls are justified in big establishments. It is absolutely necessary to have bacteriological control of employees suffering from diarrhoeal diseases.

The most important task is the teaching of food hygiene to the staff itself. Those who superintend the work in large kitchens and food plant should be given refresher courses in view of the rapid changes in the field of food hygiene and in the knowledge of diseases caused by contaminated food.

Steps should be taken to ensure that all food handlers receive suitable training. In this training visual aids should be used such as slides, films, agar-sausages, Rodac plates and swabs; these materials can be used to demonstrate the microbiological consequences of faults in hygiene. Food and other hygiene education should be encouraged at all levels and should preferably start in the primary school.

Food

As a matter of principle only bacteriologically acceptable raw materials should be used. Only 1 day's supply of highly perishable food should be held in the plant or kitchen. Frozen products should not remain out of a freezer more than 2 h. Deep frozen meat may not be thawed and refrozen and it must be used within 6 h of thawing. The preparation of food should, where possible, be done by machine to avoid direct contact with the hands. In principle, the grilling of special meat should not be done on the day before consumption. Kitchen utensils such as knives, dishes, boards and so on, that come into contact with raw materials, may not be used for finished food, unless thoroughly cleaned. Special care should be taken in the storage of prepared food. Temperatures in the range of 10° to 60° should be avoided. With automatic thermostatically controlled food-dispensers the portions of food should be kept either at 91° or below 0° (Chalke, 1965). In big kitchens the period between the end of the cooking process and the distribution of the last portion should not be more than 4 h (Munchow, 1969). Epidemiological testing is freely facilitated if a portion of every meal is kept in the refrigerator for 24–48 h after serving.

Surveillance

The statement that: "The continual surveillance of catering establishments is one of the most responsible tasks performed by the public health inspectors of the local authority" (Chalke, 1965) is fully supported. But only well trained staff can undertake these surveys and be responsible for them. Such control is not as well organized in other countries as it is in England. Some of these controls are listed in Table 3.

The public health inspector has the responsibility to ensure that a high standard of hygiene is maintained in the plant, that a healthy staff works in clean clothes, that unobjectionable raw materials are used and that the food is prepared according to hygienic guidelines. But his primary task is to give advice. Prosecution should follow only when his instructions are not followed. According to Mossel *et al.* (1968), large kitchens should be bacteriologically examined every 3–7 days by means of dabcultures, swabs and other aids.

Table 3
Hygiene in catering

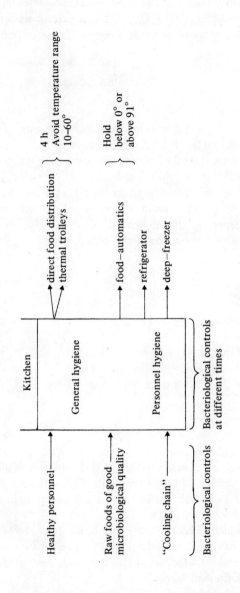

The introduction of international microbiological standards for foods would be most helpful. If such international standardization is introduced, good laboratories with trained personnel will be indispensable and great efforts will be needed to provide them (W.H.O., 1971). At present we are far from achieving this end, but this symposium should bring us a few steps nearer to our goal.

Recommendations

Kitchen

The kitchen area should be at least half that of the dining room.
Metal parts should be made of stainless steel.
Once a day there should be a thorough cleaning campaign.
Proper disinfectants should be used.
Mice, rats, flies and other vermin do not belong in the kitchen.
Sufficient lavatories should be provided [warm water, soap, hand-disinfectants, paper napkins (no towels)].
Hygiene directions should be displayed.

Staff

Clean clothes.
Personal hygiene.
Medical test.
Bacteriological control of faeces.
School food — hygiene refresher courses.
Training should make use of visual aids (slides, films, agar sausages, Rodac plates).

Food

Unobjectionable raw foods.
Only daily rations of easily spoiled food should be available.
Frozen products should not be more than 2 h without cold storage.
Deep frozen meat may not be thawed and frozen again and when thawed it must be used in 6 h.
The direct touch of cooked meat with hands should be avoided.
The grilling of special meat should not be done on the day before required.
Kitchen utensils that come in contact with raw goods may not be used uncleaned with finished food.
Special care should be taken in the storage of prepared food.
Temperatures of 10–60° should be avoided.
In food dispensers portions should be kept at 91° or below 0°.

Not more than 4 h should elapse between the end of the cooking process and distribution of the food.

Surveillance

Continual surveillance by public health inspectors.

Only well trained staff should undertake these controls.

Prosecution should only be required in instances of disobedience.

References

Anderson, E. S. (1962). Salmonella food poisoning. In *Food Poisoning*, p. 33. London: The Royal Society of Health.

Brodhage, H. (1969). Durch Bakterien verursachte Lebensmittelvergiftungen bzw.– infektionen. *Alimenta* **8**, 150.

Chalke, H. D. (1965). Food hygiene in restaurants and canteens. *Practitioner* **195**, 54.

Clean Catering: A handbook on hygiene in catering establishments. (1963). London: H.M.S.O.

Cockburn, W. C. & Vernon, E. (1969). Reporting and incidence of food poisoning. In *Bacterial Food Poisoning*, p. 7. Ed. J. Taylor. London: The Royal Society of Health.

Eschmann, K. H. (1965). Lebensmittelhygiene und Schweizerisches Lebensmittelgesetz. *Alimenta* **4**, 123.

Hobbs, B. C. (1969). Staphylococcal and *Clostridium welchii* food poisoning. In *Bacterial Food Poisoning*, p. 69. Ed. J. Taylor. London: The Royal Society of Health.

Howie J. (1969). Introduction. In *Bacterial Food Poisoning*, p. 1. Ed. J. Taylor. London: The Royal Society of Health.

Mossel, D. A. A. *et al.* (1968). Objectief Onderzoek naar de Hygienische Gesteldheid von grote Keukens. *Voeding* **29**, 105.

Munchow, S. (1969). Lebensmittelinfektionen und Gemeinschaftsverpflegung. *Alimenta* **8**, 23.

Prost, E. & Riemann, H. (1967). Food-borne salmonellosis. *Ann. Rev. Microbiol.* **21**, 405.

Seidel, G. & Muschter, W. (1970). *Die Bakteriellen Lebensmittelvergiftungen*, p. 249. Berlin: Akademie Verlag.

Taylor, J. (1969). Salmonella and salmonellosis. In *Bacterial Food Poisoning*, p. 25. Ed. J. Taylor. London: The Royal Society of Health.

The Liquid Egg (Pasteurization) Regulations (1963). Great Britain Statutory Instruments No. 1503. London: H.M.S.O.

World Health Organization. (1971). Report on a Seminar, Document EURO 0389, Regional Office of Europe, Warsaw, 1970. Copenhagen: W.H.O.

Discussion

Linderholm

Has Dr Anderhub had any outbreaks of shigella infection caused by food and not by human contact? If so were there many outbreaks and do you think that food will play a major role in outbreaks of shigellosis? What sorts of food were involved?

Anderhub

We had an outbreak of food poisoning due to *Shigella sonnei* in Central Switzerland in 1961. A butcher's wife was the carrier and 55 cases occurred.

Session 1

Bacteriology of Various Commodities in Relation to Food Poisoning

Part 2

Chairman: Professor E. Hess

Bacillus cereus in Milk and Dairy Products

F. L. DAVIES AND GRETA WILKINSON

*National Institute for Research in Dairying,
Shinfield, Reading, England*

Bacillus cereus is a common contaminant of milk, cream and other dairy products. It can on occasions multiply in such products to cause spoilage in the form of "bitty cream" or sweet curdling, though its presence, even in high numbers, is not normally considered to constitute any danger to human health. Several workers have recently shown, however, that *B. cereus* is not always a harmless micro-organism but is in fact capable of synthesizing toxic substances. The incrimination of the organism in cases of food poisoning is becoming increasingly well-documented and, in the main, meat preparations and starchy foods have been implicated with dairy products infrequently involved. The relative frequency with which *B. cereus* can occur in milk, however, and the growing popularity of new dairy products with relatively high starch contents suggest that more information is required concerning the activity of the organism in such materials. Contamination by *B. cereus* is usually in the form of heat resistant endospores which may survive processing temperatures. The spores must germinate, outgrow and multiply to be of any potential danger. We are at present studying the germination of these spores in milk derived media and have shown that various factors acting upon the spore can affect this stage of development. In addition it has become clear that the conditions of heat processing are likely to render the product more or less suitable as a germination medium for *B. cereus*. These various factors will be discussed.

IN A SYMPOSIUM concerned with the bacteriology of food commodities in relation to food poisoning it is unusual to find papers relating to *Bacillus cereus*. Indeed the microbiological programme of such meetings is frequently confined to the staphylococci, salmonellae and clostridia. While these organisms are undoubtedly the most serious and perhaps the most common agents of food poisoning outbreaks, other genera of bacteria may from time to time be implicated. Over the years there have been occasional reports of *B. cereus* induced food poisoning and of late these are becoming increasingly well-documented. Whether this represents an increase in the number of incidents, improved methods of diagnosis or a growing awareness of the problem is not clear.

We are interested in the hygiene of milk production and processing and *B. cereus* is by no means an unknown micro-organism in this field. It has in fact been said that this species can be expected to occur in the majority of farm and all bulk milks (Franklin, 1967) and its occurrence in milk has been reported repeatedly since early investigations (e.g. Lawrence & Ford, 1916). The form of the organism in milk is less clear, as the proportions of spores and vegetative cells have rarely been assessed. In raw milk the spores are the more important since they may survive pasteurization and in some instances germinate to give vegetative cells which may multiply and adversely affect the product. Spores and vegetative cells may also enter milk as post-pasteurization contaminants since non-sterile bottles are usually filled by non-aseptic processes.

The sources of *B. cereus* spores in milk are likely, therefore, to fall into 2 broad categories, firstly those entering the milk during milking or storage of the milk on the farm and secondly those entering during operations at the dairy. In a study in south-east England of the sources of various common sporeforming bacteria in milk, we have shown (unpublished data) that the overwhelming majority originate from straw bedding and soil and enter milk from inadequately washed teat surfaces; few appear to be contributed by milking machinery or auxilliary plant or by aerial contamination. It is uncertain whether this generalization is true for *B. cereus* spores since, unlike other sporeforming bacteria which reach maximum numbers in milk in the winter months when cows are housed indoors on straw bedding and other litter, they are most prevalent in August and September (Nokes, 1965) when teat surfaces are least likely to be heavily soiled. According to Billing & Cuthbert (1958), improperly washed milk cans were a primary source of *B. cereus* spores though the importance of cans must have diminished as they are largely replaced by refrigerated bulk tanks. It appears, however, that the major source of *B. cereus* spores in milk probably does differ from that of other common sporeforming contaminants. The nature of the source may be of some significance; we have some preliminary evidence that *B. cereus* spores originating from dairy sources develop more rapidly in pasteurized milk than do those derived from farm situations. Work on this is now in progress.

As a result of an extended survey of milk from several dairies, Franklin (1969) concluded that in the absence of post-pasteurization contamination, *B. cereus* "is the most important organism affecting pasteurized milk quality in warm ambient conditions". This notoriety stems from its ubiquity and spoilage potential which may manifest itself in 1 of 2 ways: (a) sweet curdling, a clotting of the milk by a rennin-like mechanism with no acid production and little off-flavour; (b) "bitty" cream, the formation of solid particles of cream which float on the surface when milk is mixed with hot beverages. The latter is thought to arise from the accumulation of spores in the cream layer of the milk. These develop into micro-colonies around which cream flakes are formed (Labots & Galesloot, 1959), possibly by a complex process involving hydrolysis of the lecithin of the fat globule membrane (Stone, 1952). In practice it is often the post-pasteurization contaminants including bacteria other than sporeforming genera which constitute the limiting factor in keeping quality. *B. cereus* can also be found in high numbers in related dairy products such as dried milk (Kim &

Goepfert, 1971) and cream (Irvine, 1959). It was shown by Davis (1969) that after incubation of cream at 15° for 24 h *B. cereus* often constituted a large part of the microflora and it is known that the resazurin and methylene blue reductase tests are readily affected by this micro-organism. Alexander & Rothwell (1970) showed that in ice cream lower numbers of *B. cereus* than of *E. coli* were required to bring about methylene blue reduction in the same times. *B. cereus* may also produce protease in raw milk. The activity of this enzyme can survive U.H.T. treatment and it has been postulated that it may be a factor in the formation of gelation during long storage of ultra heat treated milk (Harper *et al.*, 1970).

Thus *B. cereus* is a common contaminant of dairy products and it can be present in considerable numbers. We know that it can cause spoilage of these products. Is there any reason to suspect that it may be involved in food-borne illness associated with them? The documented evidence incriminating the organism in cases of food poisoning is growing and the types of food involved appear to cover a fairly wide range. Table 1 gives a number of examples of food poisoning outbreaks where *B. cereus* was the probable aetiological agent. The

Table 1

Examples of food poisoning outbreaks where B. cereus *was the probable etiological agent*

Reference	Country	Type of food	B. cereus isolated/g or/ml
Lubenau (1906)	Germany	Meatballs	Not recorded
Plazikowski (1947)	Sweden (117 outbreaks)	Meats and meat products	Not recorded
Christiansen *et al.* (1951)	Norway	"Yellow pudding dessert"	$1 \cdot 3 \times 10^7$
Pisu & Stazzi (1952) (cited in Nikodemusz *et al.*, 1962)	Italy	Chicken soup	$6 \cdot 0 \times 10^7$
Hauge (1955)	Norway	Vanilla sauce	$2 \cdot 5 \times 10^7 - 1 \cdot 1 \times 10^8$
Clarenberg & Kampel-macher (cited in Nikodemusz, 1958)	Netherlands	Mashed potatoes, meats, rice dishes, puddings, soups	$5 \times 10^5 - 2 \times 10^8$
Nikodemusz (1958)	Hungary	Vegetable soup	Not recorded
Nikodemusz *et al.* (1962)	Hungary (35 outbreaks)	Sausage, vegetable dishes, cream pastries, soups	$3 \cdot 6 \times 10^4 - 9 \cdot 5 \times 10^8$
Midura *et al.* (1970)	U.S.A.	Meatloaf	10^7
Anon (1972)	U.K.	Rice dishes	$10^6 - 10^8$

great majority of these were reported from European countries and date back to 1906, but those occurring since about 1940 have been more completely documented. The examples shown here are by no means a complete catalogue of reported incidents. The subject has been comprehensively reviewed by Goepfert *et al.* (1972), who say that "in the light of the rather high frequency of occurrence in Hungary, for example, it is quite probable that the infrequency of outbreaks in England and the United States reflects a lack of attention or examination of suspect foods rather than absence of occurrence". It can be seen that a large number of the outbreaks involved meats of various sorts and Ormay & Novotny (1968, 1970), stated that 3 out of every 5 cases occurring in Hungary between 1960–1968 involved meat or meat products. They attributed this to the fact that in Hungary, meats highly seasoned with spices which may contain high numbers of sporeformers are often inadequately cooked, particularly under conditions of mass catering. These authors ranked *B. cereus* as the third most common cause of food poisoning in Hungary over that period.

Of the remaining cases, products containing starch seem to be most frequently implicated. Usually cases have occurred where there is large scale catering and in particular where prepared food has been stored for a period prior to consumption. This is true of the one documented case in the U.S.A. (Midura *et al.*, 1970) which affected 15 members of a fraternity who ate contaminated meat loaf, and the recent cases in the U.K. which involved rice cooked and allowed to stand overnight before consumption (Anon., 1972). The large number of organisms isolated/g or /ml of incriminated foodstuffs (see Fig. 1) suggests that there is a pre-incubation period. Bonventre & Johnson (1970) suggest that *B. cereus* food poisoning is a food intoxication rather than a food infection and that a significant level of growth would be required to synthesize the necessary levels of extracellular toxin. They also reported that although *B. cereus* shares with *Clostridium perfringens* the ability to synthesize the enzyme phospholipase C, it differs in that the enzymic and toxic activities do not reside in the same molecule. Data concerning *B. cereus* enterotoxin are given by Goepfert *et al.* (1973).

From the literature it seems that dairy foods have been infrequently involved in such outbreaks and instances are confined to a few cases from cream in confectionery and certain desserts. In the light of the frequency of contamination of dairy foods by *B. cereus* it is difficult to explain why these foods should so rarely have caused food poisoning symptoms due to *B. cereus*. There are perhaps 2 possible reasons. Firstly, in accordance with the generalization made by Goepfert *et al.* (1972), laboratories may have omitted screening procedures for *B. cereus* when examining suspect foodstuffs. Secondly, since high numbers of the organism are necessary to elicit symptoms, visible spoilage of milk or associated products is likely to deter their consumption. The synthesis of toxin by *B. cereus* is known to be dependent upon the presence of certain unidentified

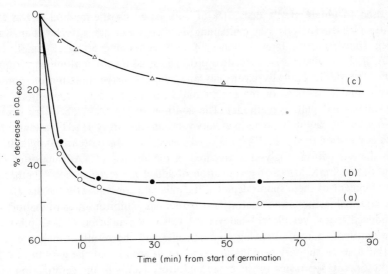

Fig. 1. Germination of spores from *B. cereus* T16 in dialysates of pasteurized milk prepared under different conditions. Dialysates prepared: (a) at 20° with the addition of chloramphenicol (50 μg/ml) to the milk; (b) at 4° with no addition; (c) at 20° with no addition.

nutritional factors present for example in beef infusion broth but absent in most other standard media (Bonventre & Johnson, 1970). It seems improbable, however, that the absence of such a factor in dairy products prevents the synthesis of food poisoning enterotoxin since Goepfert *et al.* (1973) have demonstrated that skim milk medium supports the synthesis of enterotoxin factor by *B. cereus.*

A recent development in the dairy industry is the production and sale of desserts which are essentially starch-based and contain or are topped with dairy cream. Since starchy foods have been incriminated in a number of outbreaks and the cream may provide an ample reservoir of *B. cereus,* potential spoilage problems may not, therefore, be the only hazard associated with such products.

Contamination in most cases is likely to be spores which resist most heat and chemical treatments and thus survive processing and germinate to form active vegetative cells. We are at present studying the factors controlling the *germination, outgrowth* and *multiplication* of *B. cereus* in milk and milk products. Initially we are concentrating on germination *per se,* i.e. the change from a metabolically dormant to a metabolically active spore potentially capable of outgrowth. In order to study this germination we are using mainly optical methods and estimating microscopically the change from the phase-bright to the phase-dark state or measuring spectrophotometrically the associated decrease in optical density. As milk is not an optically clear medium we have developed a method to obtain sterile dialysates of milk based on the method of Davies & White (1960). Dialysis sacs containing distilled water are autoclaved in water then transferred to large volumes of milk containing 50 μg cloramphenicol (CAP)/ml and dialysis is allowed to proceed for 24 h with continuous stirring at 20°. The sterile optically clear dialysates support the germination of spores at levels similar to those of the parent milks (as measured by conventional heat treatment and plating methods). The addition of CAP was necessary because spores in the raw milk sometimes survived our laboratory HTST pasteurization and germinated to give multiplying vegetative cells in the pasteurized milk during the dialysis period. Dialysates prepared in the absence of CAP supported only low or moderate levels of germination of added *B. cereus* spores. Whether the development of these spores depleted the milk and consequently the dialysate of an essential germination factor or whether an inhibitor of germination was produced is not yet clear. Whatever the reason the addition of CAP prevented the multiplication of surviving contaminants and the resultant dialysates were better able to support germination (Fig. 1). It seems that the growth of other sporeforming organisms might be an inhibitory factor in the germination of *B. cereus*. The germination quality of dialysates can be similarly improved by dialysing in the cold (Fig. 1) but it has been suggested (Davies & White, 1960) that those produced at 20° most closely resemble the soluble fraction of milk.

Using such a system we have started to accumulate data on other factors which affect the germination of *B. cereus* in milk, some of which may be of interest. Heat (and therefore processing temperatures) influences germination in 2 ways, firstly by a direct effect upon the spore and secondly by affecting the medium within which the spore may germinate. The first of these concerns the activation temperature and it has long been known that the germination of most spores is activated by a period of sub-lethal heat treatment (Curran & Evans, 1945). As it is the practice of some processors to increase the severity of heat treatment when spoilage problems are manifest, we carried out a small experiment to determine the effect upon *B. cereus* spores of changing the pasteurization temperature. Using a laboratory scale HTST pasteurizer (Franklin, 1965), we heated spores in water at various temperatures for 15 sec and tested their ability to germinate in a dialysate from milk which we had pasteurized at the normal temperature (71·7° for 15 sec) for commercial pasteurization. From the results in Table 2 it appears that, for the strain tested, little activation occurred within the normal pasteurization range or below but that an elevation of this temperature could significantly increase the level of germination.

With regard to the effect of temperature on the germination medium, we found that after pasteurization the ability of milk to support germination may be profoundly affected. In an experiment where spores were activated at 75° for 30 min and added to dialysates prepared from raw and pasteurized milks, germination was negligible in dialysates from the former but proceeded at a

Table 2

Germination of B. cereus *T16 spores in a dialysate of pasteurized*
milk after activation in distilled water at various temperatures

Activation temperature (°)	Per cent decrease in O.D. after 30 min at 30°
50	4·0
55	2·5
60	2·5
65	2·5
70	2·5
H.T.S.T.	
75	4·8
80	5·0
85	30·0
90	44·0

rapid rate in dialysates from the pasteurized product (Fig. 2). This effect was confirmed using whole milk rather than dialysates and also with a range of other strains of *B. cereus* isolated from pasteurized milks. In a more definitive experiment, we heat-treated milks, again using the laboratory pasteurizer over a range of temperatures for 15 sec each, prepared dialysates from them and added spores which had been activated at a single temperature. Figure 3 shows that for pasteurization temperatures above 55° the extent of germination increased markedly to a maximum over range the *c.* 65–75°, a range within which the temperature for commercial HTST pasteurization unfortunately falls. According to these data an elevated processing temperature might in fact be advantageous.

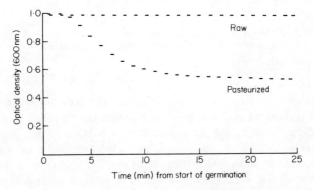

Fig. 2. Germination of spores of *B. cereus* T16 in dialysates of raw and pasteurized milk at 30°.

There are thus 2 opposed phenomena associated with heat treatment and it is the balance between them which provides an important factor controlling the ability of *B. cereus* spores to germinate in dairy products. A change in processing temperature must be considered with caution. Our current studies aim to analyse the effect of heat on milk with regard to germination and there is a considerable amount of information but by no means a complete explanation of the phenomenon.

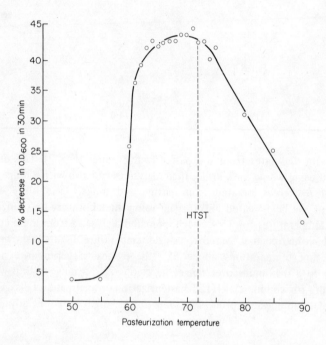

Fig. 3. Germination of spores of *B. cereus* T16 (previously activated in water at 75° for 30 min) in milks pasteurized for 15 sec over a range of temperatures.

In conclusion, we have shown that the source of spores, the presence of other sporeforming bacteria and the processing temperature may all play a role in dairy products in controlling the germination of *B. cereus* spores. We have also, in work not discussed here, established that factors such as the age of the spores and the conditions under which they are formed are important in this respect. Before a spore of *B. cereus* can grow out to form a cell potentially capable of synthesizing toxin in dairy products, it must germinate. It is obvious that there is much we still need to know concerning the factors which control this event.

References

Alexander, J. & Rothwell, J. (1970). A study of some factors affecting the methylene blue test and the effect of freezing on the bacterial content of ice cream. *J. Fd Technol.* 5, 387.

Anon. (1972). Food poisoning associated with *Bacillus cereus. Br. Med. J.* 1, 189.

Billing, E. & Cuthbert, W. A. (1958). "Bitty" cream: the occurrence and significance of *Bacillus cereus* spores in raw milk supplies. *J. appl. Bact.* 21, 65.

Bonventre, P. F. & Johnson, C. E. (1970). *Bacillus cereus* toxin. In *Microbial Toxins III. Bacterial Protein Toxins,* p. 415. Eds. T. C. Montie, S. Kadis & S. J. Ajl. London & New York: Academic Press.

Christiansen, O., Koch, S. O. & Madelung, P. (1951). Et udbrud af levnedmiddlelforgiftning forarsoget af *Bacillus cereus. Nord. VetMed.* 3, 194.

Clarenburg, A. & Kampelmacher, E. H. (1957). *Voeding* 18, 384. Cited in *Nikodemusz et al.* (1958).

Curran, H. R. & Evans, F. R. (1945). Heat activation inducing germination in the spores of thermotolerant and thermophilic aerobic bacteria. *J. Bact.* 49, 335.

Davies, D. T. & White, J. C. D. (1960). The use of ultrafiltration and dialysis in isolating the aqueous phase of milk and in determining the partition of milk constituents between the aqueous and disperse phases. *J. Dairy Res.* 27, 171.

Davis, J. G. (1969). Microbiological examination of cream. *Dairy Inds.* 34, 555.

Franklin, J. G. (1965). A simple laboratory scale HTST milk pasteurizer. *J. Dairy Res.* 32, 281.

Franklin, J. G. (1967). The incidence and significance of *Bacillus cereus* in milk. *Milk Ind.* 61, 34.

Franklin, J. G. (1969). Some bacteriological problems in the market milk industry in the U.K. *J. Soc. Dairy Technol.* 22, 100.

Goepfert, J. M., Spira, W. M. & Kim, H. U. (1972). *Bacillus cereus*: Food poisoning organism: A review. *J. Milk Fd Technol.* 35, 213.

Goepfert, J. M., Spira, W. M., Glatz, B. A. & Kim, H. U. (1973). Pathogenicity of *Bacillus cereus. In Proceedings of the 8th International Symposium. The Microbiological Safety of Food.* p. 69. London: Academic Press.

Harper, W. J., Hidalgo, J. E. & Mikolajcik, E. M. (1970). *Bacillus cereus* protease as a factor in gelation of sterilized dairy products. *XVII Int. Dairy Cong. IE,* 192.

Hauge, S. (1955). Food poisoning caused by aerobic sporeforming bacilli. *J. appl. Bact.* 18, 591.

Irvine, R. W. (1959). Churning and sweet coagulation in table cream. *Aust. J. Dairy Technol.* 14, 17.

Kim, H. U. & Goepfert, J. M. (1971). Enumeration and identification of *Bacillus cereus* in foods. I. 24 hour presumptive test medium. *Appl. Microbiol.* 22, 581.

Labots, H. & Galesloot, Th. E. (1959). Vlokvorming in de roomlag van gepasteuriseerde melk. *Ned. Melk-en Zniveltijdschr.* 13, 79.

Lawrence, J. S. & Ford, W. W. (1916). Studies on aerobic spore bearing non-pathogenic bacteria. *J. Bact.* 1, 273.

Lubenau, C. (1906). *Bacillus peptonificans* als erreger einer gastroenteritis epidemie. *Zentbl. Bakt. Abt. I.* 40, 433.

Midura, T., Gerber, M., Wood, R. & Leonard, A. R. (1970). Outbreak of food poisoning caused by *Bacillus cereus. Publ. Health Dept.* 86, 45.

Nikodemusz, I. (1958). *Bacillus cereus* als Ursache von Lebensmittelvergiftungen. *Z. Hyg. InfektKrankh.* 145, 335.

Nikodemusz, I., Bodnar, S., Bojan, M., Kiss, M., Kiss, P., Laczko, M., Molnar, E. & Papay, D. (1962). Aerobe Sporenbildner als Lebensmittelvergifter. *Zentbl. Bakt. Abt. I.* 184, 462.

Nokes, E. J. (1965). "Bitty" or broken cream. *Dairy Ind.* 30, 870.

Ormay, L. & Novotny, T. (1968). The significance of *Bacillus cereus* food poisoning in Hungary. *Proc. VI Intern. Symp. Food Microbiol.* 279.

Ormay, L. & Novotny, T. (1970). Über sogenannte unspezifischen Lebensmittelvergiftungen in Ungarn. *Zentbl. Bakt. Abt. I.* **215**, 84.

Pisu, I. & Stazzi, L. (1952). *Nuovi Ann Ig.* **1**, 1 and 14. Cited in *Nikodemusz et al.* (1958).

Plazikowski, U. (1947). Further investigations regarding the cause of food poisoning. *Congr. Int. Microbiol.* **4**, 510.

Stone, M. J. (1952). The action of the lecithinase of *Bacillus cereus* on the globule membrane of milk fat. *J. Dairy Res.* **19**, 311.

Discussion

Rehm

What amounts of *B. cereus* must be present to cause an intoxication?

Davies

The literature on this point would indicate that numbers in the order of 10^6 10^7/g would be required. Would Dr Goepfert concur?

Goepfert

One case is recorded where numbers were as low as 300,000/g but usually numbers of the order of $10^6 - 10^8$ are required.

Hobbs

The numbers of *B. cereus* in fried and boiled rice involved in U.K. outbreaks were many millions/g.

Olson

Dr Davies, how did you eliminate the activation effect on residual spores or spores normally present in milk from the activation effect on milk *per se* when you heated the milk prior to adding the test spores to such heat-treated milk?

Davies

The germination of *added B. cereus* spores was followed in dialysates of the heat-treated milks, therefore there would be no question of residual spores remaining. Furthermore, even if milks rather than dialysates were used, the numbers of residual spores would be negligible in relation to the numbers of *B. cereus* spores added for germination studies.

Hess

Increasing the temperature and time of pasteurization makes the investigation on influence of heating on spore germination very important especially with regard to competitive growth by non-sporeforming organisms which survive the normal pasteurization procedures.

Davies

Increasing the time or temperature of pasteurization can certainly affect the possibility of germination and outgrowth of surviving spores. Both the activation of the spores and the effect upon the product (germination medium) must be considered in this respect. Presumably in a situation of competition with non-sporeformers, increasing the severity of heat treatment would shift the balance in favour of the spore-forming bacteria. There are in the literature cases suggesting that associative growth of *B. cereus* with non-sporeforming bacteria, e.g. *Str. lactis,* may be of significance.

von Bocklemann

We have found it difficult to remove fat residues from plastic components of milking equipment. In such fat layers bacilli seem to grow or to get enriched. Have you any experience of this problem? In deep-fat fried meat balls the centre temperature may vary some degrees around $70 \cdot 4°$. It is likely that this variation may give different activation to spores? Have you found any psychrophilic spores?

Davies

I have no experience of bacterial growth or deposition in fat layers on plastic surfaces. This can occur of course in crevice conditions though I cannot give any information specifically relating to plastics. In the experiments I mentioned regarding the sources of spores, very few indeed derived from plant surfaces as determined by rinse techniques. Virtually all the spore contamination stemmed from teat surfaces.

The time temperature combinations used for cooking meat balls are presumably very different to the pasteurization conditions about which I talked. All I can say is that at a given temperature increasing the time of treatment can affect activation and that for a given time, altering the temperature can have a similar effect. Depending upon the penetration, altering the temperature some degrees around $70 \cdot 4^{\circ}$ presumably could have an effect on activation. We have just isolated a number of psychrophilic spore formers from milk some of which look like *B. cereus*. Whether they have any significance it is not yet possible to say.

Pathogenicity of *Bacillus cereus*

J. M. GOEPFERT, W. M. SPIRA, B. A. GLATZ AND H. U. KIM

Food Research Institute,
University of Wisconsin,
Madison, Wisconsin,
U.S.A.

The pathogenicity of *Bacillus cereus* has been recognized for over 2 decades. However, the mechanism of this pathogenicity remains unknown. Some time ago, it was suggested that phosphorylcholine generated by the lecithinase of both *Clostridium perfringens* and *B. cereus* was responsible for the symptoms of food poisoning caused by these organisms. Subsequently, it was shown that *Cl. perfringens* produces a protein enterotoxin that elicits the typical food poisoning syndrome. More recently, work in this laboratory has indicated that a protein enterotoxin is also elaborated by *B. cereus*. This protein, termed enterotoxic factor (EF) is synthesized by certain strains of *B. cereus* growing in some but not all laboratory media. The factor is produced most rapidly during exponential growth and is elaborated by the cells into the surrounding environment without lysis of the cells. Cell-free culture filtrates and partially purified derivatives thereof cause fluid accumulation in ligated ileal segments of rabbit intestine and a dermal response in albino guinea pigs. The use of mutant strains of *B. cereus* and specific antisera have shown that EF is separate and distinct from the lecithinase and haemolysin normally produced by *B. cereus*. The assay for and properties of EF are described in this paper.

THE ROLE OF *Bacillus cereus* as a food poisoning agent has been recognized for over 20 years (Hauge, 1950, 1955). However, with the exception of studies by Nygren (1962) and Dack *et al.* (1954), and animal feeding trials by Nikodemusz and his colleagues (1965, 1966), little work has been done on the mechanism of pathogenicity of *B. cereus*. Within recent years, it has been demonstrated that the symptoms of *Clostridium perfringens* food poisoning are due to a protein enterotoxin elaborated in the intestine by sporulating cells of toxigenic strains (Hauschild, 1971; Hauschild *et al.*, 1970). Therefore, the illness caused by *Cl. perfringens* may be termed an "intoxication" rather than an infection. Despite some very significant differences, *B. cereus* and *Cl. perfringens* have much in common. Both are Gram positive, sporeforming, rod shaped, soil inhabitants. Both may cause food poisoning and each species produces a variety of extracellular, biologically active entities. Therefore, it seems plausible that each is pathogenic by a similar if not identical mechanism.

One of the most common methods used to demonstrate enteropathogenicity is the ligated ileal loop procedure. Enteropathogenic organisms, both invasive and toxigenic types, will elicit fluid accumulation in a ligated ileal segment of sensitive test animals. Although mice, chickens, dogs, and pigs have been used, the rabbit is the most popular test animal. This paper will describe the results obtained with cultures of *B. cereus* and *Bacillus thuringiensis* using the rabbit ileal loop test.

The loop test

The surgical procedure employed has been described in the literature (De & Chatterje, 1953; Duncan *et al.,* 1968). Only slight modifications are necessary for application of the test to *B. cereus.* The most important modification involves the age (or size) of the test animal. After the rabbits (New Zealand White in this case) reach 1200 g, they become somewhat refractory to the enterotoxic substances and results are erratic. For this reason, it is desirable to use rabbits weighing from 500–900 g.

When 3 or more of the test loops contain active enterotoxic material, there is little chance that the animals will survive an 18–24 h holding period. Because of this, and after experimental verification, a holding time of 7 h was established (Spira & Goepfert, 1972). Seven hours is sufficient time for loop responses to be significant if the culture is toxigenic. Since the absolute volume of fluid accumulated is not used as a measure of virulence even sub-maximum values of fluid accumulation are adequate for qualitative purposes. A positive loop response after 7 h is defined as that in which the ratio of loop fluid volume (ml) to the loop length (cm) exceeds 0·2.

It has also been found that there is a significant variation in loop response to the same test preparation between rabbits. Consequently, when assaying a given preparation it is advisable to perform the tests in at least 2 and preferably 3 different rabbits. Although the influence of nutrition on the synthesis of enterotoxic material has not been studied in depth, a brief survey of various media for the ability to support toxin production has been conducted. Each medium was tested at 2 temperatures, i.e. 32° and 37° with both shaking and static incubation. The best combination of incubation conditions appeared to be shake cultures at 32°. Toxin production occurred best in brain heart infusion (BHI) broth, with trypticase soy broth, fresh beef infusion and skim milk being progressively less effective. Toxin production was not detected by the loop test when cells were grown in nutrient broth.

The appearance of toxic activity in whole cultures of *B. cereus* is a function of the age of the culture as well as cell numbers. Toxic activity of BHI-grown cultures is first detectable when the cell population reaches 10^6-10^7 cells/ml. Toxin is synthesized continuously until the culture enters the stationary phase when either synthesis ceases or the rate of destruction equals the rate of synthesis. Whole cultures remain fully active for 24 h at 32° with the activity decreasing upon further incubation.

Thus, at present, the loop assay for determining the enteropathogenicity of *B. cereus* strains is performed as follows. The test strain is grown with shaking in BHI broth at 32° for 18 h. Two ml of this culture are injected into ligated ileal sections of several New Zealand White rabbits (500–900 g). The rabbits have been given water but not food during the 48 h preceding surgery. The rabbits are held at room temperature for 7 h, killed and the small intestine excised. The fluid accumulation is measured and an average fluid volume: loop length is computed.

Nature and production of the toxic substances

As indicated above, the enterotoxic material(s) is synthesized and released by logarithmically growing cells. Sterile culture filtrates of BHI-grown cells are almost as active as whole cultures. Most of the data presented below have been compiled from experiments conducted primarily on strain B-4ac, which was obtained from Dr D. A. A. Mossel and was originally isolated from a food poisoning outbreak.

This strain synthesized toxin over a temperature range of 18–44°. Lower temperatures were not tested and the organism did not grow at 45°. Other strains in our possession are able to grow at 49° but toxin synthesis at this temperature has not been assessed. Although toxin is synthesized quite readily in BHI broth, refractile or heat resistant spores are not formed to any significant degree in this medium. It appears that toxin is synthesized during logarithmic growth and that synthesis ceases at an early stage of sporulation.

So far, 25 strains of *B. cereus* have been tested in ileal loops and 20 have been positive. In addition, 10 strains of *B. thuringiensis* have been surveyed and 8 of these have elicited accumulation of fluid in loops.

The toxin is sensitive to trypsin and pronase, and is completely inactivated in 60 min at 37° in the presence of 0·01% of either enzyme. Toxin is destroyed in 30 min at 56° but remains fully active after 30 min at 45°. It is precipitated by 60% saturation with ammonium sulphate and by cold ethanol. Toxin activity is non-dialyzable but diminishes upon concentration of culture filtrates. Antisera prepared in rabbits against crude culture filtrates neutralize the toxic activity of filtrates from the homologous culture and 4 of 5 heterologous cultures thus far tested. One of 5 strains produced toxin that was not neutralized by anti B-4ac serum, indicating that multiple serological forms of toxin exist in a similar fashion to *Clostridium botulinum* and *Staphylococcus aureus*.

One of the most pressing questions is the relation of the toxin to some of the known extracellular products of *B. cereus* such as lecithinase, haemolysin and lethal toxin. Experiments were performed to elucidate whether the enterotoxin was in fact separate and distinct from these other entities. The results are summarized as follows.

Lecithinase (from *B. cereus*) was purchased from General Biochemicals, Chagrin Falls, Ohio. Direct inoculation of 2 ml of lecithinase (10 mg/ml) into the rabbit loops failed to elicit fluid accumulation. Antiserum prepared against this lecithinase did not neutralize the ability of any of the 5 *B. cereus* culture filtrates tested to accumulate fluid in the loops. The lecithinase activity of the filtrates was neutralized by this serum. Lecithinase activity of the culture filtrates was not altered by trypsinization or by heating to 56° for 30 min while loop activity was destroyed by both treatments. Lecithinase-negative mutants were obtained from investigators in Great Britain and on the European continent and these cultures were as loop active as lecithinase-positive strains.

The role of haemolysin in causing loop fluid accumulation was similarly ruled out. Haemolysin-negative mutants were obtained that showed loop activity. Haemolysin was adsorbed from active culture filtrates by mixing in rabbit erythrocytes at 4° (a temperature too low to permit enzymic lysis), holding for 60 min and subsequently removing by centrifugation. Culture filtrates treated in this fashion remained fully active. In addition, haemolytic activity of culture filtrates was destroyed by heating to 45° for 30 min while loop activity was not affected.

The lack of relationship between enterotoxin and the lethal toxin described by Johnson & Bonventre (1967) has not been so easily established. Loop active but non-lethal culture filtrates have not been found among the strains tested to date. However, we have 2 cultures that produce demonstrable lethal toxin but do not evoke a loop response. Ten-fold concentration of the culture filtrate of one of these organisms did not produce a loop-active preparation in the only trial to date. Whether this is a reflection of a differential sensitivity of the 2 assays or is evidence that the 2 activities are distinct has not been determined. Consequently, the non-identity of enterotoxin and the Johnson & Bonventre lethal toxin has not yet been unequivocally established.

Dermal reaction to culture filtrates

While the ileal loop test is satisfactory for the qualitative demonstration of potential pathogenicity, it is cumbersome for quantitative determinations of activity and for purification work. Therefore, an alternate procedure for demonstrating enterotoxic activity of *B. cereus* was sought. Increased skin capillary permeability (Craig, 1965) and erythemal reaction (Hauschild, 1970) to intradermal inoculation of culture filtrates of *Vibrio cholerae* and *Cl. perfringens* have been described. To evaluate the potential usefulness of these reactions for testing *B. cereus* strains, 0·05 ml of loop-active culture filtrates were inoculated intradermally into the shaved back and flanks of albino guinea pigs. The results of these trials showed that *B. cereus* filtrates did not elicit increased permeability of skin capillaries. An erythemal response was obtained but this was found to be a nonspecific reaction unrelated to the enterotoxic activity. Loop active preparations caused a dermonecrotic reaction which was shown to be separate and distinct from the reactions caused by haemolysin and lecithinase. Skin necrosis was prevented by preincubation of the test material with anti-serum prepared against B-4ac culture filtrate. Only filtrates from strains of *B. cereus* that were loop-active evoked a dermonecrotic response. Dilution experiments have demonstrated that the skin test is about 40 times more sensitive than the loop test and, to a degree, more simple to perform.

Relationship of B. cereus *to food poisoning*

As yet, we have not performed feeding trials with human volunteers and thus have not proven unequivocally that the toxin causing fluid accumulation in rabbit ileal loops is the factor responsible for food poisoning symptoms in human beings. The case for the toxin's role in food poisoning is circumstantially supported by the fact that, to the authors' knowledge, all organisms able to cause fluid accumulation in rabbit ileal loops are also able to evoke gastro-intestinal symptoms in man.

Several pieces of information merit further, brief discussion. First, the demonstration that *B. thuringiensis* has the ability to induce fluid accumulation may be of considerable importance to food microbiologists because of the possible use of these organisms for biological control of insects on certain crops. Application of these organisms may have to be re-evaluated if it is proved that the loop reaction observed is related to human food poisoning.

Secondly, experiments in our laboratory have indicated that toxin synthesis does not occur in the ileal loop of rabbits even though there is a significant increase in cell number during the test period. This would suggest that for symptoms to be observed in man, the afflicted would have to ingest toxin that was synthesized by *B. cereus* in the food product prior to consumption. It would then follow that large numbers (e.g. $>10^6$) of *B. cereus* would develop in the food before there was a potentially dangerous situation. In virtually all of the outbreaks of *B. cereus* poisoning that have been described in the literature a high number of cells or spores of this organism has been found (Goepfert *et al.,* 1972).

It is obvious that the surface of the problem has only been scratched. Much more remains to be done. Work on the purification, properties, serology, pharmacology, and genetics of *B. cereus* toxin is in progress.

References

Craig, J. P. (1965). A permeability factor (toxin) found in cholera stools and culture filtrates and its neutralization by convalescent cholera sera. *Nature (Lond.)* **207**, 614.

Dack, G. M., Sugiyama, H., Owens, F. J. & Kisner, J. B. (1954). Failure to produce illness in human volunteers fed *Bacillus cereus* and *Clostridium perfringens. J. Infect. Dis.* **94**, 34.

De, S. N. & Chatterje, D. N. (1953). An experimental study of the mechanism of action of *Vibrio cholerae* on the intestinal mucous membrane. *J. Path. Bact.* **66**, 559.

Duncan, C. L., Sugiyama, H. & Strong, D. H. (1968). Rabbit ileal loop response to strains of *Clostridium perfringens. J. Bact.* **95**, 1560.

Goepfert, J. M., Spira, W. M. & Kim, H. U. (1972). *Bacillus cereus*: Food poisoning organism. A review. *J. Milk Fd Technol.* **35**, 213.

Hauge, S. (1950). Matforgiftninger fremkalt av *Bacillus cereus. Nord. hyg. Tidskr.* **31**, 189.

Hauge, S. (1955). Food poisoning caused by aerobic spore-forming bacilli. *J. appl. Bact.* **18**, 591.

Hauschild, A. H. W. (1970). Erythemal activity of the cellular enteropathogenic factor of *Clostridium perfringens* type A. *Can. J. Microbiol.* **16**, 651.

Hauschild, A. H. W. (1971). *Clostridium perfringens* enterotoxin. *J. Milk Fd Technol.* **34**, 596.

Hauschild, A. H W., Niilo, L. & Dorward, W. J. (1970). Enteropathogenic factors of food-poisoning *Clostridium perfringens* type A. *Can. J. Microbiol.* **16**, 331.

Johnson, C. E. & Bonventre, P. F. (1967). Lethal toxin of *Bacillus cereus.* I. Relationships and nature of toxin hemolysin, and phospholipase. *J. Bact.* **94**, 306.

Nikodemusz, I. (1965). Die Reproduzierbarkeit der von *Bacillus cereus* virusacteten Lebensmittelvergiftungen bei Katzen. *Zentbl. Bakt.* **196**, 81.

Nikodemusz, I. (1966). Comparaison du pouvoir pathogene experimental de *Bacillus cereus* et de *Bacillus laterosporous* administeres par voie orale. *Annls Inst. Pasteur (Lille),* **17**, 229.

Nygren, B. (1962). Phospholipase C-producing bacteria and food poisoning. *Acta Pathol. Microbiol. Scand.* Suppl. **160**, 1.

Spira, W. M. & Goepfert, J. M. (1972). *Bacillus cereus*-induced fluid accumulation in rabbit ileal loops. *Appl. Microbiol.* **24** (In press).

Discussion

Hobbs

Incubation periods in *B. cereus* outbreaks in the U.K. have been as short as 2 h (rice outbreaks) whereas in Norway incubation periods of 15 h are cited (cornflower outbreaks). I should like to question the figures from Hungary in relation to *B. cereus* outbreaks associated with meat products and query the *Cl. perfringens* figures.

Goepfert

I'm quite surprised by the 2 h incubation period characterizing the fried rice outbreak. It would seem to suggest that perhaps other agents or toxic factors were present in addition to *B. cereus.* Any comment I might make concerning the validity of the Hungarian outbreak of *B. cereus* would be pure speculation. I am not aware of the incidence of *Cl. perfringens* food poisoning in Hungary.

Riemann

Dr Goepfert, did you state that the loop-positive principle is produced only by cells committed to sporulation?

Goepfert

No! I said that the factor is produced by logarithmically growing cells and seems to be synthesized before the cells are committed to sporulation. The factor is excreted by log phase cells and appears to be retained within stationary phase or sporulating cells. Its role in the cell is not known.

Microbiological Quality Assurance for Weaning Formulae

D. A. A. Mossel

The Catholic University,
Louvain,
Belgium

G. A. Harrewijn

Central Institute for Nutrition and Food Research TNO,
Zeist,
The Netherlands

F. J. van Sprang

Wilhelmina Kinderziekenhuis,
University of Utrecht,
The Netherlands

UNICEF'S efforts to provide child protection are concentrated mostly on the supply of highly nutritious, dried formulae foods to be manufactured in developing areas, if possible from locally available staple ingredients. In view of the low immunity of young infants to enteric infections, careful sanitary monitoring of this project is required. Following Sir Graham Wilson's principle, assurance of good microbiological quality is based primarily on the formulation, checking and implementation of an adequate code of good manufacturing practice. The weakest link in the system is the reconstitution phase, which is often beyond the control of local health authorities. None the less, epidemiological evidence suggests that errors committed at this stage are infrequent because there are recommendations to: (i) boil the reconstituted formulae before feeding; and (ii) discard all left-overs. The assurance of prevention is thought to be effective and only limited laboratory testing is required. For the examination of the dried formulae in the field, three simple presence or absence tests (total count, *Enterobacteriaceae* and *Staphylococcus aureus*) are recommended. The tests for water used for reconstitution follow the W.H.O. recommendations, with an additional test for *Pseudomonas aeruginosa* by enrichment in nitrofurantoin broth at 42° followed by streaking on to glycerol mannitol acetamide cetrimide agar, also incubated at 42°. This method is satisfactory for the rapid detection of regular as well as apyocyanogenic types which frequently occur.

Introduction

THE IMMUNITY of young infants against various infective agents is absent or impaired. Those that are under-nourished and/or otherwise sick are in an even worse situation (Scrimshaw, 1970; Selvaraj & Bhat, 1972; Mathews *et al.*, 1972). Such children are therefore particularly liable to food-borne bacterial, viral and

77

helminthic enteric infections. These diseases cannot be controlled effectively by vaccination, as may be possible in some other diseases threatening youthful populations. Hence, the only mode of control is to feed such infants exclusively with microbiologically wholesome commodities. When such children are breast-fed, there is, as a rule, almost no risk of enteric infections (Bullen & Willis, 1971). However, at the moment breast-feeding is abolished, for one reason or another, risks are magnified. Appropriately therefore, UNICEF has concentrated on weaning foods in its efforts to provide child protection through nutrition. These foods should follow breast feeding at the age of 4–12 months (Jelliffe, 1971; Buffa, 1971; Viteri & Bressani, 1972). Rightly also UNICEF has chosen dried formulae, particularly mixtures of dried ingredients that can for the greater part be manufactured from locally available crops in developing countries (Barja et al., 1971; Pak & Barja, 1971; Sadre et al., 1972). This approach, however, well justified from the viewpoints of economy and ecology, is nevertheless fraught with risks of contamination; developing areas invariably suffer from low standards of general sanitation for both educational and economic reasons. From the inception the UNICEF Programme has paid careful attention to microbiological hazards. This paper will review the achievements over about 10 years in the development of a tight system to ensure good microbiological quality for dried weaning foods manufactured for UNICEF.

The Dried Formulae

Weaning foods of the type under consideration invariably contain cereals and pulses as the main ingredients. These require an enzymic predigestion because infants cannot tolerate great amounts of starch. The UNICEF manufacturing project recommends the use of thermotolerant amylase preparations. If this were replaced by a mesophilic predigestion step, the so-called pre-process proliferation of various bacteria would be almost inevitable. Many of the metabolites formed when this occurs, e.g. staphyloenterotoxins (Denny et al., 1971), and pressor amines (Mossel, 1968), are thermostable and they would not be inactivated by terminal pasteurization. An additional advantage in the use of a thermotolerant amylase is that the starch digestion involves heating the mixture to 60–85° for a minimum of 40 min, which results in an intensive pasteurization of the local raw materials which are often highly contaminated.

The subsequent steps in manufacture are mostly the conventional roller dehydration processes used for dairy products. The apparatus used for these purposes has a good sanitary history and requires no complex handling. The only hazards encountered in the UNICEF process are : (i) recontamination during the ultimate phase, wherein the dried cereal mixture is blended with skim milk powder, vitamins and some other ingredients; and (ii) infestation of the plant by insects and rats. It appeared possible to overcome these problems by: (i) thorough sanitation of the blending area; and (ii) proofing the premises against rats and mice and providing an effective system of metal screens against flies. Obviously, similar risks must be eliminated from the packaging and storage operations, but here again there is certainly no lack of technological knowledge. In summary, the manufacture of dried weaning foods is a classical example of preventive quality assurance (Recommended International Code of Practice, 1969) and hence a faithful application of Sir Graham Wilson's recommendation of 1955 (Wilson, 1955). Monitoring the process is therefore relatively easy for the product is so intensively heat-processed that the only hazard to be controlled is post-process recontamination.

Since this may stem from enteric as well as dermal foci, the tests to be applied to the end products are obvious: the Presence or Absence (P.A.) test for *Enterobacteriaceae* (Mossel, 1967) and another for *Staphylococcus aureus* (Giolitti & Cantoni, 1966; Asperger, 1971; Ovejero del Agua & Fernandez, 1971). In addition to these health checks, the technologists responsible for UNICEF plants have often requested somewhat more elaborate examinations to provide guidance in improving the mode of processing when disappointing results are obtained. This aspect will be dealt with later.

Reconstitution

However safe the dried formulae may be, the infants may nevertheless be exposed to massive infective challenge doses of organisms, namely when the water used for reconstitution is not of impeccable microbiological and chemical quality (Scardino, 1971) and particularly when the rehydrated formulae are kept at a temperature allowing microbial proliferation for any significant period (Huhtanen *et al.*, 1972). These faults still occur frequently in the preparation of infant formulae in much more advanced areas (Mossel & Weijers, 1957; Kepler & Fiedler, 1970; Stockhausen, 1971). They have been fully acknowledged by UNICEF from the beginning and attempts have been made to control such hazards by educational campaigns coupled with the launching of dried weaning food.

The wholesomeness of public water supplies cannot be taken for granted in many areas served by UNICEF. Here again there is little need for repetitive analytical work. Where the regular simple examination for *E. coli* and Lancefield group D streptococci indicates a potential lack of safety, chlorination must be applied and systematically monitored where possible by determination of the chlorine decontamination efficiency (log. count before—log. count subsequent to chlorination) (Perkins & Mossel, 1971). Clearly, the most difficult and as yet unresolved aspect of quality assurance is the avoidance of microbial proliferation subsequent to reconstitution, because this rests with the mothers. This danger was overcome in former times (Marriott & Davidson, 1923; Field, 1924) by supplying only acidified formulae, where mishandling would at worst lead to spoilage by lactic acid bacteria, yeasts and occasionally moulds, but virtually never to the growth of pathogenic bacteria.

According to Table 1, UNICEF weaning foods have pH values around neutrality and hence could lead to food-borne enteric disease if mishandled after reconstitution. So far, however, the epidemiological record of the formula has been impeccable, most probably because the reconstituted products are thoroughly cooked before being fed to infants.

Should this situation change, the inclusion of a suitable antimicrobial preservative may be considered. However, no physiologically acceptable synthetic substance giving sufficient protection against Gram positive as well as Gram negative enteric pathogens are available (Mossel, 1971). The use of non-toxic factors inhibitory to Gram positive bacteria only, e.g. tylosin, could

Table 1

Properties of a typical UNICEF weaning formula: "Superamine" as manufactured in the U.A.R.

Ingredients (%): chick pea flour, 38; hard wheat flour, 28; lentil flour, 18; dried skim milk, 10; sucrose, 5; nutritional additions, 1 pH = 6·5–6·6	
Moisture (%)	4·0–5·0
Water activity (a_w)	<0·40
Protein (%)	20·5–21·0*
Total lysine (%)	1·0–1·2
Available lysine (%)	0·9–1·1
Methionine	0·03
Lipids (%)	3·0–3·5
Carbohydrates (%)	
Total, calc. as starch	57–58
Digestible, calc. as starch	52–54
Dextrins	12–14
Reducing sugars, calc. as maltose	14–15
Glucose	1 –3
Calories/100 g	410–420
P.E.R.	0·3–0·4
N.P.U.	64–70
Vitamin A (I.U./kg)	20,000
Vitamin D_2 (I.U./kg)	4,000†
Aneurin (mg/kg)	4
Riboflavin (mg/kg)	8
Pyridoxin (mg/kg)	5
Nicotinicacid (mg/kg)	50
Ascorbic acid (mg/kg)	500

* Nutritionally considered as rather elevated; † nutritionally considered as suboptimal.

increase rather than decrease the hazards by clearing the way for the growth of enteric pathogens of the *Enterobacteriaceae* group which are many and widespread in developing areas of the world. Inclusion in the formulae of sufficient citric or tartaric acid to reduce the pH of the liquid food to below 4·5 might then be reconsidered, demonstrating the wisdom of some paediatric Stone Age principles (Marriott & Davidson, 1923; Field, 1924).

Microbiological Standards

Dried formulae

As in all instances of microbiological quality assurance. the use of microbiological standards cannot achieve quality control, but they are most useful in assessing whether the applied quality assurance is adequate. In this context microbiological standards for dried weaning foods have been laid down by the present authors at the request of UNICEF and in consultation with W.H.O. The specifications, readily and regularly attainable where good sanitary practices are followed, are summarized in Table 2. The ability to attain such levels has been demonstrated elsewhere (Powers *et al.*, 1971; Naguib *et al.*, 1972).

Table 2

Microbiological specification for UNICEF dried weaning formulae

Class	Group	Organisms	Limits	
			Aimed at‡	Maximum ever tolerated
I*	1	*Salmonella, Arizona* / *Edwardsiella* / *Shigella*	absent in § 15 g	
	2	E. coli	absent in 10 g	1 g
	3	Staph. aureus	absent in 1 g	
	4	B. cereus, per 1 g	$<10^2$	$<10^3$
	5	Mould spores, per 1 g	$<10^2$	$<10^2$
	6	*Enterobacteriaceae*	absent in 1g	0.1g
II†	7	Total aerobic count, per 1 g	$<10^4$	$<10^5$
	8	Aerobic spore count. Virtually equals aerobic count, per 1 g		
	9	Lancefield D streptococci, per 1 g	$<10^2$	$<10^3$
	10	*Clostridium* species, per 1 g	<10	$<10^2$

*Pathogenic or sometimes pathogenic organisms; † indicator organisms; ‡ not to be encountered in more than 2 out of 10 samples; § "absent in" = not found in the prescribed aliquot.

As clarified later, the entire battery of 10 tests is not carried out regularly. In fact, tests 6 and 3 are the cardinal ones. Tests 1 and 2 serve mainly to assess the seriousness of post-process recontamination with *Enterobacteriaceae* (Mossel, 1967; Drion & Mossel, 1972), should this group be detected. Test 4 is done only when required by epidemiological considerations while examinations 7 to 10 are carried out to suggest to the technologists where to search for areas of microbial build-up in the processing line. Finally, test 5 serves to preclude the germination of mould spores due to water migration into the dried product leading to foci with an a_w value permitting mould growth (Mossel, 1971). Two sets of limits are given in Table 2. The first column represents the desired values, the second the maximum tolerable counts.

Clearly there is no tolerance for pathogenic organisms such as *Salmonella* or *Staph. aureus*, but only for groups of organisms that serve as indicators. Generally, not more than 2 out of 10 samples tested should show counts in excess of those recorded in the first column, and none should exceed the limits of the second column (Mossel *et al.*, 1973).

Water

There are at least 3 reasons why strict standards must be applied to water used to reconstitute weaning food formulae. The first is that, unlike the weaning formulae, water may be used without any prior heat treatment for dilution or cooling of the rehydrated mix. Also, organisms which are generally of no significance in drinking water can be hazardous in this instance, because they may proliferate rapidly in the reconstituted formulae. Finally, water serves as a component of the food provided for a very vulnerable class of the population, as outlined in the Introduction.

For these reasons we apply 1 or 2 supplementary microbiological criteria to those regularly used for drinking water (W.H.O., 1971) to the water supply authorized for use in conjunction with UNICEF weaning foods. The first is a general test for *Enterobacteriaceae* so that lactose-negative types will not be overlooked by testing solely for *E. coli* and for the coli-aerogenes group of bacteria (Mossel, 1967). Lactose-negative types including pathogens occasionally even outnumber the lactose-positive types in water (Seligmann & Reitler, 1965; Boring *et al.*, 1971); they will be detected when a search is made for the entire *Enterobacteriaceae* group. We consider it to be also essential to test for *Pseudomonas aeruginosa*, because its pathogenicity for debilitated infants when absorbed by the oral route seems to be well established (Kubota & Liu, 1971; Weber *et al.*, 1971). Fortunately, a simple, rapid and reliable method is available (Thorn *et al.*, 1971; Mossel & Indacochea, 1971), which detects apyocyanogenic types as well as pigmented strains of *Ps. aeruginosa (vide infra)*.

Review of Available Methodology

Dried formulae

It is always a sound principle to reduce to a minimum the number of tests to be applied to food samples, because it allows the resources of a laboratory to be applied to the testing of more samples. This is especially important in the examination of dried foods, where the contamination is unusually sporadic (Silverstolpe *et al.*, 1961) due to its mode of origin—fortuitous contamination followed by a rather unpredictable proliferation mechanism.

A more pressing reason in this particular instance is that much of this testing must be carried out in or near the manufacturing plants, as both qualified manpower and laboratory facilities are limited in developing regions. We have already noted that, for these reasons, the local bacteriological testing has been limited to the 1 or 2 cardinal tests summarized in Table 2. In addition we have thought it wise to go one step further and attempt to replace a few commonly used plate count tests by the much simpler dilution count or "titre" method of the French Public Health Veterinarian School (Névot *et al.*, 1953; Pantaléon & Bavdeau, 1958). This required some preliminary research into the feasibility of this type of procedure for the present purpose.

Some 30 samples of UNICEF dried weaning foods and similar materials were subjected to comparative testing by the generally applied methodology and by dilution counts (most probable number technique). For this purpose the following techniques were used:

Total counts. 1 ml aliquots of serial dilutions in peptone saline were inoculated in triplicate into 9 ml of tryptone soya peptone broth with 0·005% resazurin added and incubated at $31 \pm 1°$ for 2 days. Tubes in which the medium had become colourless were considered positive (Mossel *et al.*, 1972).

Total spore counts. Same procedure, but with the dilutions heated for 1 min at 80° to eliminate virtually all non-spore-bearing cells (Mossel, 1967).

Counts of Enterobacteriaceae. 1 ml aliquots of serial decimal dilutions in peptone saline were kept at 20–25° for 2 h to allow sublethally impaired cells to be restored to fully robust cells (Mossel & Ratto, 1970; Rosenthal *et al.*, 1972). These were then inoculated in triplicate into 9 ml aliquots of buffered brilliant green bile glucose broth and incubated overnight at $30 \pm 1°$; gas formation, shown by Durham tubes pasteurized with the medium, was considered as the criterion for positive results. The medium was always pretested for the absence of intrinsic inhibitory properties towards non-impaired low populations of *Enterobacteriaceae* (Mossel & Harrewijn, 1972).

Counts of Lancefield Group D streptococci. The procedure was the same as for total counts (above) but using crystal violet azide ("streptosel") broth (Pike, 1945; Schaub *et al.*, 1958) with 0·005% resazurin (Mossel *et al.*, 1973).

The results obtained are summarized in Table 3 and analysed in Table 4. They show that the dilution count technique of the French School gave results that agreed very well indeed with results obtained from conventional techniques. Actually there was a trend towards slightly higher counts with the P.A. tests, so that falsely negative conclusions are unlikely to be a problem.

Table 3

A review of typical results obtained in comparative tests of dried weaning formulae by plate (P) and dilution (D) methods (all data expressed in log$_{10}$ values)

Sample code	Total count		Aerobic spores		Enterobacteriaceae		Lancefield D streptococci	
	P	D	P	D	P	D	P	D
T 6	2·5	2·7	1·8	2·0	1·0	1·0	<2·0	<0·5
T 7	2·7	3·4	2·0	1·6	1·0	1·4	<2·0	<2·0
T 8	3·7	4·7	2·3	2·9	3·0	<4·0	3·0	3·0
T 9	2·7	2·7	2·0	2·7	<1·0	1·4	2·3	2·0
T 10	2·7	3·2	2·8	2·7	1·8	1·2	<2·0	1·0
T 11	2·8	3·0	1·7	2·2	1·9	1·6	2·0	1·3
T 12	3·0	2·7	2·3	2·7	1·0	0·9	2·0	2·0
T 13	3·9	3·4	2·0	2·2	–	–	2·9	3·0
T 14	2·8	2·7	2·0	1·5	<1·0	0·6	2·8	3·0
T 15	3·6	3·7	2·3	3·7	2·0	1·6	<2·0	2·4
T 16	2·7	4·0	2·5	4·0	1·9	1·6	2·5	2·4
T 17	2·5	2·9	2·3	3·0	1·0	1·0	<2·0	1·4
T 18	2·7	3·0	2·0	3·2	2·0	1·6	<2·0	1·6
T 19	2·7	3·6	2·5	3·2	2·3	2·2	2·0	1·4
T 20	3·0	3·7	2·5	3·0	1·0	1·0	2·0	3·0
T 21	2·5	3·4	2·5	4·0	<1·0	1·4	<2·0	1·2
T 22	2·9	3·0	2·6	3·4	1·7	1·4	2·7	2·4
T 23	3·6	4·0	2·3	3·0	<1·0	<0·5	2·8	2·3
T 24	3·9	3·7	3·0	3·4	1·5	3·0	2·8	<3·0
T 25	3·3	3·5	3·0	3·0	<1	<0·5	2·3	1·9
T 26	3·6	4·0	2·8	2·5	1	<0·5	2·9	2·7
S 10	4·3	4·2	2·5	2·0	<2	1·6	2·5	1·6
S 11	3·7	3·6	2·9	2·6	2·3	1·6	2·3	3·0

Water

The methods currently used for the sanitary surveillance of water supplies are mostly dilution techniques (W.H.O., 1971). The total count can also, in this instance, be replaced by a simple dilution technique as shown by the results obtained with dried formulae, summarized in the first column of Table 3. The P.A. test for *Ps. aeruginosa* is also simple. One and 10 ml aliquots enriched overnight at 42° in peptone water containing 100 μg/ml of nitrofurantoin

Table 4

Evaluation of the results of plating (P) versus dilution (D) methods

	Aerobic colony count	Aerobic spore count	Entero-bacteriaceae	Lancefield D streptococci
Number of samples yielding results fit for comparison	38	38	26	26
Percent of samples with				
$P > D^1$	10	15	46	39
$P = D^2$	38	19	39	46
$P < D^3$	52	66	15	15
Percent of samples with				
$P \ll D^4$	10	20	7	7
$D \ll P^5$	0	0	0	0

[1] $P > D; P - D > 0\cdot2$ (log colony forming units)
[2] $P = D; P - D \leqslant 0\cdot2$
[3] $P < D; D - P > 0\cdot2$
[4] $P \ll D; D - P \geqslant 1\cdot0$
[5] $D \ll P; P - D \geqslant 1\cdot0$

(Thorn *et al.*, 1971) and positive tubes are streaked on to plates of glycerol mannitol acetamide cetrimide (GMAC) agar (Mossel & Indacochea, 1971). The GMAC plates are incubated overnight at $42°$ and prolific growth surrounded by red halos is considered to be positive evidence of the presence of *Ps. aeruginosa* in the water sample under examination (Mossel & Indacochea, 1971). This procedure detects apyocyanogenic strains in addition to regular ones, which is useful as the former occur more frequently than is often recognized (Taylor & Whitby, 1964; Gilardi, 1968; Azuma & Witter, 1970).

References

Asperger, H. (1971). Zum Nachweis Koagulase-positiver Staphylokokken. II. Mitteilung: Koagulase-positive Staphylokokken in Trockenmilchprodukten. *Öst. Milch u. fettw.* **26**, 41.

Azuma, Y. & Witter, L. D. (1970). Pyocyanine degradation by apyocyanogenic strains of *Pseudomonas aeruginosa. Can. J. Microbiol.* **16**, 395.

Barja, I., Munoz, P., Solimano, G., Vallejos E., Undurraga, O. & Tagle, M. A. (1971). Formula de garbanzo (Cicer arietinum) en la alimentacion del lactante sano. *Arch. Latino-Amer. Nutricion* **21**, 485.

Boring, J. R., Martin, W. T. & Elliott, L. M. (1971). Isolation of *Salmonella typhimurium* from municipal water, Riverside, California, 1965. *Am. J. Epidem.* **93**, 49.

Buffa, A. (1971). Food Technology and Development: Processing low-cost nutritious native foods for world's hungry children. *Food Engng* **43**, No. 11, 79; No. 12, 61.

Bullen, C. L. & Willis, A. T. (1971). Resistance of the breast-fed infant to gastroenteritis. *Br. Med. J.* **3**, 338.

Denny, C. B., Humber, J. Y. & Bohrer, C. W. (1971). Effect of toxin concentration on the heat inactivation of staphylococcal enterotoxin A in beef bouillon and in phosphate buffer. *Appl. Microbiol.* **21**, 1064.

Drion, E. F. & Mossel, D. A. A. (1972). Mathematical-ecological aspects of the examination for *Enterobacteriaceae* of foods processed for safety. *J. appl. Bact.* **35**, 233.

Field, M. C. (1924). A clinical study of lactic acid milk as a routine feeding for sick and healthy infants. *Archs Pediat.* **41**, 541.

Gilardi, G. L. (1968). Diagnostic criteria for differentiation of pseudomonads pathogenic for man. *Appl. Microbiol.* **16**, 1497.

Giolitti, G. & Cantoni, C. (1966). A medium for the isolation of staphylococci from foodstuffs. *J. appl. Bact.* **29**, 395.

Huhtanen, C. N., Brazis, A. R., Arledge, W. L., Cook, E. W., Donnelly, C. B., Ginn, R. E., Jezeski, J. J., Pusch, D., Randolph, H. E. & Sing, E. L. (1972). Effects of time of holding dilutions on counts of bacteria from raw milk. *J. Milk Fd Technol.* **35**, 126.

Jelliffe, D. B. (1971). Approaches to village-level infant feeding. VI. The essential characteristics of weaning foods. *Envir. Child Hlth* **17**, 171.

Kepler, H. & Fiedler, K. (1970). Untersuchungen zur Bedeutung der Milchküche für die Keimübertragung in Kindereinrichtungen. *Z. ges. Hyg.* **16**, 584.

Kubota, Y. & Liu, P. V. (1971). An enterotoxin of *Pseudomonas aeruginosa. J. infect. Dis.* **123**, 97.

Marriott, W. M. & Davidson, L. T. (1923). Acidified whole milk as a routine infant food. *J. Am. med. Ass.* **81**, 2007.

Mathews, J. D., Whittingham, S., Mackay, I. R. & Malcolm, L. A. (1972). Protein supplementation and enhanced antibody-producing capacity in New Guinean school-children. *Lancet ii,* 675.

Mossel, D. A. A. & Weijers, H. A. (1957). Uitkomsten, verkregen bij bacteriologisch onderzoek van vrouwenmelk van diverse herkomst en de betekenis daarvan voor de pediatrische praktijk. *Maandschr. Kindergeneesk.* **25**, 37.

Mossel, D. A. A. (1967). Ecological principles and methodological aspects of the examination of foods and feeds for indicator microorganisms. *J. Ass. off. analyt. Chem.* **50**, 91.

Mossel, D. A. A. (1968). Bacterial toxins of uncertain oral pathogenicity. In: *The Safety of Foods: International Symposium, University of Puerto Rico, 1968,* Eds J. C. Ayres *et al.* Westport, Conn., AVI Pub. Co., p. 168.

Mossel, D. A. A. & Ratto, M. A. (1970). Rapid detection of sublethally impaired cells of *Enterobacteriaceae* in dried foods. *Appl. Microbiol.* **20**, 273.

Mossel, D. A. A. (1971). Physiological and metabolic attributes of microbial groups associated with foods. *J. appl. Bact.* **34**, 95.

Mossel, D. A. A. & Indacochea, L. (1971). A new cetrimide medium for the detection of *Pseudomonas aeruginosa. J. med. Microbiol.* **4**, 380.

Mossel, D. A. A. (1971). Ecological essentials of antimicrobial food preservation. In *Microbes and Biological Productivity. 21st Symposium of the Soc. Gen. Microbiol.,* Eds D. E. Hughes & A. H. Rose, pp. 177–195. Cambridge: The University Press.

Mossel, D. A. A. & Harrewijn, G. A. (1972). Les défaillances dans certains cas des milieux d'isolement des *Enterobacteriaceae* des aliments et des médicaments secs. *Alimenta* **11**, 29.

Mossel, D. A. A., Shennan, J. L. & Vega, C. (1973). The bacteriological condition of animal feeds: a survey to aid in determining product standards for proteinaceous feed ingredients. *J. Sci. Fd Agric.* **24**, 499.

Naguib, K., Abd-el-Ghani, S. & Taha, S. M. (1972). Spray dried milk of the Sakha plant. II. Identification of predominating micro-organisms. *J. Milk Fd Technol.* **35**, 45.

Névot, A., Pantaléon, J. & Rosset, R. (1953). Contrôle bactériologique systématique des préparations des viandes. *Bull. Acad. Vét. France* **26**, 331.

Ovejero del Agua, S. & Suarez Fernandez, G. (1971). Récherche de staphylocoques pathogènes dans du lait en poudre. *Le Lait* **51**, 294.

Pak, N. & Barja, I. (1971). Mezclas de alimentos de adecuado valor proteico recomendables para la alimentacion del lactante y preescolar. *Arch. Latino-Amer. Nutricion* **21**, 473.

Pantaléon, J. & Bavdeau, H. (1958). Le contrôle bactériologique des viandes et des produits carnés dans la pratique journalière: organisation, méthodes, résultats. *Bull. Soc. Sci. Hygiène Aliment.* **46**, 137.

Perkins, F. T. & Mossel, D. A. A. (1971). The testing of disinfectants. *Alimenta* **10**, 67.
Pike, R. M. (1945). The isolation of hemolytic streptococci from throat swabs. Experiments with sodium azide and crystal violet in enrichment broth. *Am. J. Hyg.* **41**, 211.
Powers, E. M., Ay, C., El-Bisi, H. M. & Rowley, D. B. (1971). Bacteriology of dehydrated space foods. *Appl. Microbiol.* **22**, 441.
Recommended International Code of Practice (1969). *General Principles of Food Hygiene.* Rome: F.A.O.
Rosenthal, L. J., Martin, S. E., Pariza, M. W. & Iandolo, J. J. (1972). Ribosome synthesis in thermally shocked cells of *Staphylococcus aureus. J. Bact.* **109**, 243.
Sadre, M., Payan, R., Donoso, G. & Hedayat, H. (1972). Protein food mixture for Iran. *J. Am. diet. Ass.* **60**, 131.
Scardino, P. T. (1971). Bacterial diarrhea in eastern Nicaragua. *South. Med. J.* **64**, 823.
Schaub, I. G., Mazeika, I., Lee, R., Dunn, M. T., Lachaine, R. & Price W. H. (1958). Ecologic studies of rheumatic fever and rheumatic heart disease. I. Procedure for isolating beta haemolytic streptococci. *Am. J. Hyg.* **67**, 46.
Scrimshaw, N. S. (1970). Synergism of malnutrition and infection. Evidence from field studies in Guatemala. *J. Am. med. Ass.* **212**, 1685.
Seligmann, R. & Reitler, R. (1965). Enteropathogens in water with low *Escherichia coli* titer. *J. Am. Wat. Wks Ass.* **57**, 1572.
Selvaraj, R. J. & Bhat, K. S. (1972). Metabolic and bactericidal activities of leukocytes in protein-calorie malnutrition. *Am. J. clin. Nutrition* **25**, 166.
Silverstolpe, L., Plazikowski, U., Kjellander, J. & Vahlne, G. (1961). An epidemic among infants caused by *Salmonella muenchen. J. appl. Bact.* **24**, 134.
Stockhausen, H. B. von (1971). Welchen Platz hat die Frischmilch heute in der Ernährung des jungen Säuglings? *Ernährungsumschau* **18**, 101.
Taylor, J. J. & Whitby, J. L. (1964). *Pseudomonas pyocyanea* and the arginine dihydrolase system. *J. clin. Path.* **17**, 122.
Thorn, A. R., Stephens, M. E., Gillespie, W. A. & Alder, V. G. (1971). Nitrofurantoin media for the isolation of *Pseudomonas aeruginosa. J. appl. Bact.* **34**, 611.
Viteri, F. E. & Bressani, R. (1972). The quality of new sources of protein and their suitability for weanlings and young children. *Bull. Wld Hlth Org.* **46**, 827.
Weber, G., Werner, H. P. & Matschnigg, H. (1971). *Pseudomonas aeruginosa* im Trinkwasser als Todesursache bei Neugeborenen. *Zentbl. Bakt. Abt. I.* **216**, 210.
Wilson, G. S. (1955). Symposium on food microbiology and public health: general conclusion. *J. appl. Bact.* **18**, 629.
World Health Organization (1971). *International Standards for Drinking-Water*, 3rd ed. Geneva.

Discussion

de Groote

Why a standard for *B. cereus* and not for *Cl. perfringens*, but only for total numbers of clostridia?

Mossel

In order to save labour and money we have contented ourselves with a total clostridial count rather than apply the more complicated procedure for *Cl. perfringens*. So far most results have been negative, i.e. $<10/g$ for total clostridia; hence there have been $\leqslant 10/g$ *Cl. perfringens*.

Jarvis

Do you routinely screen the raw materials in the Superamine mix for the presence of mycotoxin residues and what levels of mycotoxins do you accept as satisfactory?

Mossel

We screen the raw materials for the absence of visible mould spoilage and, in addition, do some spot checks on viable mould counts. Although this should afford sufficient protection against mycotoxin risks we have also examined occasional samples for aflatoxins; as expected, the results were negative.

Foster

What size sample do you recommend for salmonella testing and why do you test for Group D streptococci?

Mossel

We examine 25 g aliquots for the absence of *Salmonella, Arizona, Edwardsiella* and *Shigella* in combination with the customary systematic testing for *Enterobacteriaceae*; this secures more than ample consumer protection [Drion, E. F. & Mossel, D. A. A. (1972). *J. appl. Bact.* **35**, 233] Examination for Group D streptococci is done for the following purposes: (1) verification of proper terminal processing; (2) checking on the absence of significant microbial proliferation both before and after heat processing. (Note: testing for pathogens and Lancefield Group D streptococci is done in special instances only, cf. Table 2 in paper).

Semi-Preserved Foods: General Microbiology and Food Poisoning

V. BARTL

*Hygiene Laboratories,
District Hygienic Station,
Prague,
Czechoslovakia*

The term semi-preserved is used for perishable foods where the shelf-life is prolonged by physical, physico-chemical or biological means. Only some such foods are commonly described as semi-preserved and opinions differ as to what is and what is not a semi-preserved food. Three typical kinds of semi-preserved food are reviewed, the pasteurized meat products, with and without brine, the fish preserves and sweetened condensed milk. Each of them has its characteristic microbiological association and spoilage micro-organisms. Microbiological standards, a help in safety control, are discussed. Food poisoning due to these types of foods is rather rare. Typical examples of intoxications are given.

SEMI-PRESERVED foods are products which, by application of physical, chemical or biological methods, have acquired a marked increase in shelf-life compared to the non-preserved, raw product (Sanker, 1970). This means that pasteurized milk as well as sauerkraut are semi-preserved foods, although not generally described as such.

For convenience semi-preservation will be defined as a process sufficient to ensure that food will neither spoil nor present a microbial health hazard after incubation for from 1 to 10 days, depending on the process used for preservation, at 18–20°. In Belgium the "Sous-commission des semi-conserves" divided these foods into various groups, according to the type of packing and storage conditions that these products need. Unfortunately the definition of semi-preserved foods was omitted (Report, 1970).

It must be emphasized that semi-preservation produces, microbiologically, new types of food, which due to their intrinsic parameters (Mossel & Ingram, 1955) have a different microflora from that of the raw material.

In commercial as well as scientific practice, pasteurized, cured meat products, fish preserves and, in our country, sweetened condensed milk are designated as semi-preserved foods. This paper will be confined to these three types of food.

Pasteurized Meat Products

The semi-preserved foods which are most thoroughly studied are the pasteurized meat products, i.e. hams, frankfurters and some other products. Their technology includes soaking in brines and pasteurization at 80–105°. They are packed in cans, often under vacuum. The interior of the raw meat is not always sterile; for example, 20% of raw meat for ham production and 31% for shoulder production were found by Cervenka (1966) to be contaminated. The meat is either dry or wet salted, and there is no microbiological difference between these 2 methods (Juchniewicz, 1961).

The salt content and low temperature of brine favour the growth of lactobacilli, micrococci and vibrios. The pseudomonads are excluded by salt concentrations of about 5% (Ingram & Kitchell, 1967), but salt favours growth of micrococci and lactobacilli (Goldman et al., 1963). Lactobacilli lower the pH and reduce nitrates to nitrites (Buttiaux, 1963; Leistner, 1958a) and they can influence the development of flavours and of off-flavours (Sharpe, 1962a, b). The micrococci also may influence the flavour, but this has not been fully proved (Leistner, 1958a; Giolitti et al., 1971a). The micrococci, which may constitute as much as 40% of all micro-organisms present, withstand a brine concentration of 15% (Kitchell, 1962). Pohja (1962) divided them into 6 groups, of which he specified only Staphylococcus lactis and Staph. saprophyticus. A great many types of micro-organisms, but few gram-negative rods, were found in brine by Leistner (1959). Experimental attempts to produce better flavours by means of pure cultures were unsuccessful (Michailova, 1971), as it was difficult to determine the influence of single strains in normal production practise (Pohja, 1958). It is much easier to control the quality of ham by regulating technology or changing additives. Coagulase-negative staphylococci and halophilic vibrios are usually present (Ingram, 1962; Riberio, 1964).

There are more bacteria in old brines and the enzymic activity promotes undesirable changes. Growth of Leuconostoc spp. leads to slime formation and yeasts can form a pellicle (Leistner, 1958b). The yeasts present are mainly Debaryomyces kloeckeri, D. hanseni, D. nicotianae, Candida rugosa and some others. They can influence the flavour of the product and its spoilage (Leistner & Bem, 1970). The activity of the bacteria changes during ripening of the meat, especially when smoking or some other heat treatment raises the temperature (Diebel et al., 1961).

The aerobic plate counts of cured meats in Czechoslovakia are of the order of 10^4/g, with maximum counts of about 9×10^4/g. In Danish products similar results were obtained (Zeuthen, 1970). In Italian hams, which are only salted, the plate count was $1 \cdot 8 \times 10^6$/g after 60 days at 2° and the maximum count was $2 \cdot 5 \times 10^6$/g, with micrococci and coryneform bacteria predominating (Giolliti et al., 1971b). High counts were, surprisingly, found in cooked ham and hamburgers (10^6/g), and in roast pork (10^5/g) (Walton & Lewis, 1971; Baran et al., 1970). 47·5% of Yugoslav semi-preserved meats were sterile and relatively low counts were recorded in the remainder. This is shown in Table 1.

Table 1

*Aerobic plate counts in non-sterile Yugoslav hams
(Kendereski, 1970)*

Count/g	No. of samples	%
Less than 100	562	42·0
100–1000	438	32·8
1000–10,000	204	15·3
10,000–50,000	89	6·6
50,000 and more	45	3·3

Bacillus subtilis, B. cereus, B. coagulans, B. pumilis and *B. megaterium* were found in most cans. From a similar product in the USSR, *B. cereus* was isolated from 6·1–9·2% of cans (Pivovarov *et al.*, 1970).

When these foods are analysed quantitatively, micrococci, coagulase-negative staphylococci and lactic acid bacteria predominate (Kitchell, 1962). The lactic acid bacteria grow at temperatures below 10° and include representatives of the genera *Streptococcus, Leuconostoc, Pediococcus* and *Lactobacillus* (Cavett, 1963; Reuter, 1967). Enterococci are also present (Zlamalova, 1964) and they can act antagonistically towards some *Clostridium, Bacillus* and *Lactobacillus* spp. (Kafel & Ayres, 1969). Lactobacilli predominate also in the Swedish sausage *isterbad* which is prepared without any heat treatment and is fermented microbiologically (Ostlund, 1970).

The aerobic plate count is not usually a good indicator of shelf-life because the lactobacilli, which form 20–30% of the microflora (Mol *et al.*, 1971) are not usually included in the routine plate count. Their rate of growth does not differ greatly during the first 4 weeks in various meat products kept at 5° (Kempton & Bobier, 1970).

Spoilage micro-organisms

Spoilage of semi-preserved meat products is quite unlike the spoilage of raw meat. Aerobic spore-formers are involved only as secondary microflora. The relatively high salt concentration inhibits not only the halosensitive bacteria, but it is likely that it can influence the microbial metabolism of amino acids, and perhaps suppress lipolytic and proteolytic activities (Dainty, 1971). To this must be added the effect of acid produced by the lactobacilli. It was found that, among bacteria from cured meat, those which were sensitive to acidity were resistant to salt, and those insensitive to salt were sensitive to acidity (Ingram & Kitchell, 1967). Further, the mutagenic effect of nitrite and the inhibitory effect of nitric acid cannot be ignored. The bacterial count in semi-preserved meat products can be 10^6/g without causing deterioration, but a count of 10^7/g seems to be critical and at this point organoleptic changes start.

Spoilage is strongly influenced by the storage temperature: increasing temperature makes spoilage more rapid. Pasteurized crab cake mix may be cited as an example. The shelf-life is stated to be 27 h at 30°, 4 days at 18° and 6 months at 2° (Loaharanu & Lopez, 1970; Vrban, 1971). The chief spoilage organisms at 20° are micrococci and lactobacilli, and, in products with higher salt contents, *Leuconostoc* spp.; at 30° they are coagulase-negative staphylococci and lactobacilli, and a sour or fishy flavour is often produced (Cavett, 1962; Ingram & Dainty, 1971). The micrococci, or enterococci if present, are suppressed during spoilage and the number of lactobacilli, especially the unclassified streptobacteria, increases (Mol *et al.*, 1971; Allen & Foster, 1960; Gardner, 1968). The latter can produce a sweet-sour flavour (Ristic *et al.*, 1967). Coryneforms, spore-forming bacilli and clostridia, most frequently *B. licheniformis, B. coagulans, B. cereus* and *Cl. sporogenes,* can take part in further spoilage (Mol & Timmers, 1970; Pěgřímková *et al.*, 1970; Pivovarov *et al.*, 1970; Eddy & Ingram, 1956; Verhoeven, 1950; Popovic, 1965; Kafel 1964; Maleszewski, 1964). Spoilage due to *Alcaligenes, Achromobacter* and *Brevibacterium* spp. has been reported (Gianelli & Maggi, 1962). A special kind of spoilage of frankfurters in brine, the so-called weakening, has been observed. It was caused by the proteolytic activity of spore-forming bacilli, and could be prevented by better technology. On the whole, bad technology is always followed by a poor microbiological situation.

Food poisoning micro-organisms

For salmonellae a pH of 4·4 is critical and pasteurization temperatures are sufficient to kill them. Exposure to salt before pasteurization increases the heat resistance of salmonellae (Cotterill & Glauert, 1969). Nitrites retard their growth, but the concentrations required to ensure total inhibition are so high that they cannot be used in cured meats (Goepfert & Chung, 1970). In inoculated cans of corned beef *Salmonella typhi* grew well at 37° and at room temperature (Hobbs, B. C. 1970). Salmonellae can survive low temperatures in brines for quite long periods (Ingram & Kitchell, 1967). The growth of salmonellae can be influenced by the composition of meat and an antagonistic effect of short-chained fatty acids was reported by Khan & Katamay (1969).

Cl. botulinum can grow in media containing up to 7% salt; types A and B did not grow in 10% salt and type E was inhibited by 4–5% salt (Spencer, 1966). Nitrites, low pH, and low brine and storage temperatures are further inhibitory factors. To this must be added pasteurization, which impairs spores and/or makes them sensitive to brines. Brine salts alone, or pasteurization alone, is not sufficient for spore inactivation. All these factors interact, for example nitrites are more inhibitory at low pH than at higher levels, but salt is more inhibitory at higher pH. Nitrates are unimportant, as they vanish during heat treatment.

Various studies have been made of the growth of *Cl. botulinum*. It grew in ham with 3·5% salt, but not with 4·6% salt and 60 p/m nitrite (Ajmal, 1968). In laboratory tests, type E was inhibited by 5·0% and one strain by 4·5% salt (Segner *et al.*, 1966). The outgrowth of spores was retarded in jelly with 4·5% salt and a pH of 5·7–5·8 (Blanche-Koelensmid & van Rhee, 1968). Delay in toxin production is directly related to the concentration of nitrite (Pivnick *et al.*, 1967; Roberts & Ingram, 1966; Steinke & Foster, 1951). On the other hand, nitrite up to 2000 p/m had no effect on heated spores. How the nitrite works is still not known (Pivnick, 1970). Packing in plastic pouches had no effect on the growth of *Cl. botulinum* (Pivnick & Bird, 1965), but nitrite concentration had a direct effect in such pouches (Pivnick *et al.*, 1967). The temperature was important, as toxin was produced in inoculated packs at 13° and 23° but not at 3° (Warnecke *et al.*, 1967). Toxic cans did not always swell nor did swollen cans always contain toxin (Anellis *et al.*, 1965). Duncan (1970) reviewed thoroughly all the relevant work on spores and concluded that "inhibition of the initiation of spore germination is an important factor in the stability of the preservative system".

In fully appertized foods, the death of spores during heat treatment is of particular interest, but with semi-preserved foods, pasteurization is not adequate to kill the spores and intrinsic factors affecting the further life of spores are extremely important. Therefore Pivnick (1970) suggested expressing the safety of cured meats, with respect to botulism, by the following equation:

$$P = D + I$$

where P = protection against spores of *Cl. botulinum* that are present in the raw product

   D = destruction of spores during heating

   I = inhibition of growth of spores that survive heating.

This equation is an expression of the complicated dynamic factors in semi-preserved foods.

Staph. aureus may be present in brines. It can withstand salt up to 15% and occasionally even to 20% (Ingram & Kitchell, 1967). It is killed by pasteurization temperatures, if they are maintained for long enough (Silliker *et al.*, 1962). This does not mean that they will be absent indefinitely. Further handling and manipulation of the product may lead to re-infection and *Staph. aureus* can outgrow the other microflora—the lactobacilli are important, as they act antagonistically—and present a renewed health hazard (Minor & Marth, 1971; Eddy & Ingram, 1962). The staphylococci are unable to invade the inner parts of the meat (Elmossalami & Wassee, 1971) and are slightly inhibited in vacuum-packs (Christiansen & Foster, 1965), but they can produce toxin even under these conditions. Higher temperatures favour toxin production (Genigeorgis *et al.*, 1969).

Enterotoxin formed before pasteurization is not inactivated by the heat treatment. When a pasteurized product was inoculated, toxin was present for as long as 4 months (Bajlozov, 1966). Toxin is produced in the presence of 9·2% salt (Baird-Parker, 1971).

The growth of *Cl. welchii* was inhibited in a germination medium containing nitrite in the range 0·01 to 0·2%, depending upon the pH and other factors (Labbe & Duncan, 1970; Perigo & Roberts, 1968). However, *Cl. welchii* survives the usual concentrations of curing salts and when inoculated into the raw material spores were present in the smoked products (Gough & Alford, 1965). In a laboratory medium *Cl. welchii* grew in 7% salt, and growth was strongly influenced by E_h (Mead, 1969). Spores can survive for a long time in the product, as long as 4 years (Del Vecchio *et al.*, 1969). They occur in the raw product or contaminate the meat during processing and handling.

The counts of *Cl. welchii* in products vary greatly. *Cl. welchii* was isolated by Hall & Angelotti (1965) from 43·1% of meat dishes in the U.S.A., and in Poland it was frequently found in semi-preserved meat products (Słuzewski, 1969). In Czechoslovakia it is isolated rarely.

Food poisoning

There were 2 salmonella outbreaks associated with semi-preserved meats in 1968 in the U.K. (Hobbs, B. C., 1970) and in Germany *S. enteritidis* caused poisoning of 2 adults and 1 dog in 1971 (Alterauge, 1971).

Cl. botulinum intoxications are attributed mostly to home-made products. In Germany, there were 434 outbreaks, mostly type B, in the years 1898–1948 with 1294 persons involved. In the years 1947–1960, there were 400 outbreaks with 49 deaths. A typical outbreak was caused by home-cured ham, which intoxicated 5 members of 1 family; type B toxin was found in the ham, but the organism was not isolated (Muller-Prasuhn & Bach, 1971). In the U.S.S.R. there were 2 outbreaks in 1961 from ham in brine and red beet in brine (Sakaguchi, 1969). In Poland there were as many as 3420 cases in 1959–1969, 80% of which involved home-made, inadequately heated preserves (Anusz, 1971). An outbreak associated with pasteurized tomato juice of pH 3·8–4·5 occurred in the U.S.S.R. The cans were not hermetically sealed and there was mould on the juice. Nine persons were involved and 2 died. *Cl. botulinum* was isolated from the soil where the tomatoes were grown (Popugaila *et al.*, 1972).

In England, staphylococcal toxin in ham caused death. The ham was kept 2 days at room temperature before consumption and *Staph. aureus* producing type

A toxin was isolated from the ham and from stools (Anonymous, 1971). In the U.S.A., in the state of Wisconsin alone, at least 3 outbreaks involving hams, salami and cream-filled coffee cakes occurred. A further outbreak involved 1300 persons who had eaten ham and there were other outbreaks in which frankfurters were thought to be the food involved (Bergdoll, 1970a, b).

One outbreak of Cl. welchii food poisoning from ham has been reported (Parry, 1963).

In our country there are no confirmed reports of food poisoning due to semi-preserved meat products although in 1 outbreak involving 60 persons ham was suspected to be the cause.

Standards

Standards used in Czechoslovakia limit the plate count in freshly-produced products to 10^3/g. In products pasteurized at higher temperatures a maximum of 5×10^2/g is recommended. Bacteria posing a health hazard must be absent. Mossel & Ratto (1970) suggested pre-incubation of the products at 17° for a period of from 5 to 10 days. The pre-incubated food should have aerobic and anaerobic counts of less than 10^6/g, or less than 10^4/g in products subjected to a higher pasteurization temperature. These authors make other valuable recommendations. A maximum count of 10^4/g was proposed by the IAMS Food Hygiene Committee for non-sterile hams preserved in cans (Ingram et al., 1955).

Semi-Preserved Fish Products

In Czechoslovakia, cold and heated marinades, marinades with mayonnaise and with remoulade are produced. Remoulade is mayonnaise mixed with pickles. These products are packed in cans, glass and plastics.

The raw fish, after thawing, passes through a pickling bath of acid pH. The type of acid used determines, to some extent, the inhibition effect of the given pH value (Mossel & Bruin, 1960). The Pseudomonas, Bacillus and Proteus spp. are suppressed by these factors and also by inhibitory agents produced by lactobacilli in the pickling bath (Price & Lee, 1970).

The microflora of cold marinades consists of obligate or facultative halophilic bacteria. The plate count is generally 10^3/g, but higher counts are reached after passing through the subsequent technological processes where contamination occurs, and then 10^4/g is the usual population count. There are 3 main groups present: micrococci, which come with the raw product and possibly help in producing flavour; pediococci, which appear after 2–3 weeks of storage and give a higher acidity to the pickle; and, last but not least, lactobacilli, which do not come with the fish, but are present in the pickle, and on the utensils and surfaces (Lerche, 1960; Erichsen, 1967a, 1970). In addition yeasts occur in numbers higher than the bacterial counts. Enterococci, coliforms and *Bacillus* spp. are present in low concentrations (Bartl, 1966).

The above mentioned 3 groups of bacteria take part in the ripening of fish in the pickling bath, as without some bacterial activity the characteristic odours and flavours are not produced (Hobbs, G., 1970). Some of the lactobacilli are capable of producing diacetyl (Meyer, 1965). In Japan, yeasts, moulds or bacteria are sometimes added to produce desirable flavours (Amano, 1962). The ripening process is a degradation of proteins, in which fish tissue enzymes are most active at first, to be superceded later by enzymes produced by microorganisms (Voskresensky, 1965). Salt at the same time inhibits the hydrolysis and regulates the process to within desired limits.

The marinades with mayonnaise have a microflora similar to that of various types of salads. Bacterial counts are rather high (usually 10^5/g), the count of moulds and yeasts is about 10^4/g (Christiansen & King, 1971), and the composition of the microflora is more varied, as the pH is nearly neutral. *Pseudomonas, Achromobacter, Flavobacterium, Serratia* and *Bacillus* spp. are present.

The microflora of fried marinades is influenced by the heat treatment. The plate count is less than 10^3/g, and cocci and bacilli are present. Coliforms, yeasts and moulds (when present) are indicative of contamination (Skolnikova, 1968).

Spoilage micro-organisms

Spoilage is much influenced by storage temperature, being most rapid at 30° (Buttiaux, 1953). Lactobacilli are very important in the spoilage of cold marinades. Their numbers reached 4×10^8/g in 13 days when stored at 20°, but only 2×10^5/g when stored at 9° (Holtzapffel & Mossel, 1968). In the anaerobic pickle these bacteria can obtain their energy by decarboxylation of amino acids (arginine, glutamic acid and tyrosine). CO_2 is produced in this process (Meyer, 1956a, b). *Betabacterium buchneri*, which was shown to carry out this reaction, was isolated from spoiled preserves and was probably the cause of spoilage (Meyer, 1956b). The amines produced neutralize part of the acetic acid, causing the rise in pH and encouraging further bacterial growth (Hobbs, G., 1970). *Betabact. buchneri, Betabact. breve, Streptobacterium plantarum* and *Streptobact. leichmani* were isolated from blown cans, likewise some bacteria of the *B. subtilis–B. mesentericus* group (Lerche, 1960). This author reached the unusual conclusion that cans blown by the action of lactobacilli were spoiled but at the same time edible.

Bacilli are found in spoiled cooked marinades, and occasionally moulds and rope organisms in fried marinades (Meyer, 1965). Large quantities of yeasts are present in spoiled marinades with mayonnaise. *Pichia membranaefaciens, P. fermentans, Saccharomyces exiguus,* various types of *Torulopsis* and *Candida tropicalis* were isolated by Muzikar *et al.* (1970) and Muzikar (1972). These products thus have a spoilage microflora similar to salads (Kurtzman *et al.*, 1971) and sour vegetable preserves. In these products *Saccharomyces acidifaciens, P. membranaefaciens,* the mould *Moniliella acetobutans* and lactobacilli were the main spoilage organisms (Dakin & Radwell, 1971).

The addition of hexamethylenetetramine as a preservative is legal in Sweden, but not in other European countries. Preservation tests with 20–40 p/m tylosin lactate gave encouraging results (Erichsen, 1967*b*).

Food poisoning micro-organisms

Salmonella spp. can be present, but their growth is inhibited by low temperatures and various intrinsic factors.

Cl. botulinum is capable of multiplying in pickles of pH 4·5–4·8; and toxin production in pickled herring with an even lower pH of 4·0–4·2 was reported (Sakaguchi, 1969). In laboratory tests type E toxin was produced at pH 5·3 (Meyer, 1958). Low temperatures delay toxin production but holding at 5–10° does not guarantee complete absence of toxin formation. Refrigeration at 3° or lower is recommended (Sakaguchi, 1969). Botulinum toxin once formed is stable at chilling temperatures and at acid pH.

Staph. aureus does not occur in cold marinades, but it is generally present in marinades with mayonnaise at levels below 10^2/g, that is, in much the same numbers as in salads (Christiansen & King, 1971; Holtzapffel & Mossel, 1968).

Little attention has been given to *Cl. welchii* in semi-preserved fish products. It was found that spores survive for some days, but that low temperatures block their germination (Holtzapffel and Mossel, 1968).

Bacillus spp. occur in warm marinades but they can reach high and dangerous numbers only when the product is badly mishandled.

Food poisoning

Outbreaks associated with salmonellae are rare. One due to pickled herring and involving 4 persons was reported in the U.S.S.R. *Salmonella mission* was isolated and poor technology was responsible (Korovina & Artischeva, 1972). Two outbreaks caused by ingestion of marinated herring infected with *S. typhimurium,* were reported in Germany (Seidel & Muschter, 1967).

Botulism caused by fish preserves was nearly unknown before the Second World War, but since that time outbreaks have been reported with increasing frequency. In Scandinavian countries, pickled meat and fish products have been responsible for 50% of outbreaks. Fifty-seven per cent of outbreaks in the U.S.S.R. from 1958 to 1964 were associated with home-made fish products (Riemann, 1969). Three outbreaks of botulism were reported from salted herring in mackerel in Denmark (Jensen & Hahnemann, 1959), one type A outbreak from imported pickled fish in England (Mackay-Scollay, 1958), and one of the same type from a vinegar-cured herring in Germany (Wasmuth, 1948). Nine outbreaks of type E botulism from lactic-fermented *izushi* were reported in Japan (Nakamura *et al.*, 1956). From 1951 to 1960, 166 outbreaks with 56 deaths were caused by the ingestion of this food in Japan (Kawabata & Sakaguchi, 1963). In Northern Canada, type E botulism resulted from the consumption of fermented fish by Eskimoes (Dolman, 1960). Home-made marinated fungi were associated with 3 outbreaks in the U.S.S.R., and another outbreak involved 5 persons with 1 death. Type E spores were isolated from the fungi (Makarkin, 1970; Litvinov & Peskov, 1969). The same type was involved in an outbreak following the ingestion of tunny fish salad (Johnston *et al.*, 1963).

There are no reports of staphylococcal intoxications associated with cold marinades. However, marinades with mayonnaise are involved quite often and in Czechoslovakia such an outbreak involved 25 persons.

It is of interest that mild outbreaks attributed to yeasts have been reported by Muzikar *et al.* (1970) from marinated herring containing 10^5–10^6 *Pichia fermentans*/g.

B. cereus was isolated from fried fish in tomato sauce which caused intoxication. The population of 5×10^4/g was surprisingly low (Smykal & Rokoszewska, 1965).

Standards

Cold marinades should have, by Czechoslovakian standards, less than 10^3 anaerobes/g in mayonnaise products and 5×10^3/g in other types. Aerobic plate counts should be less than 10^5/g, and coliforms, *Proteus* and enterococci less than 20/g. Our standards meet the requirements published in England by Shewan (1970, 1971)—plate count less than 10^5/g at $35°$, the coliform count less than 200/g, *E. coli* less than 100/g and *Staph. aureus* less than 100/g. In fried marinades a limit of 5×10^3/g for the plate count and no coliform bacilli in 10 g was suggested by Skolnikova (1968).

In Czechoslovakia it is fairly common to standardize microbiological limits in foods, and it is generally accepted to be useful in maintaining good sanitation and in minimizing health risks. It has helped also to introduce better technology, as for example the use in pickling tanks of liquid spice extracts in place of pulverized spices.

Sweetened Condensed Milk

The microbiology of a different type of semi-preserved food, sweetened condensed milk, will now be mentioned briefly. This food does not appear to have been involved in food-poisoning. It is a product of standard chemical and microbiological quality in our country. The plate count is less than 10^3/g in the newly-canned product, coliforms are usually absent and the counts of enterococci are less than 10^2/g.

When testing the shelf-life, we were surprised by the increasing viable counts shown in Fig. 1. The maximum count is reached in 20–38 days when it begins to fall. After 11–12 months the count is about 10^4/g. The organisms are unspecified and harmless micrococci which form a nearly pure culture in the cans. They probably come from the raw milk; they are found in pasteurized milk, and isolations from sweetened condensed milk have been reported previously (Terplan & Gedek, 1965; Asperger, 1965; Jarchovska *et al.*, 1970; Bartl *et al.*, 1970). Micrococci find conditions favourable in the cans and the population develops normally. Organoleptic or chemical changes or swelling do not occur. It had one effect—the population exceeded the microbiological standard which had been set. This harmless coccus caused great concern before the problem was solved. It is interesting that *Staph. aureus* inoculated in this product grows in the same way (Lukásová, 1972).

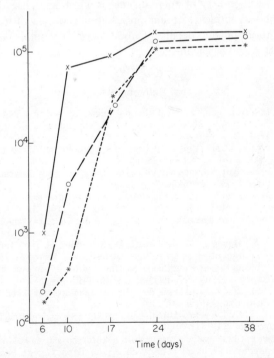

Fig. 1. Mean counts in 1 g of sweet condensed milk. x———x, Blood agar, o————o, M-P agar; *————* micrococci.

This example is presented to show that semi-preservation totally changes the intrinsic factors of a food, produces a new type of food, and introduces new and often unexpected problems.

The great variety of semi-preserved foods makes comparison of published results difficult, and furthermore the microflora in these foods changes from the beginning of processing to the end of storage. There is still much to do before we can hope to understand the dynamics of the microbiological associations in such foods.

Recommendations

Research should be carried out on saprophytic microflora in semi-preserved foods. This could lead to: improved technology; organoleptically better products; prolonged shelf-life; products of greater nutritive value; conversion of less valuable or less attractive foods to acceptable foods.

The succession of various groups of saprophytic microflora and the metabolic products which enable or cause the bacterial changes and the dynamics of these changes are still not perfectly understood. The fate, viability and degree of pathogenicity of micro-organisms which are a hazard to health as well as the inhibitory influence of semi-preservation is worthy of study.

The publication of full details of food poisoning outbreaks from semi-preserved foods should be encouraged.

References

Anon. (1971) Death from staphylococcal food poisoning. *Br. med. J.* **4**, 244.

Ajmal, M. (1968) *Clostridium botulinum* type E: Growth and toxin production in food. *J. appl. Bact.* **31**. 124.

Allen, J. R. & Foster, E. M. (1960). Spoilage of vacuum-packed sliced processed meats during refrigeration storage. *Fd Res.* **25**, 19.

Alterauge, W. (1971). Mit Salmonellen infiziertes Hackfleisch als Ursache einer Lebensmittelvergiftung, *Archs Lebensmittelhyg.* **22**, 69.

Amano, K. (1962). The influence of fermentation on the nutritive value of fish with special reference to fermented fish products of South-east Asia. In *FAO International Conference on Fish in Nutrition.* Eds. E. Heen & R. Kreuzer. London: Fishing New Books.

Anellis, A., Grecz, N., Huber, D. A., Berkowitz, D., Schneider, M. D. & Simon, M. (1965). Radiation sterilization of bacon for military feeding. *Appl. Microbiol.* **13**, 37.

Anusz, Z. (1971). Předběžná epidemiol. analýza otrav z potravin, zpusobenych klobasovým jedem v Polsku, 1959–1969. *Polski Tygad. lek.* **26**, 1491.

Asperger, H. (1965). Zur Klassifikation der Mikrokokken-Stafylokokken aus milchwirtschaftlicher Sicht. *Oster. Milch.* **20**, 23.

Baird-Parker, A. C. (1971). Factors affecting the production of bacterial food poisoning toxins. *J. appl. Bact.* **34**, 181.

Bajlozov, D. (1966). Obrazovanije i ustojčivost někotorych stafylo-kokovych toksinov i fermentov v mjasnych i rybnych konzervach. *Epid. mikrob. i infekc. bolezni.* **3**, 51.

Baran, W. L., Kraft, A. A. & Walker, H. W. (1970). Effect of carbon dioxide and vacuum packaging on color and bacterial count of meat. *J. Milk Fd Technol.* **33**, 77.

Bartl, V. (1966). *Mikrobiologie potravin* (Food Microbiology). UVUPP: Praha.

Bartl, V., Jarchovská, H., Kopečná, M. & Prikryl, F. (1970). Changes in the composition of microflora of condensed sweetened milk during storage. Presented at the *Symposium on Microbiology of Semi-preserved Foods*, Liblice, Czechoslovakia, 5–7 October, 1970. (To be published.)

Bergdoll, M. S. (1970a). The staphylococcal enterotoxins in semi-preserved foods. Presented at the *Symposium on the Microbiology of Semi-preserved Foods, Liblice, Czechoslovakia, 5–7 October, 1970*. (To be published.)

Bergdoll, M. S. (1970b). Personal communication.

Blanche-Koelensmid, W. A. A. & van Rhee, R. (1968). Intrinsic factors in meat products counterattacking botulogenic conditions. *Antonie van Leeuwenhoek* **34,** 287.

Buttiaux, R. (1953). Quelques problemes d'ordre microbiologique dans les semiconserves de viande. *Proc. Symposium on Cured and Frozen Fish Technology*. SIK-Göteborg. Publ. No. 100.

Buttiaux, R. (1963). Les bactéries nitrifiantes des saumures de viandes. *Revue fr. ind. alim.* **18,** 191.

Cavett, J. J. (1962). The microbiology of vacuum packed sliced bacon. *J. appl. Bact.* **25,** 282.

Cavett, J. J. (1963). A diagnostic key for identifying lactic acid bacteria of vacuum packed bacon. *J. appl. Bact.* **26,** 453.

Christiansen, L. N. & Foster, E. M. (1965). Effect of vacuum packaging on the growth of *Clostridium botulinum* and *Staphylococcus aureus* in cured meats. *Appl. Microbiol.* **13,** 1023.

Christiansen, L. N. & King, N. S. (1971). The microbial content of some salads and sandwiches at retail outlets. *J. Milk Fd Technol.* **34,** 289.

Cotterill, O. J. & Glauert, J. (1969). Thermal resistance of salmonellae in egg yolk products containing sugar or salt. *Poult. Sci.* **48,** 1156.

Cervenka, J. (1966). Mikrobiologicky obraz suroviny pro vyrobu pasterovanych šunek a plecí. *Prum. Potrav.* **17,** 16.

Dainty, R. H. (1971). The control and evaluation of spoilage. *J. Fd Technol.* **6,** 209.

Dakin, J. C. & Radwell, J. Y. (1971). Lactobacilli causing spoilage of acetic acid preserves. *J. appl. Bact.* **34,** 541.

Del Vecchio, V., D'Arca, S. U. & D'Arca, S. A. (1969). Indagini sulla persistenza nel tempo del *Cl. perfringens* in un campione di carve naturalmente contaminato conservata a +4°, e sulla termoresistenza delle sue spore. *Nuovi ann. ig. e microbiol.* **20,** 477.

Deibel, R. H., Niven, C. F., Jr. & Wilson, G. D. (1961). Microbiology of meat curing. III. Some microbiological and related technological aspects in the manufacture of fermented sausages. *Appl. Microbiol.* **9,** 156.

Dolman, C. E. (1960). Type E botulism. A hazard of the North. *Arctic* **13,** 230.

Duncan, C. L. (1970). Arrest of growth from spores in semi-preserved foods. *J. appl. Bact.* **33,** 60.

Eddy, B. P. & Ingram, M. (1956). A salt tolerant denitrifying *Bacillus* strain which "blows" canned bacon. *J. appl. Bact.* **19,** 62.

Eddy, B. P. & Ingram, M. (1962). The occurrence and growth of staphylococci on packed bacon, with special reference to *Staphylococcus aureus. J. appl. Bact.* **25,** 237.

Elmossalami, E. & Wassee, N. (1971). Penetration of some micro-organisms in meat. *Zentbl. VetMed.* B. **18,** 329.

Erichsen, I. (1967a). The microflora of semi-preserved fish products. III. Principal groups of bacteria occurring in tidbits. *Antonie van Leeuwenhoek* **33,** 107.

Erichsen, I. (1967b). The effect of tylosin lactate on the shelf-life of semi-preserved herring fillets ("Tidbits"). *J. Fd Technol.* **2,** 61.

Erichsen, I. (1970). The microflora of semi-preserved herring products. Presented at the *Symposium on the Microbiology of Semi-preserved Foods. Liblice, Czechoslovakia 5–7 October, 1970*. (To be published.)

Gardner, G. A. (1968). Effect of pasteurization or added sulphite on the microbiology of stored vacuum packed baconburgers. *J. appl. Bact.* **31,** 462.

Genigeorgis, C., Riemann, H. & Sadler, W. W. (1969). Production of enterotoxin B in cured meats. *J. Fd Sci.* **34,** 62.

Gianelli, F. & Maggi, E. (1962). Studiu microbiologico della considetta "moliga" dei prosciutti. *Nuova veterin.* **38,** 121.

Giolitti, G., Cantoni, C., Bianchi, M. A., Renon, P. & Beretta, G. (1971a). Microbiological and biochemical changes in dry hams during ripening. *Arch. vet. Ital.* **22,** 61.

Giolitti, G., Cantoni, C. A., Bianchi, M. A. & Renon, P. (1971b). Microbiology and chemical changes in raw hams of Italian type. *J. appl. Bact.* **34,** 51.

Goepfert, J. M. & Chung, K. C. (1970). Behaviour of salmonella during the manufacture and storage of a fermented sausage product. *J. Milk Fd Technol.* **33,** 185.

Goldman, M., Deibel, R. H. & Niven, C. F., Jr. (1963). Interrelationship between temperature and sodium chloride on growth of lactic acid bacteria isolated from meat-curing brines. *J. Bact.* **85,** 1017.

Gough, B. J. & Alford, J. A. (1965). Effect of curing agents on the growth and survival of food poisoning strains of *Clostridium perfringens. J. Fd Sci.* **30,** 1025.

Hall, H. E. & Angelotti, R. (1965). *Clostridium perfringens* in meat and meat products. *Appl. Microbiol.* **13,** 352.

Hobbs, B. C. (1970). Microbial intoxications from semi-preserved foods. Presented at the *Symposium on the Microbiology of Semi-preserved Foods. Liblice, Czechoslovakia, 5–7 October, 1970.* (To be published.)

Hobbs, G. (1970). The microbiology of semi-preserved and fermented fish products. Presented at the *Symposium on the Microbiology of Semi-preserved Foods. Liblice, Czechoslovakia, 5–7 October, 1970.* (To be published.)

Holtzapffel, D. & Mossel, D. A. A. (1968). The survival of pathogenic bacteria in, and the microbial spoilage of, salads, containing meat, fish and vegetables. *J. Fd Technol.* **3,** 223.

Ingram, M. (1962). Microbiological principles in prepacking meats. *J. appl. Bact.* **25,** 259.

Ingram, M., Cheftel, M. H., Clarenburg, A., Lerche, A., Buttiaux, R. & Mossel, D. A. A. (1955). Suggestions regarding bacteriological norms for non-sterile hams preserved in cans. *Annls Inst. Pasteur, Lille* **7,** 254.

Ingram, M. & Dainty, R. H. (1971). Changes caused by microbes in spoilage of meats. *J. appl. Bact.* **34,** 21.

Ingram, M. & Kitchell, A. G. (1967). Salt as a preservative for foods. *J. Fd Technol.* **2,** 1.

Jarchovska, H., Brodsky, F. & Hartmanova, J. (1970). Celkovy počet zarodku ve slazeném kondensovaném mléce. *Prům. Potravin* **21,** 298.

Jensen, B. B. & Hahnemann, F. (1959). Botulism. Reports of two similar epidemics. *Ugeskr. Laeg.* **121,** 1363.

Johnston, R. W., Feldman, J. & Sullivan, R. (1963). Botulism from canned tuna fish. *Publ. Hlth Rep., Wash.* **78,** 561.

Juchniewicz, J. (1961). Badanie roznic jakosiowych w szynce mielonej w zaleznosci od spodobu peklowania (suche, mokre). *Gospodarka miesna* **4,** 20.

Kafel, S. (1964). Wpyw najszcściej wystupujacych zakazen bakteryjnych na trwalosc konserw mieasnych pasteryzowanych. *Med. Weteryn.* **20,** 158.

Kafel, S. & Ayres, J. C. (1969). The antagonism of enterococci on other bacteria in canned hams. *J. appl. Bact.* **32,** 217.

Kawabata, T. & Sakaguchi, G. (1963). The problem of type E botulism in Japan. In *Microbiological Quality of Foods.* Eds. L. W. Slanetz, C. O. Chichester, A. R. Gaufin & Z. J. Ordal. New York: Academic Press.

Kempton, A. G. & Bobier, S. R. (1970). Bacterial growth in refrigerated vacuum-packed luncheon meats. *Can. J. Microbiol.* **16,** 287.

Kendereski, S. (1970). Degree and nature of bacteriological contamination of semi-conserves of meat. Presented at the *Symposium on the Microbiology of Semi-preserved Foods. Liblice, Czechoslovakia, 5–7 October, 1970.* (To be published.)

Khan, M. & Katamay, M. (1969). Antagonistic effect of fatty acids against salmonella in meat and bone meal. *Appl. Microbiol.* **17,** 402.

Kitchell, A. G. (1962). Micrococci and coagulase-negative staphylococci in cured meats and meat products. *J. appl. Bact.* **25**, 416.

Korovina, K. E. & Artischeva, L. J. (1972). A case of poisoning with a pickled herring. *Gigiena i sanitarija (Moskva)* **37**, 97.

Kurtzman, C. P., Rogers, R. & Hesseltine, C. W. (1971). Microbiological spoilage of mayonnaise and salad dressings. *Appl. Microbiol.* **21**, 870.

Labbe, R. G. & Duncan, C. L. (1970). Growth from spores of *Clostridium perfringens* in the presence of sodium nitrite. *Appl. Microbiol.* **19**, 353.

Leistner, L. (1958*a*). Bakterielle Vorgange bei der Pokelung von Fleisch. *Fleischwirtschaft* **10**, 226.

Leistner, L. (1958*b*). Bakterielle Vorgange bei der Pokelung von Fleisch. *Fleischwirtschaft* **10**, 530.

Leistner, L. (1959). Die Keimarten und Keimzahlen von Pökellaken. *Fleischwirtschaft* **11**, 726.

Leistner, L., Bem, Z. (1970). Vorkommen und Bedeutung von Häfen bei Pökelfleischwaren. *Fleischwirtschaft* **50**, 350.

Lerche, M. (1960). Bombage-Ursachen in Fischpräserven. Berl. Munch. tierärzttl. *Wochenschrift* **73**, 12.

Litvinov, A. P. & Peskov, V. G. (1969). O slučajach botulisma ot upotreblenija konservirovanych gribov domašnego prigotovlenija. *Gigiena i sanitarija (Moskva)* **34**, 100.

Loaharanu, P. & Lopez, A. (1970). Bacteriological and shelf-life characteristics of canned, pasteurized crab cake mix. *Appl. Microbiol.* **19**, 734.

Lukásová, J. (1972). Vliv doby skladování na přežívání patogenních stafylokoku v slazených kondensovaných mléčnych vyrobeich. *Prům Protravin* **23**, 168.

Mackay-Scollay, E. M. (1958). Two cases of botulism. *J. path. Bact.* **75**, 482.

Makarkin, A. I. (1970). A case of botulism caused by marinated fungi. *Vopr. pitanija* **4**, 80.

Maleszewski, J. (1964). Zmiany mikroflory resztkovyw w szynkach mielonych pdczas skadowania w temperature 5–6° C. *Rocz. Panstw. Zakl. Higieny.* **15**, 1964.

Mead, G. C. (1969). Combined effect of salt concentration and redox potential of the medium on the initiation of vegetative growth of *Cl. welchii*. *J. appl. Bact.* **32**, 468.

Meyer, V. (1956*a*). Aminosäuren-Dekarboxylose durch Beta-bakterien als Ursache von Bombagen. *Arch. Fisch. Wiss.* **7**, 260.

Meyer, V. (1956*b*). Probleme des Verderbens von Fischkonzerven in Dosen. II. Aminosauredecarboxylose durch Organismen der Beta-Bakterium buchneri Gruppe als Ursache bombierter Marinaden. *Veoffentl. Inst. Meeresforsch. Bremerhaven.* **4**, 1.

Meyer, V. (1958). Das Verhalten von Botulismuskeimen in Geleeheringen. *Veroffentl. Inst. Meeresforsch. Bremerhaven.* **5**, 195.

Meyer, V. (1965). Marinades. In *Fish as Food*, Vol. 3. Ed. G. Borgstrom. London: Academic Press.

Michailova, A. (1971). Kurzere Pokelzeiten bei Schinken durch Bakterienkulturen. *Fleischwirtschaft* **25**, 42.

Minor, T. E. & Marth, E. H. (1971). *Staphylococcus aureus* and staphylococcal food intoxications. *J. Milk Fd Technol.* **34**, 557.

Mol, J. H. H., Hietbrink, J. E. A., Mollen, H. W. M. & Tinteren, J. van. (1971). Observation on the microflora of vacum packed sliced cooked meat products. *J. appl., Bact.* **34**, 377.

Mol, J. H. H. & Timmers, C. A. (1970). Assessment of the stability of pasteurized comminuted meat products. *J. appl. Bact.* **33**, 233.

Mossel, D. A. A. & Bruin, A. S. de. (1960). The survival of *Enterobacteriaceae* in acid liquid foods stored at different temperatures. *Annls Inst. Pasteur, Lille* **11**, 65.

Mossel, D. A. A. & Ingram, M. (1955). The physiology of the microbial spoilage of foods. *J. appl. Bact.* **18**, 232.

Mossel, D. A. A. & Ratto, M. A. (1970). Wholesomeness of some types of semi-preserved foods. Presented at the *Symposium on the Microbiology of Semi-preserved Foods, Liblice, Czechoslovakia, 5–7 October, 1970.* (To be published.)

Müller-Prasuhn, G. & Bach, R. (1971). Ein hausgemachter Schinke als Ursache von Botulismus. *Archs Lebensmittelhyg.* 22, 179.

Muzikar, V., Emberger, O., Cihova, H., Cerna, J., Neskusil, J. & Pěgřímková, J. (1970). The hygienic importance of yeasts in semi-preserved marinated fish. Presented at the *Symposium on the Microbiology of Semi-preserved Foods, Liblice, Czechoslovakia, 5–7 October, 1970.* (To be published.)

Muzikar, V. (1972). Mikrobiální kontaminace čerstvých rybích marinovaných polokonzerv. *Průmysl potravin,* 23, 213–214.

Nakamura, Y., Iida, H., Saeki, K., Kanazawa, K. & Karashimada, I. (1956). Type E botulism in Hokkaido, Japan. *Jap. J. med. Sci. Biol.* 9, 45.

Östlund, K. (1970). Effect of pasteurization on the Swedish sausage *isterbad. Nord. VetMed.* 22, 634.

Parry, W. H. (1963). Outbreak of *Clostridium welchii* food poisoning. *Br. med. J.* 2, 1616.

Pěgřímková, J., Emberger, O., Cihova, H., Cerná, J. & Bartl, V. (1970). Microbiology of semi-preserved frankfurters. Presented at the *Symposium on the Microbiology of Semi-preserved Foods, Liblice, Czechoslovakia, 5–7 October, 1970.* (To be published.)

Perigo, J. A. & Roberts, T. A. (1968). Inhibition of clostridia by nitrite. *J. Fd Technol.* 3, 91.

Pivnick, H. & Bird, H. (1965). Toxinogenesis by *Clostridium botulinum* type A and E in perishable cooked meats vacuum-packed in plastic pouches. *Fd Technol.* 19, 132.

Pivnick, H., Rubin, L. J., Barnett, H. W., Nordin, H. R., Ferguson, P. A. & Perrin, C. H. (1967). Effect of sodium nitrite and temperature on toxinogenesis by *Clostridium botulinum* in perishable cooked meats vacuum packed in air-impermeable plastic pouches. *Fd Technol.* 21, 204.

Pivnick, H. (1970). The inhibition of heat-damaged spores of *Clostridium botulinum* by sodium chloride and sodium nitrite. Presented at the *Symposium on the Microbiology of Semi-preserved Foods, Liblice, Czechoslovakia, 5–7 October, 1970.* (To be published.)

Pivovarov, J. P., Akimov, A. M., Volkova, R. S. Kuchto, A. I. & Sidorenko, A. I. (1970). The occurrence of *Bacillus cereus* in food products. *Aktual. vopros. gigieny, Medgiz, Moskva,* p. 103.

Pohja, M. S. (1958). The differentiation of micrococci isolated from brines and cured meats. *Conserva.* 7, 119.

Pohja, M. S. (1962). Numerical taxonomy of micrococci of fermented meat origin. *J. appl. Bact.* 25, 341.

Popovic, P. (1965). Bombaže removki u limenkama izazyane aerobnim bacilima. *Tehn. mesa.* 6, 210.

Popugaila, V. M., Zhuravskaya, S. P., Yarishewskaja, E. E., Sukhanova, V. I., Zhukova, L. N. & Koneshina, I. A. (1972). A case of botulism caused by preserved tomato juice. *Gigiena i sanitarija.* 37, 97.

Price, R. J. & Lee, J. S. (1970). Inhibition of *Pseudomonas* species by H_2O_2-producing Lactobacilli. *J. Milk Fd Technol.* 33, 13.

Report (1970). Sous-Commission des semi-conserves. A report of the "Commission de Microbiologie-Societe belge de Zymologie pure et appliquee". President: A. M. Piette. Bruxelles, Belgium. Personal communication.

Reuter, G. (1967). Atypische Streptobakterien als dominierende Flora in reifender und gelagerter Rohwurst. *Fleischwirtschaft* 47, 397.

Riberio, A. M. (1964). Les vibrio halophiles des saumures de jambon. *Annls Inst. Pasteur, Lille* 15, 99.

Riemann, H. (1969). Botulism–types A, B and F. In *Food-borne Infections and Intoxications.* London: Academic Press.

Ristic, M., Tadic, Z. & Table, M. (1967) Uzrocnici kvara sunki u limenkama. *Techn. mesa* 8, 34.

Roberts, T. A. & Ingram, M. (1966). The effect of sodium chloride, potassium nitrate and sodium nitrite on the recovery of heated bacterial spores. *J. Fd Technol.* 1, 147.

Sakaguchi, G. (1969). Botulism type E. In *Food-borne Infections and Intoxications.* Ed. H. Riemann. London: Academic Press.

Sanker, U. (1970). Public health aspects of faulty production of semi-preserved foods. Presented at the *Symposium on the Microbiology of Semi-preserved Foods, Liblice, Czechoslovakia, 5–7 October, 1970.* (To be published.)

Segner, W. P., Schmidt, C. F. & Boltz, J. K. (1966). Effect of sodium chloride and pH on the outgrowth of spores of type E *Clostridium botulinum* at optimal and suboptimal temperatures. *Appl. Microbiol.* **14**, 49.

Seidel, G. & Muschter, W. (1967). *Die bakteriellen Lebensmittelver-fiftungen.* Berlin: Akademie Verlag.

Silliker, J. H., Jansen, C. E., Voegeli, M. M. & Chmura, N. W. (1962). Studies on the fate of staphylococci during the processing of hams. *J. Fd Sci.* **27**, 50.

Sharpe, M. E. (1962*a*). Les lactobacillus dans les produits carnes. *Annls Inst. Pasteur, Lille* **13**, 153.

Sharpe, M. E. (1962*b*). Lactobacilli in meat products. *Fd Manuf.* **37**, 582.

Shewan, J. M. (1970). Bacteriological standards for fish and fishery products. *Chemy Ind.* p. 193.

Shewan, J. M. (1971). The microbiology of fish and fishery products—a progress report. *J. appl. Bact.* **34**, 299.

Skolnikova, S. S. (1968). Microbial control in the production of fried fish products. *Ryb. Choz.* **44**, 62.

Słuzewski, R. (1969). Badania nad drobnoustrojami rodzaju Clostridium za szcregalnym uwzglednieniem *Clostridium perfringens* w produkcii konzerw miesnych. *Pol. arch. wet.* **12**, 129.

Smykal, B. & Rokoszewska, J. (1965). Zatrucia pokarmowe ryba zakazone *Bacillus cereus. Rocz. pan. zaki. Hyg.* **16**, 325.

Spencer, R. (1966). Factors in cured meat and fish products affecting spore germination, growth and toxin production. In *Botulism 1966.* Eds. M. Ingram & T. A. Roberts. London: Chapman & Hall.

Steinke, P. K. W. & Foster, E. M. (1951). Botulinum toxin formation in liver sausage. *Fd Res.* **16**, 472.

Terplan, G. & Gedek, W. (1965). Zur Differenzierung von aus Rinderutern isolierten Micrococcaceae. *Arch. Lebensm. Hyg.* **16**, 203.

Verhoeven, W. (1950). On a spore-forming bacterium causing the swelling of cans containing cured ham. *Antonie van Leeuwenhoek* **16**, 269.

Voskresensky, N. A. (1965). Salting of herring. In *Fish as Food,* Vol. 3. Ed. G. Borgstrom. New York: Academic Press.

Vrban, N. (1971). Microbiological testing of pasteurized meat preserves under various storage conditions. *VetMed. nauki* **8**, 79.

Walton, J. R. & Lewis, L. E. (1971). Contamination of fresh and cooked meats by antibiotic-resistant coliform bacteria. *Lancet ii,* 255.

Warneke, M., Carpenter, J. A. & Saffle, R. L. (1967). A study of *Clostridium botulinum* type E in vacuum packed meat. *Fd Technol., Champaign* **21**, 115A.

Wasmuth, W. (1948). Uber Botulismus. *Dtsch med. Wschr.* **73**, 636.

Zeuthen, P. (1970). Some observations on the bacterial growth in vacuum packed sliced ham in relation to various processing factors. Presented at the *Symposium on the Microbiology of Semi-preserved Foods, Liblice, Czechoslovakia, 5–7 October, 1970.* (To be published.)

Zlamalova, J. (1964). Mikroflora pasterovanych suek. *Prům. Potravin* **15**, 263.

Discussion

Elliott

Dr Bartl has reviewed an area that is of deep concern to all food microbiologists—namely the tendency, in response to consumer demand, for food of lower salt and higher moisture contents, higher pH, etc. These are on the borderline between stability and hazard or spoilage. We in the U.S. Meat and

Poultry Inspection Programme constantly evaluate requests for new products of this kind. To make the best judgments, we need additional data showing the interrelationships among the various factors that control microbial growth—principally the growth of pathogens.

Jarvis

Have you any information regarding the presence of nitrite in the "brined" hams implicated in outbreaks of botulism in Germany and Russia?

Bartl

In Germany nitrites are used, but I am not certain if they are also used in the U.S.S.R.

Roberts

Botulism associated with home-cured hams is not unusual. Sebald has reported some 50–60 cases in France over the last 10–15 years.

Jarvis

I think the total number of outbreaks reported by Sebald was 66 and some 131 persons were affected. Not all outbreaks were caused by hams, but the majority were. Sebald has told me that she has not detected nitrite in hams implicated in some of her reported outbreaks. From an epidemiological point of view the comparative number of outbreaks of botulism in the U.K. and U.S.A. on the one hand and France on the other is interesting.

Survival of *Staphylococcus aureus* and *Salmonella newport* in Dried Foods, as Influenced by Water Activity and Oxygen

J. H. B. CHRISTIAN AND BETTY J. STEWART

Commonwealth Scientific and Industrial Research Organization, Division of Food Research, North Ryde, New South Wales, Australia

Survival of *Staphylococcus aureus* and *Salmonella newport* was studied after storage for up to 27 weeks in 4 dried foods; cake mix, skim milk, onion soup mix and flummery, a gelatin-based dessert. Storage was at 6 levels of water activity (a_w) in the range 0 to 0·53, in air and in vacuum. A liquid inoculum was added to equilibrated foods, and re-equilibration and storage took place at 25°. In all foods, staphylococci survived the equilibration period better than did salmonellae. Subsequent destruction rates of both organisms were minimal in the range of 0 to 0·22 a_w, and increased sharply as the a_w level increased to 0·53. They were generally much higher for staphylococci than for salmonellae. The high destruction rates reported for freeze-dried salmonellae at 0 a_w in some laboratory media were not observed. Storage in air was much more destructive than storage in vacuum and this effect was generally greatest at high a_w levels. Survival in nitrogen with 1·5% oxygen was little better than storage in air. Above 0·22 a_w, survival of both organisms was influenced by the pH of the food, being greatest in cake mix (pH 6·8) and least in dessert (pH 3·1). Staphylococci were particularly susceptible to acidity. In air at the a_w of these foods as purchased, a 99% decrease in viable staphylococci and salmonellae required 27 weeks and 10 weeks respectively in cake mix (0·32 a_w) 18 weeks and 9 weeks in skim milk (0·22 a_w), both more than 27 weeks in onion soup (0·14 a_w), and 1 week and 2 weeks in dessert (0·42 a_w).

THE SURVIVAL of pathogens in human foods is of concern when the food is likely to be consumed without cooking or when there is a possibility that pathogens may be transferred to already prepared foods. Many dry foods on occasions do contain pathogens, and the survival of such cells in the relatively dry state is known to be influenced profoundly by the water activity (a_w) of the substrate (Scott, 1958). The IAMS Symposium on The Microbiology of Dried Foods (Kampelmacher *et al.*, 1969) covered a wide field, but very little attention was given to the effect of a_w on survival. It was partly for this reason that the work described here was undertaken.

There are few published data on the water relations of bacterial survival in dry foods for human consumption. The study by Higginbottom (1953) of the microbiology of dried milk remains a classic in this field. Most recent work on the water relations of bacterial survival has involved non-food substrates, heat treatments, or animal feeds of relatively high a_w.

This investigation is to some extent an extrapolation to foods of the studies by Scott (1958) of residual water and the survival of bacteria freeze dried in laboratory media or in solutions of defined composition. The same bacterial strains, a_w levels and storage temperature were used.

Materials and Methods

Test organisms

The organisms studied were the strains of *Staph. aureus* and *S. newport* used previously by Scott (1958) in freeze-drying and storage experiments.

Foods

Four dry packaged foods in common use were chosen: cake mix, skim milk, onion soup mix, and flummery, a gelatin-based dessert. Some properties are listed in Table 1. The pH values were obtained with a glass electrode in a thick aqueous slurry of the food. The a_w levels were determined by measuring the relative humidity of air in equilibrium with the food.

Table 1

Characteristics of 4 foods used in storage experiments

Food	Ingredients	pH	ERH (%)	Aerobic plate count (per g)*
Cake mix	Sugar, flour, shortening, non-fat milk solids, leavening salt, glucose, dried egg albumen, vegetable gum, artificial flavour and colouring	6·8	32	$1·4 \times 10^3$
Skim milk powder	Protein (36%), lactose (51%), fat (1%), minerals (8%), moisture (4%)	6·2	22	$1·3 \times 10^3$
French onion soup	Not given	5·4	14	$1·6 \times 10^4$
Dessert (lime-flavoured flummery)	Sugar, gelatin, cornflour, citric acid, flavouring and colouring	3·1	42	$1·0 \times 10^1$

* < 10 coliforms/g.

Experimental procedure

0·5 g amounts of dried food were weighed into small test tubes (12 x 75 mm) and the tubes placed in desiccators over dry P_2O_5 or over saturated solutions (containing some solid phase) of the required a_w (Robinson & Stokes, 1955). The desiccators were evacuated, and the foods equilibrated to that a_w level for 12 days. The levels used were 0·00, 0·11, 0·22, 0·33, 0·43 and 0·53.

Bacteria were grown in brain-heart broth at $30°$ for 17 h with aeration. Cells were harvested, washed in 0·5% (W/V) peptone water, and suspended in peptone water to give, when the cell suspensions were combined, c. 3×10^5 *Staph. aureus* and 5×10^5 *S. newport* in each 0·04 ml drop.

Each equilibrated food sample received 0·04 ml of the mixed inoculum. This was stirred into the food with a sterile glass rod which remained in the small tube. Each tube was then placed in a larger test tube (19 x 150 mm) which was sealed under vacuum, in air, or at an intermediate oxygen concentration. The a_w in the larger tube and the re-equilibration of the inoculated food was controlled by c. 1 ml of the appropriate saturated solution plus solid phase, or by P_2O_5, in the bottom of the tube. Reduced oxygen levels were prepared by mixing nitrogen and oxygen in metal foil bags. The large tubes were evacuated and filled with the gas mixture 3 times before sealing. The oxygen concentration in the gas mixture was checked by gas chromatography before and after gassing the tubes.

A comparative experiment was performed in which inoculated food samples were freeze-dried before equilibration to the desired a_w levels. An inoculated slurry (c. 40% solids) of the food was prepared, 0·5 g samples were weighed into 12 x 75 mm test tubes and shell-frozen. After overnight storage at $-20°$, the samples were freeze-dried for 6 h. Viable counts were carried out before and after freezing and after drying. As described in detail above the small tubes were placed in larger tubes containing saturated equilibration solutions or P_2O_5, which were then sealed in air or under vacuum.

Tubes were stored at $25°$. At intervals up to 27 weeks, single tubes of each treatment were removed, opened and the food sample transferred by pasteur pipette to 10 ml peptone water (0·5%) with repeated rinsing of the tube and stirring rod.

Dilutions were made in 0·5% peptone water, followed by surface plating onto brain-heart agar (BHA) and brilliant green agar (BGA). After 48 h incubation of BHA plates at $30°$, colonies of *Staph. aureus* and *S. newport* could be readily distinguished and differential viable counts performed. The very small number of organisms present before inoculation did not interfere. The BGA plates were incubated for 48 h at $37°$ and gave an independent estimate of *S. newport*.

When the number of viable cells had declined to less than 10^3/tube, the 10-tube MPN method was used to estimate survivors. Initial inoculation was made into tubes of brain heart broth. Positive cultures were streaked onto Baird

Parker medium for *Staph. aureus* detection, and inoculated into tetrathionate broth followed by streaking onto BGA for detection of *S. newport*.

To prepare onion dip, 42 g of dry onion soup mix was stirred thoroughly into 280 g of sour cream. The pH was 4·5 and the a_w value was 0·97. After inoculation with a mixture of staphylococci and salmonellae, portions were incubated at 20° and 3·3° and samples taken at intervals for plating and counting on BHA, BGA, and BHA + 10% NaCl.

Results

Recovery of stored bacteria

Higher viable counts of the *Salmonella* inoculum were invariably obtained by direct plating onto BHA rather than onto BGA. In contrast, samples of all inoculated foods at all stages of the experiments gave similar counts of salmonellae on these 2 media. Baird-Parker medium was not used routinely for viable counts of staphylococci as it proved to be inhibitory. MPN determinations of salmonellae were much lower when dilutions were inoculated directly into tetrathionate broth than when brain-heart broth was the initial medium.

Survival during equilibration

Although foods were equilibrated to the desired a_w levels before inoculation, these levels were substantially increased by the addition of inoculum. It seems likely from survival curves at low a_w levels that re-equilibration was achieved within 1–2 weeks, as by this time the rapid initial rate of destruction which occurred in some foods had ceased. *S. newport* always showed this rapid loss of viability which ranged from 0·5 log units in onion soup to 2·0 log units in the dessert. *Staph. aureus* was much less affected in all foods (e.g. Fig. 1).

To attempt the separation of effects of initial drying from the later stages of equilibration, survival of bacteria freeze-dried in skim milk was followed at 0·00, 0·22 and 0·53 a_w. The reductions in viability during the first 2 weeks were similar to those observed with cells dried at 25° throughout. However, most of these losses occurred during storage at 25° after freezing and freeze-drying were completed.

Effect of a_w level

In a particular food, the survival patterns of staphylococci and salmonellae were generally similar, but with steeper destruction curves for staphylococci (Fig. 1). In all foods, survival was greatest at some a_w in the range of 0·00–0·22 (Table 2). Without exception, survival decreased as a_w was increased incrementally from

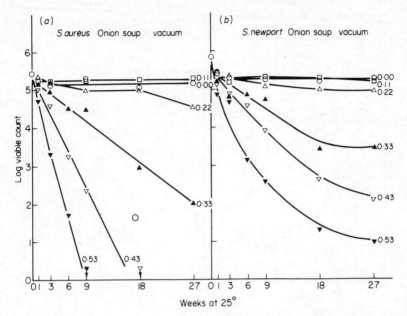

Fig. 1. Effect of a_w value on survival during storage at $25°$ in vacuum of bacteria in dried onion soup. (*a*) *Staphylococcus aureus*; (*b*) *Salmonella newport*.

Table 2

Optimum a_w *for storage at* $25°$

Organism	Atmosphere	Cake mix	Skim milk	Onion soup	Dessert
Staphylococcus	Air	0·11	0·11	0·00–0·11	0·11
aureus	Vacuum	0·00	0·00	0·00–0·11	0·22
Salmonella	Air	0·11	0·11	0·00–0·11	0·00
newport	Vacuum	0·00	0·00	0·00–0·11	0·11

0·22–0·53. Survival at 0·00 a_w was generally close to optimal, except in three situations where it was lower at this a_w than at 0·33 a_w. These all involved storage in air and were obtained with both organisms in the dessert and with staphylococci in cake mix (Fig. 2).

The most marked effects of a_w were recorded with *Staph. aureus* in onion soup stored in air. At 0·00 a_w, 30% of cells remained viable for 18 weeks, while at 0·53 a_w, only 0·001% survived for 10 days. The influence of a_w on survival was least for bacteria stored in cake mix (Fig. 2).

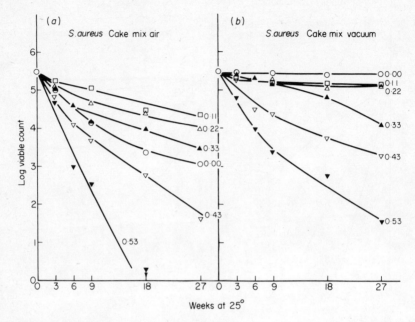

Fig. 2. Effect of a_w value on survival during storage at 25° of *Staphylococcus aureus* in cake mix. (*a*) Storage in air; (*b*) storage in vacuum.

Effect of oxygen

Survival was always increased by removal of air from the inoculated food before storage. The increase was most marked at higher a_w levels and, as noted above, in several treatments at 0·00 a_w. The adverse effect of air was generally greater on the staphylococci and was evident in both greater initial destruction rates at low a_w levels, and much steeper destruction curves at the higher levels of a_w tested (Fig. 2).

The data of Fig. 3 establish that most of the destructive effect of air could be obtained by storage in 1·5% oxygen in nitrogen. Several experiments performed in atmospheres of lower oxygen content (*c.* 0·2%) did not give reproducible results but there was no evidence that any appreciable concentration threshold existed for the deleterious effects of oxygen on bacterial survival in dry foods.

Effect of Food

Marked differences were observed in the effects of the 4 foods on bacterial survival. Destruction in the initial days of storage was much greater in the dessert

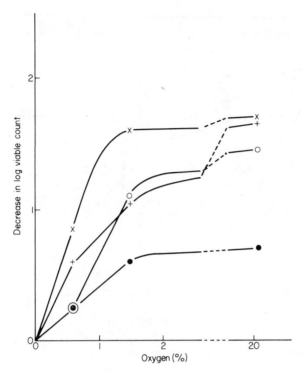

Fig. 3. Effect of oxygen concentration on the loss of viability during storage for 6 weeks at 25° and 0·43 a_w relative to the loss occurring in vacuum. *Staphylococcus aureus* in onion soup (×), and in skim milk (○). *Salmonella newport* in onion soup (+), and in skim milk (●).

(Fig. 4*b*) than in other foods. Overall survival at the higher a_w levels appeared by inspection to be highest in cake mix (Fig. 2), followed by skim milk (Fig. 4*a*), onion soup (Fig. 1), and dessert (Fig. 4*b*).

The above list of foods is also one of decreasing pH. To show the relation between pH and survival, the reductions in viable counts in 9 weeks at 0·43 a_w, in air and in vacuum, are given for both organisms in Fig. 5. Reduction in pH increased greatly the destruction of staphylococci during storage under these conditions, but had no consistent effect upon salmonellae. Skim milk was substantially more destructive to both organisms in air, but not in vacuum, than might have been expected from its pH value.

Another attempt at a quantitative assessment of the survival data is shown in Table 3, where the time required at 0·53 a_w for a 99·9% reduction in viability is recorded. Here destruction of both organisms was generally greater as pH decreased, but in this assessment onion soup was more destructive to staphylococci in both air and vacuum, than the more acid dessert.

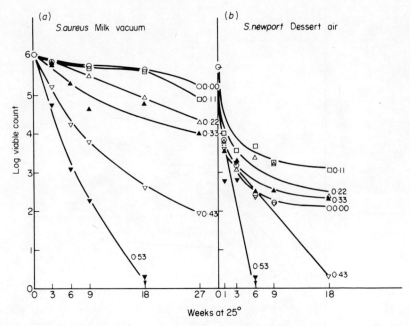

Fig. 4. Effect of a_w value on survival during storage at 25° of (a) *Staphylococcus aureus* in skim milk in vacuum; and (b) *Salmonella newport* in dessert in air.

Survival in onion dip

Effects of 2 temperatures on bacterial survival in an onion soup–sour cream dip are shown in Fig. 6. At 20°, the viability of staphylococci was contant for 2 days but decreased by more than 99% in the ensuing 3 days. The salmonella count declined steadily by less than 90% over the 5-day period. At 3·3°, destruction was very slow, with salmonellae again being the more resistant organisms.

Lest lactic acid bacteria, contributed by the sour cream, should interfere with development of staphylococcal colonies on BHA, dilutions were also plated on BHA + 10% NaCl. In fact, lactic acid bacteria did not interfere with counts on BHA, but the ability of staphylococci to grow on the salt medium decreased before any marked reduction in colony formation was recorded on BHA.

Discussion

The water relations patterns for survival of *Staph. aureus* and *S. newport* present in dried foods are basically similar to those reported by Scott (1958) for the same organisms freeze-dried in papain digest broth. Storage in cake mix under

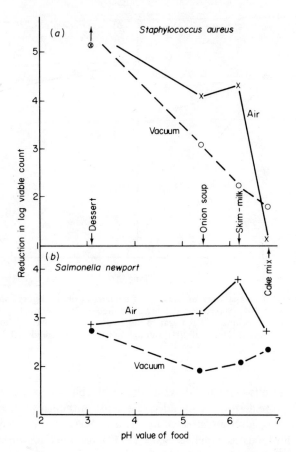

Fig. 5. Relation of pH value of foods to the reduction in viability of bacteria during storage for 9 weeks at 25° and 0·43 a_w in air and in vacuum. (*a*) *Staphylococcus aureus*; (*b*) *Salmonella newport*.

Table 3

Time in weeks for 99·9% decrease in viable count at 0·53 a_w and 25°

Organism	Atmosphere	Cake mix	Skim milk	Onion soup	Dessert
Staphylococcus	Air	8	3	1	2
aureus	Vacuum	18	7	4	5
Salmonella	Air	7	3	2	1
newport	Vacuum	14	6	7	2

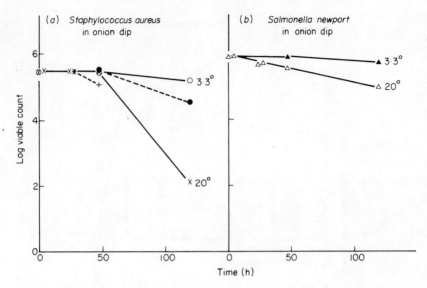

Fig. 6. Effect of temperature on survival of bacteria in onion dip. (a) *Staphylococcus aureus*: x, stored at 20°, counted on BH agar; +, stored at 20°, counted on BH agar with salt; ○, stored at 3·3°, counted on BH agar; ●, stored at 3·3°, counted on BH agar with salt; (b) *Salmonella newport*: △, stored at 20°, counted on BH agar; ▲, stored at 3·3°, counted on BH agar.

vacuum gave results very like those in the broth, but the other 3 foods were progressively more destructive with decrease in pH. Although survival at $0·00\,a_w$, in particular, was adversely affected by the presence of air, in no case was the destruction in foods as rapid as Scott (1958) recorded for broth or water suspensions of salmonellae freeze-dried and stored in air. Thus even very acid and destructive foods provide substantial protection against this effect.

Skim milk appears to be the only one of these 4 foods in which bacterial survival has been studied previously as a function of a_w. The survival data presented here for *Staph. aureus* in milk in air agree closely with those of Higginbottom (1953) for a *Micrococcus* sp. treated similarly. In both instances, optimal survival was obtained at about $0·1\,a_w$ and the destruction curves at $0·0\,a_w$ fell between those recorded at 0·3 and $0·4\,a_w$.

At a more quantitative level, 1% survival of *S. newport* after 9 weeks in skim milk at $0·22\,a_w$ is in accord with data of LiCari & Potter (1970). They recorded that after 8 weeks at 25°, survival of *S. thompson* was 3% if the milk was contaminated before spray drying and 0·9% if contaminated after drying.

The sets of survival curves in skim milk (Fig. 4a) were very similar to those in cake mix (Fig. 2) but were steeper at the higher a_w levels. Data from onion soup and dessert experiments did not conform to this simple pattern. In onion soup (Fig. 1), there was a much greater tendency, especially with staphylococci, for death rates to be exponential. The typical response in dessert was a rapid and substantial fall in viability followed, at low a_w levels, by a persistent survival of 0·1–0·01% of initial cells. The influence of pH is being studied further, using a range of foods, each adjusted to several levels of pH.

The only property common to all the foods which appears related to bacterial survival under the conditions studied is pH. If this is a basic factor, then it can be deduced from Fig. 5 that skim milk contains an additional destructive agent which is active at high a_w levels but only in air. Table 3 suggests that onion soup contains an ingredient active against staphylococci, but not salmonella, in both air and vacuum. Johnson & Vaughn (1969) reported that the threshold concentration of rehydrated onion powder at which detectable death of *S. typhimurium* occurred was 1% (w/v). The soup mix may have contained protective ingredients. Its onion content and other constituents are not known.

The observation that the lethal effects of a 1·5% oxygen atmosphere were comparable with those of air is of some practical significance. In commercial vacuum (or nitrogen) packing, the residual oxygen concentration after evacuation (and flushing) is likely to be greater than 1·5%, so that the protection afforded to bacteria by vacuum in all these experiments is unlikely to exist in practice.

From the initial a_w levels of the foods (Table 1) and the experimental data for survival during storage in air, the approximate times required for a 99% decrease in bacterial count in each food as purchased can be derived. For staphylococci, these are 1 week in dessert, 18 weeks in skim milk, 27 weeks in cake mix, and more than 27 weeks in onion soup. The periods for salmonellae are 2 weeks, 9 weeks, 10 weeks and more than 27 weeks, respectively.

One illustration of a possible hazard of survival in an unheated, reconstituted dried food is given by the onion dip. Growth of staphylococci and salmonellae was not supported by the mixture during 5 days incubation at 20°, in spite of the high a_w. The data obtained with salt agar suggest that this may be due in part to sublethal injury on prolonged contact with the rather acid dip. However, in spite of the destructive nature of dried onion soup in the storage trials at various much lower a_w levels, little loss of viability occurred with either organisms when the dip was held for 2 days at 20° or 5 days at 3·3°.

The data show that a_w has a major effect on the survival of pathogens in dried foods, and that constituents of the food, oxygen tension, and particularly pH may drastically modify this effect. In practical terms, the most important factors appear to be a_w and pH. It may be possible, when more data have been obtained, to make useful predictions of bacterial survival from a knowledge of these 2 parameters in a particular food.

References

Higginbottom, C. (1953). The effect of storage at different relative humidities on the survival of micro-organisms in milk powder and in pure cultures dried in milk. *J. Dairy Res.* **20**, 65.

Johnson, M. G. & Vaughn, R. H. (1969). Death of *Salmonella typhimurium* and *Escherichia coli* in the presence of freshly reconstituted dehydrated garlic and onion. *Appl. Microbiol.* **17**, 903.

Kampelmacher, E. H., Ingram, M. & Mossel, D. A. A. (Eds.) (1969). *Proc. 6th Int. Symp. Food Microbiol., Bilthoven, June 1968.*

LiCari, J. J. & Potter, N. N. (1970). Salmonella survival during spray drying and subsequent of skim-milk powder. III. Effects of storage temperature on salmonella and dried milk properties. *J. Dairy Sci.* **53**, 877.

Robinson, R. A. & Stokes, R. H. (1955). *Electrolyte Solutions.* London: Butterworths.

Scott, W. J. (1958). The effect of residual water on the survival of dried bacteria during storage. *J. gen. Microbiol.* **19**, 624.

Discussion

Skovgaard

Enzymic reactions in bacterial cells are decreased when the water activity is diminished. This may explain the effect of the water activity mentioned on the survival. Could you maybe give an explanation for the effect of oxygen and pH?

Christian

An important chemical reaction that may result in death in the relatively dry state is non-enzymic browning, which occurs, for example, between sugars and amino acids. This is accelerated at low pH. I don't recall whether oxygen is involved.

Ingram

Following the hypothesis that browning reactions may be involved, do you notice any browning in foods you tested?

Christian

It was not obvious.

Ingram

At low water activities fat oxidation reactions tend to become much more important than browning reactions which could explain the marked effect of oxygen at low water activities.

Christian

This is quite likely at the very lowest a_w levels tested, where destruction rates seem to be sensitive to factors less important under other conditions.

Hess

Destruction rates for staphylococci were stated in the summary to be much higher than the rates for salmonellae. Can you explain why?

Christian

No. In spite of the general belief to the contrary, staphylococci appear to be more sensitive than salmonellae to a number of physical agents when other environmental factors, for example pH, are suboptimal.

Nelson

Stressed micro-organisms, as the survivors of these treatments would be, are more sensitive to recovery conditions than are "normal" organisms. Gram positive organisms tend to be recovered better at slightly less acid conditions after stress than are Gram negative organisms. This could help explain the better apparent survival of salmonellae than of staphylococci in the more acid, low a_w foods.

Christian

This is true. No attempt was made to buffer cells against low pH during rehydration.

Beuchat

The question may be answered when Dr Christian collects data from experiments which he now has in progress. He has eliminated the variability of different chemical compounds which are present in the 3 food substrates studied here. He will be able to contrast Gram negative versus Gram positive survival at various pH values and a_w levels since he is using only 1 substrate.

Means and Methods

H. RIEMANN

Department of Epidemiology and Preventive Medicine,
University of California,
Davis, California,
U.S.A.

A critical review of the effectiveness of meat and poultry inspection in the U.S.A. indicates lack of efficiency and the necessity to find methods to evaluate the system. As many of the food-borne disease agents are shared by man and food animals, the aim is to produce pathogen-free livestock and poultry and ensure the wholesomeness of food. There should be provision for sufficient laboratory capacity to obtain optimum performance and epidemiology should be used as a diagnostic tool by the Meat Inspection Service.

ON MARCH 27, 1972, the American Public Health Association and 6 other organizations filed a public interest law suit against the Secretary of Agriculture and the administrators of the United States Department of Agriculture's Consumer and Marketing Service. The suit charges that the Secretary has violated his statutory duty under the Wholesome Meat Act and the Wholesome Poultry Products Act that meat and poultry products are not adulterated or misbranded when delivered to the consumer. It is stated that the USDA has repeatedly refused to attach instruments for safe handling and cooking of these products even though the USDA admits that they often contain salmonella and other bacteria that can cause food poisoning.

This incident illustrates problems inherent in the consumer protection system that covers an important part of the food supply—and the problems are not unique to the United States.

Meat inspection as it is practised today—and it is practised in much the same fashion as when it was instituted 100 years ago—has had a very important function in reducing food-borne diseases. However, while the inspection system has not changed in principle, the disease pattern in livestock has changed and some of the most important food-borne disease agents today elude the efforts of the meat inspector. Thanks to past successes it is generally taken for granted that inspection serves its intended purpose but there have apparently been only a few studies devoted to critical evaluation of inspection effectiveness. Three examples are given below. The purposes of meat and poultry inspection are: (1) to prevent micro-organisms or other disease agents present in animals brought to slaughter from causing food-borne diseases; (2) to prevent contamination of carcasses with disease agents (sanitation); and (3) to keep filth and other foreign material out of the food, that is, to insure general wholesomeness.

Sadler *et al.* (1966) studied the prevalence of bacterial agents and the histopathology in passed and condemned poultry carcasses and livers. Salmonellae were found in the caecal area of 5·8% of condemned chicken fryers and in 1% of passed carcasses; this difference was statistically significant. The corresponding figures for turkeys were 16·5% and 12·7%, and for chicken hens 1·2% and 1·7%, which revealed no significant differences between condemned and passed carcasses. Examination of livers from passed and condemned carcasses for a variety of bacteria showed no significant difference between passed livers and condemned livers. With respect to the effectiveness of sanitation inspection even less information is available. Huff (1972) examined 10 years of inspection records from rendering plants where laboratory data for salmonella contamination of the final product were available. Although the level of contamination declined over the period involved, no inspection criteria were found to have predictive value with respect to contamination.

The last of the above mentioned functions of meat inspection, judgment of wholesomeness, was studied by Brant *et al.* (1967) for poultry. It was concluded from this study that the appearance factors with respect to eviscerated and uneviscerated poultry, used by consumers and laymen for evaluation of wholesomeness, were quite different from those employed by veterinary inspectors.

These studies do not show that inspection is without value but they do indicate that: (1) the present inspection system is not sufficient; and (2) there is a need to develop methods to evaluate the effectiveness of inspection.

There is probably no disagreement that inspection—which applies to the left part of Fig. 1—is important in the microbiological safety of foods. Many of the food-borne disease agents are shared by man and food animals; the latter are the natural reservoir for some of the agents which are most prevalent today, salmonellae and probably also enteropathogenic *Clostridium perfringens* type A.

The question is how to augment inspection and this was discussed at the 1971 National Conference on Food Protection (American Public Health Association, 1971) which recommended that programs be directed towards the reduction and eventual elimination of *Salmonella* and other pathogens from livestock and poultry. The recommendations included: (1) elimination of animal pathogens from animal by-products and feed no later than 1976; (2) establishment of a surveillance and reporting system of outbreaks of salmonellosis in livestock and poultry; (3) development of an eradication program for pullorum disease and fowl typhoid involving all chickens and poultry breeding flocks in the United States; (4) establishment of poultry breeding flocks and livestock herds free from the 10 most prevalent salmonella serotypes; and (5) research to find ways

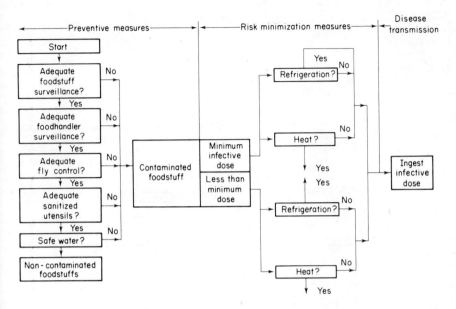

Fig. 1. Methods of food protection (from APHA, 1971).

to assure a salmonella-free environment including housing, husbandry practices and waste disposal.

Some of these recommendations have in fact already been put into practice on a local level (Sadler, 1972). The role of the meat inspection service in efforts of this type has not been established, but there is a great potential. Meat and poultry inspection is already an invaluable part of the effort to eradicate animal diseases (Van Houweling, 1965). The market cattle identification program which consists of sampling at the slaughterhouse and a traceback system forms a basis for tuberculosis and brucellosis control. It has also proven its value in epidemiological studies of anaplasmosis and has been suggested as a basis for monitoring environmental pollutants. Market swine identification programs have been established as a basis for brucellosis control and have been used in epidemiological studies of pseudorabies.

There is no doubt that these programs could be equally valuable in epidemiological studies and control of *Salmonella* and *Cl. perfringens* infections. Neither is there any doubt that the meat inspector, if freed from some of his routine duties, could play a more active role which in a sense would mean that part of the meat inspection would be moved from the slaughter establishment to the auction yard and the producer. Such use of the meat inspection system would save expense and provide a more meaningful activity and probably higher

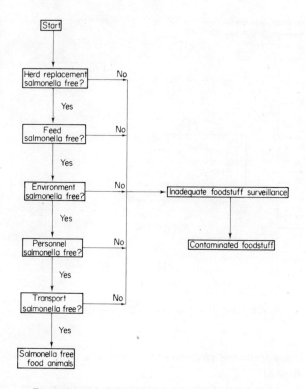

Fig. 2. Outline of expanded meat inspection system.

prestige for the veterinary meat inspector. An outline of an expanded meat inspection is shown in Fig. 2.

Sufficient laboratory capacity is an important prerequisite for the successful control of food animal infections and of the risk of post-mortem contamination. In some countries all or a majority of slaughterhouses have laboratories for bacteriological examination; in other cases service is given by central laboratories. However, laboratory examinations are costly and serious consideration must be given to ensuring optimum laboratory performance. Two factors must be considered when a laboratory procedure is chosen: (1) the specificity of the method; and (2) the sensitivity of the method. The first is a measure of the risk of getting false positives, the second of getting false negatives. Often too much emphasis is placed on the sensitivity of the method *per se*. In practice sensitivity cannot be separated from the sampling variance which is often of considerable magnitude. It is often better to use a less sensitive method if it is simpler, and to increase the number of samples. Efforts should be devoted to the development, automation and acceptance of such methods.

As already indicated meat inspection could play an important role in epidemiological studies related to infection of animals with agents of food-borne disease. While much effort has been devoted to the control of *Salmonella* in food products, relatively little attention has been paid to the epidemiology in populations of food animals and work done generally falls into the category of descriptive epidemiology. Yet little precise information has been gained with respect to important questions such as why salmonella prevalence in hogs is low in some areas and high in others and why transportation sometimes increases faecal shedding of salmonella. There is a need for the application of epidemiology as a diagnostic discipline in these areas and it would be natural for the meat inspection service to be at the center of such activities.

References

American Public Health Association. (1971). *Proceedings of the 1971 National Conference on Food Protection*. U.S. Government Printing Office, 1712-0134. Washington, D.C.

Brant, A. W., Sadler, W. W. & Lewis, H. (1967). Consumer and trained panels versus the veterinary inspector on wholesomeness of poultry carcasses. *Poult. Sci.* **46**, 450.

Huff, I. (1972). Evaluation of inspection effectiveness. A study of the California animal by-products inspection records. Research report for the Master of Preventive Veterinary Medicine degree, University of California, Davis. To be published.

Sadler, W. W. (1972). Personal communication.

Sadler, W. W., Corstvet, R. E., Adler, H. E. & Yamamoto, R. (1966). Relative prevalence of bacterial agents and histopathology in passed and condemned poultry carcasses and livers. *J. Fd Sci.* **31**, 773.

Van Houweling, C. D. (1965). The value of meat and poultry inspection to disease control and eradication. *Proc. IVth Symp. World Association of Veterinary Food Hygienists*, p. 404.

Discussion

Linderholm

I think we, or I, can agree to the principle in some but not all of Prof. Riemann's remarks about the necessity of reviewing some inspection systems. But I should like to add that the opinion is not new and the problem has been observed and discussed internationally for several hours. As 2 examples I would mention that the World Association of Veterinary Food Hygienists has held 2 symposia (Round Table Conferences) with participants from more than 20 countries during the past years. One was on Meat Inspection and Meat Hygiene and the other on Poultry Inspection and Poultry Meat Hygiene.

Resolutions stressing the necessity of solving the problems touched upon by Prof Riemann were made at these conferences. Sweden has some disease control programmes for food producing animals and their environment which have been set up from the public health point of view. As examples salmonella-control programmes for pigs, cattle and poultry and leukosis-control programmes for cattle and poultry can be mentioned. We know, however, too little about the harmfulness of leukosis to the consumer and more investigation must be done. It seems to me that such control programmes are more advanced and that the authorities have more power over the handling of the food-producing animals already on the farm if the veterinary services are more involved in the food control administration than they are in some countries.

Elliott

The suit by the American Public Health Association would require that we put a warning label on raw meat and poultry. This singling out of a specific commodity among many that may carry salmonellae seems to us undesirable. We, however, are now increasing our consumer health programme since for food infection to occur there usually must be mistreatment at the serving level. Our sanitation programme has been constantly improving. Whereas it is difficult to demonstrate that improvements in general sanitation will also improve the salmonella picture we have high hopes that this will happen. Statistically valid data to show the effects of programmes such as this require large numbers of samples, usually more than a laboratory system can handle.

Our Agricultural Research Service conducts experiments designed to improve processing methods. Our Animal Health Services are continuously studying methods for animal disease control. Salmonella control, as Dr Riemann says, requires improvement in each step of production from farm to consumer. The Meat and Poultry Inspection Programme of the USDA has authority from the time animals are received for slaughter until the food is delivered to the retail store. As long as carrier animals are received for slaughter, salmonella-free meats will be impossible to produce. Dr Williams' paper this morning was of great interest; he has apparently developed procedures that will reduce salmonella incidence dramatically in pork. We need more of this kind of work.

Insalata

Is there presently a salmonella method quick enough (specific and selective) to allow valid statistical sampling?

Riemann

Different methods have been suggested for more rapid detection of salmonellae. It is conceivable that one or more of these methods may be acceptable but to my knowledge none has been sufficiently tested. The sensitivity of the method is not too critical if the method is not biased and permits a large number of samples to be examined.

Session 2

Epidemiology of Food-Borne Infection in Man and Animals

Part 1: Man
Chairman: Sir Graham Wilson

Food Poisoning in England and Wales

BETTY C. HOBBS

Food Hygiene Laboratory,
Central Public Health Laboratory,
Colindale Avenue,
London NW9 5HT, England

The Public Health Laboratory Service has recorded food poisoning incidents since 1941 and these reports together with those submitted to the Department of Health and Social Security form the annual statistics for food poisoning. Facts are presented about the ratio of one agent to another, the foods involved in relation to agents and the causes of some outstanding outbreaks. During the years 1968–1970 there was an average of 6000 incidents each year including general and family outbreaks and sporadic cases; the reported figure of *c*. 10,000 cases annually is thought to be a gross under estimate of the true number. Salmonellae were responsible for the largest proportion of incidents; reported infections due to *Salmonella typhimurium* were falling while those due to other serotypes of salmonellae were rising. *Clostridium welchii* and *Staphylococcus aureus* were the next most common food poisoning agents. In 80–90% of incidents meat and poultry were the food vehicles for all types of food poisoning. Isolations of salmonellae from surveillance studies on various foods and feeds of animal origin and veterinary reports on salmonellae isolated from dead animals are presented in tabular form in the annual reports; the serotypes are compared with those causing human infections. The data for 1968–1970 indicate that as usual animals and animal products were the main sources and vehicles of salmonella food poisoning. Spread of salmonellae from raw to cooked food (meat and poultry) by hands, surfaces and equipment, and sometimes survival of a few organisms after cooking, followed by multiplication under poor conditions of storage are considered to be mainly responsible for salmonellosis. Inadequate cooling and cold storage facilities for bulks of cooked meat and poultry are the faults leading to *Cl. welchii* food poisoning. Contact of hands, frequently contaminated with staphylococci, with cooked meats and poultry followed by lengthy storage at ambient temperatures continually give rise to staphylococcal food poisoning. One outbreak only of *Vibrio parahaemolyticus* has been reported in the U.K. and it was not clear whether cross contamination from raw to cooked crabs or survival after cooking was responsible for the presence of the vibrio in cooked crab meat. Gross multiplication of *Bacillus cereus* in cooked rice is causing food poisoning; the fault is lengthy non-refrigerated storage after cooking.

THE ASSESSMENT of the significance of food poisoning in relation to morbidity and mortality in any country requires statistics on incidence, proportion of agents in relation to incidents, vehicles of infection and other factors; without such data it would be difficult to know where to place the greatest emphasis in surveillance and preventive studies. To obtain the necessary statistical data there must be statutory obligation to report food-borne disease and a department assigned to collate the data which become available from a good reporting system and conscientious public health workers. The prevention of food-borne disease requires not only a knowledge of the statistics but also an awareness of the characteristics of the organisms responsible for gastro-intestinal disease, their ecology in the environment and their behaviour in foods.

Statistics

For the past 25 years the Public Health Laboratory Service and the Department of Health and Social Security have collected and analysed data from reported incidents of food poisoning. The results are published annually in a report called *Food Poisoning in England and Wales*; formerly they appeared in the Monthly Bulletin of the Ministry of Health and Public Health Laboratory Service which is no longer published. The 1968 figures are given in Vernon (1970); the 1969 to 1971 reports are due for publication. Facts are presented about the proportion of incidents due to the different bacterial agents in relation to the food vehicle of infection. Details are given about the causes of some outstanding outbreaks. The results of surveillance studies for salmonellae in various foods and animal feeds are given in tabular form and compared with serotype studies of salmonellosis in man. Information is also given on the salmonella serotypes isolated from dead animals taken to veterinary investigation stations for post-mortem examination.

Incidents and Agents

As in other countries the predominant organisms causing food poisoning are those of the salmonella, staphylococcal and clostridial groups. Lately, but as yet unrecorded in the statistics, *Bacillus cereus* has been isolated in a number of incidents. The first outbreak known to be caused by *Vibrio parahaemolyticus* in the United Kingdom has recently been investigated. The numbers of incidents reported annually for the years 1968–1970 were *c.* 5000–6000 including general and family outbreaks and sporadic cases. The cases were estimated to be *c.* 10,000–12,000 each year, but this figure should be multiplied many times to include all those which are unreported. Salmonella was the commonest agent of food poisoning, but the incidents included a large number (greater than 50%) of sporadic cases; these so-called sporadic cases may be parts of outbreaks due to widely distributed food vehicles of infection. Fewer sporadic cases are reported from staphylococcal and *Clostridium welchii* food poisoning; these agents are usually responsible for outbreaks affecting many people. The figures for incidents alone, therefore, can be misleading unless cases are considered also. Most outbreaks of *Cl. welchii* food poisoning involve an average of at least 30 persons, based on the 1967–1969 figures; those for staphylococcal food poisoning seem to average about 16 in number. Table 1 shows the number of incidents and cases according to the agents of food poisoning for 1968–1970 as given by the Public Health Laboratory Service. Table 2 shows the Chief Medical Officer's statistics for the years 1968–1970; the reasons for the discrepancy between the 2 sets of figures is not altogether clear but it is assumed to be due to irregularities in reporting incidents and other data. Another interesting fact is that whereas 80–90% of *Cl. welchii* outbreaks and about 70% of staphylococcal outbreaks are traced to their food source, only *c.* 24% of general and family outbreaks and 1% of sporadic cases of salmonellosis are followed through successfully.

Table 1

Bacterial food poisoning in England and Wales 1968–1970–data from Public Health Laboratory Records

| Year | Causal agents | | | | All causes | |
	Salmonella	*Staphylococcus*	*Cl. welchii*	Others and unrecorded causes	Incidents	Cases*
1968	3796(5948)	63(309)	60(1840)	1165(2414)	5084	(10,475)
1969	4820(7322)	17(397)	47(1534)	Not recorded	4884	(9,253)
1970	5225(6852)	29(523)	32(1263)	Not recorded	5286	(8,638)

Incidents (general and family outbreaks and sporadic cases) and total cases (in brackets) according to causal agents.

* Include 361 symptomless excretors of *S. typhimurium* and 880 symptomless excretors of other salmonellae.

Table 2

Bacterial food poisoning in England and Wales 1968–1970–data from the Annual Reports of the Chief Medical Officer, Department of Health and Social Security

| Year | Causal agents | | | | All causes | |
	Salmonella	*Staphylococcus*	*Cl. welchii*	Others and unrecorded causes	Incidents	Cases*
1968	1919(3236)	72(394)	67(1441)	1196(2039)	3254	(7110)
1969	2761(4449)	51(208)	69(1243)	1581(2698)	4462	(8598)
1970*	2956(5018)	56(370)	83(945)	1755(2474)	4241	(8088)

Incidents (general and family outbreaks and sporadic cases) and total cases (in brackets) according to causal agents.

* England only.

Food Vehicles

The food vehicles responsible for the various types of food poisoning differ according to the agent. Staphylococci from the fingers of food handlers contaminate the cooked foods, chickens, semi-preserved meats and cold meats of any kind. The spores of *Cl. welchii* and *B. cereus* in food and environment survive cooking and grow out when food is cooled slowly and stored unchilled. There is a similar phenomenon in preserved foods contaminated with *Cl. botulinum* in those countries which experience this form of food poisoning. *V. parahaemolyticus* may be eaten directly in raw seafoods or it may be transferred from the raw to the cooked product by cross-contamination in ways similar to those described for salmonellae in meat and poultry. Salmonellae from raw meats and poultry contaminate the environment and are transferred to cooked foods by means of fingers, surfaces, utensils, equipment and other environmental factors.

Inadequate thawing and cooking may allow survival of salmonellae, in poultry, for example. This, followed by careless storage, may encourage the development of a dose large enough to cause illness. All these organisms causing food poisoning must grow in food before there are sufficient numbers or a sufficient volume of toxin to cause symptoms. It is possible that *V. parahaemolyticus* may survive in the flesh inside the claws and body of crabs given too short a period of boiling. Experiments suggest that this may be so. Cooking, except under pressure, can never guarantee the destruction of all the spores of *Cl. welchii, Cl. botulinum* or even *B. cereus. Cl. welchii* and *B. cereus* are ubiquitous organisms and it would be impossible to ensure their exclusion from food or from the environment.

Table 3 gives the food vehicles and causal agents for 280 general and family outbreaks investigated in the years 1968–1970. In 87·5% of the incidents meat and poultry proved to be the vehicles of infection and Table 4 gives the categories of meat and poultry meals involved, according to the causal agent. It will be seen that a high proportion of incidents caused by salmonellae and *Cl. welchii* were due to fresh meat and poultry cooked and eaten cold; *Cl. welchii* was the predominant agent associated with precooked and reheated meat and poultry and also pies cooked at home and in canteens; staphylococci accumulated in cold cooked processed food subjected to handling and poor storage conditions. Salt in cures will also play a part in the growth of staphylococci by depressing the development of the competing microflora.

Surveillance

The source of the staphylococci causing food poisoning is almost certainly human, from the nose and skin of the food handlers. The main focus of

Table 3

General and family bacterial food poisoning outbreaks in England and Wales 1968-1970*—food vehicles and causal agents

Food vehicle	Causal agent				
	Salmonella	Staphylococcus	Cl. welchii	Not discovered	All agents
Meat and poultry	78	44	113	10	245
Sea food		4	0	6	10
Milk (unpasteurized)	17	0	0	0	17
Dairy products (trifle, cream cakes, egg and egg products)	2	5	0	1	8
All foods	97	53	113	17	280

* Public Health Laboratory Service records.

Table 4

General and family bacterial food poisoning outbreaks in England and Wales 1968-1970*—food vehicles (meat and poultry) and causal agents

Food vehicle	Causal agent				
	Salmonella	Staphylococcus	Cl. welchii	Not discovered	All agents
Cold cooked meat and poultry					
Processed (including canned)	8	22	0	2	32
Fresh	56	14	37	2	109
Reheated cooked meat and poultry	11	3	53	5	72
Meat pies					
Manufactured	3	2	0	0	5
Home and canteen	0	3	23	1	27
All foods	78	44	113	10	245

* Public Health Laboratory Service records.

salmonella organisms, except for *Salmonella typhi* and *S. paratyphi* bacilli, is the animal intestine. The cycle of spread of salmonellae amongst animal excreters must be infinite, but it may be initiated by contaminated feedstuffs and enhanced by the conditions under which animals are reared, transported, marketed and slaughtered. Many published works testify to this chain of infections and they are reviewed by Williams & Hobbs (1973). From the gut of slaughtered animals and poultry to the flesh prepared for human consumption is a short step, and the microbes from raw foods are readily distributed around the factory, retail shop and kitchen. Cross contamination from raw to cooked foods will be a day-to-day occurrence. Table 5 gives the rates of contamination with salmonellae of some foods.

Table 5
Surveillance studies—Salmonella in poultry and sausages

Food	No. of samples examined	Salmonella found in		
		No.	%	Range %
Poultry (raw, frozen)	101	36	35·6	3–62
Sausages				
Producer				
A Large	252	98	38·9	
B Large	135	3	2·2	
C Medium	52	6	11·5	
D Medium	50	0		0–44
E Small	30	0		
F Small	30	0		
Prawns (cooked frozen)				
Malaysian	300	4	1·3	
Other	400	0		0–1·3

Every food microbiologist should seek the origin of the bacteria which are found in the foods for which he is responsible. It is also essential to study the behaviour of these organisms in foods as well as in the human and animal body.

Ecology

Characteristics and spread

(a) Cl. welchii *and* B. cereus

The characteristics and ecology of food poisoning agents provide the keys to their control and to the prevention of gastro-enteritis. An overall list of notes on personal hygiene will do little or nothing to prevent *Cl. welchii* and *B. cereus* food poisoning, and it is even doubtful whether it would have much effect on the occurrence of salmonellosis. The spores of *Cl. welchii* will survive in food after most normal cooking procedures and outgrowths of cells from the spores is facilitated by the heat shock of cooking. Long slow cooling and warm storage will encourage the multiplication of this organism, which has a short generation time and an optimum growth temperature between 43–47°. Control measures must seek to prevent the multiplication of the organism in masses of food required to be eaten cold, still warm or reheated some hours after preparation; proper cooling facilities should be provided in canteens preparing meals for large numbers of people.

There is a similar sequence of events before *B. cereus* food poisoning can occur—survival of the spores after cooking and outgrowth and multiplication of cells in the slowly cooling unchilled foodstuffs. There are differences between the organisms in that *B. cereus* spores readily in foods whereas *Cl. welchii* does not, and *B. cereus* forms a toxin in the foodstuff (Goepfert *et al.*, 1973) whereas *Cl. welchii* appears to release its toxin at the moment of sporulation in the intestine (Strong *et al.*, 1971; Hauschild *et al.*, 1971). Both *Cl. welchii* and *B. cereus* are ubiquitous organisms, and, except by reducing their incidence in the environment of food preparation by strict attention to cleanliness and anti-dust measures, it would be impossible to eliminate them altogether. Thus their presence in foods in small numbers must be accepted but their development into large numbers cannot be tolerated. The time between cooking and eating must be short and if cold dishes are required rapid cooling and cold storage facilities are essential. When reheating is essential, the mass of food must be seen to boil through to the centre. *Cl. welchii* spores rarely if at all in cooked foodstuffs but the non-sporing vegetative cells which are swallowed appear to sporulate readily in the intestine; masses of spores will pass out into the sewage system with the numerous fluid stools caused by diarrhoea and perpetuation of the organism is thus assured.

Although *B. cereus* spores readily in foodstuffs, it is not always found in the stools of those who suffer an acute attack of vomiting (presumably due to the toxin) only 2 h after eating food heavily populated with this organism. Because of the active sporulation in food the spores will spread readily in the environment. This organism gives rise to spoilage problems in dairy plants, in bakeries and also in hatcheries where coagulation of the yolk of the fertilized eggs robs the developing embryo of nourishment. *Cl. welchii* commonly causes food poisoning by its growth in cooked meat and poultry dishes, whereas *B. cereus* causes food poisoning by its growth in cooked cereal products. Both are harmless in small numbers, both have a high survival rate and *Cl. welchii* has a short generation time.

Table 6 compares the characteristics of these 2 organisms with regard to their ability to cause food poisoning.

(b) Staphylococcus aureus

Staphylococci which produce enterotoxin in food appear to originate almost entirely from man. The organism remains viable on the skin and it is difficult to eliminate by ordinary washing methods. Hot water and scrubbing will sometimes lead to an increase in the numbers of staphylococci on the hands. *Staph. aureus* from boils, carbuncles, whitlows and other staphylococcal infections linger on

Table 6
Characteristics in relation to food poisoning

	Clostridium welchii	Bacillus cereus
Prevalence	Common Food, faeces, soil, dust	Common Cereals, soil, dust
Growth	Anaerobic growth in meat and poultry	Aerobic growth in rice and other cereals
Generation time	10–12 min	27 min
Optimum temperature Minimum Maximum	43–$47°$ 15–$20°$ $50°$	$30°$ 10–$12°$ $48°$
Sporulation	Poor in food Good in intestine	Good in food
Heat resistance D_{100}	Less than 1 min to 17 min	1 min to 7·5 min
Mechanism	Toxin liberated in intestine	Toxin produced in food
Incubation	12–18 h	2 h (15 h)
Symptoms	Diarrhoea and pain (vomiting)	Vomiting (diarrhoea and pain)
Dose required	Large numbers of organisms	Large numbers of organisms plus toxin
Stool	Large numbers of organisms	Variable number of organisms
Prevention	Stop growth in food	Stop growth in food

the skin long after the lesions have healed. Staphylococci from the fingers of food handlers will inevitably find their way into cooked foods unless instructions not to touch cooked foods with the fingers are more stringent. In small numbers they are harmless and again the main line of control is the prevention of growth by the proper use of cold storage facilities. Papers from this laboratory (Simkovicova & Gilbert, 1971; Gilbert et al., 1972; Gilbert & Wieneke, 1973) describe some of the methods for the detection of staphylococcal enterotoxin in culture filtrates of Staph. aureus and in samples of food implicated in food poisoning outbreaks. These papers also confirm the findings of various workers that staphylococci incriminated in food poisoning are most likely (greater than 90%) to belong to phage-group III or groups I/III; most of the strains produce enterotoxin A or enterotoxin A and B. However, unpublished data from the same workers on a large number of cultures isolated from a wide variety of routine food samples and from wound infections among hospital patients indicate that it is almost impossible to predict whether a phage-group III or any other strain produces enterotoxin A, B or C.

(c) Salmonella

The cycle of events for salmonella food poisoning is more complex. Except for host specific serotypes and *S. typhimurium*, this cycle can be described simply as feedstuffs–animal–foodstuffs–man, with an inner circle of perpetuation because the imperfect treatment of animal remains rendered into feedstuffs cycles the salmonellae back again to animals. Excretion of even small numbers will passage the organisms throughout overcrowded intensive rearing systems. When animals and birds are huddled together in transport, markets and slaughterhouses, the rates of excretion will be increased by tension factors common to all living creatures under such conditions. The host specific types will spread more rapidly because they more often cause illness and increase enormously in the intestine of the infected animal; the liquid faeces will be heavily contaminated and cause a profuse environmental scatter. The more exotic serotypes picked up from feedstuffs may be carried in a transient manner and in small numbers sufficient, however, to contaminate carcasses when methods of slaughter and dressing are not hygienic. Many of the serotypes which do not readily initiate acute infection in animals are invasive for man and cause illness. The aerosol spread of salmonellae must be profuse when poultry are slaughtered and the only way to ensure freedom from salmonellae of the dressed carcass is to ensure freedom from salmonellae of the live bird.

Countries which legislate for the elimination of salmonellae from feedstuffs, for example Denmark, may still have a fair incidence of *S. typhimurium* in animals, birds and man but the incidence of other serotypes is much reduced and the figures for outbreaks are low.

The significance of contamination of food after cooking is often not appreciated and the fact that raw meat and poultry are to be cooked is considered to be a panacea for selling salmonella-contaminated raw foods. Competitive flora repress the growth of salmonellae in raw meat, but small numbers of salmonellae picked up after cooking, when the spoilage organisms are destroyed will grow relatively undeterred by competition with other organisms. Sterilization of the environment between preparation of the raw food and removal of the cooked food from the oven is impossible. Complete separation in time and space of raw preparation and the manipulation of foods after cooking is required, otherwise inhibition of growth is the only safeguard. Table 7 compares the characteristics of salmonellae and staphylococci in relation to food poisoning.

(d) Vibrio parahaemolyticus

Food poisoning from *V. parahaemolyticus* in seafoods both raw and cooked is described frequently in the East; Sakazaki and his colleagues have contributed many papers on the subject (Sakazaki, 1969; Nickelson & Vanderzant, 1971). More recently outbreaks due to *V. parahaemolyticus* have been described in

Table 7
Characteristics in relation to food poisoning

	Salmonella	*Staphylococci*
Prevalence	Common Mammals (animals) Faeces	Common Man Skin
Growth	Meat and poultry Milk and dairy products Infective dose	Cold cooked foods Toxin in food
Incubation	12–24 h	2–4 h
Symptoms	Diarrhoea, pain, vomiting, fever	Vomiting (diarrhoea, pain)
Stool	Large numbers	Small numbers
Prevention	Stop growth	Stop growth

Australia (Battey *et al.*, 1970), the U.S.A. (Sumner *et al.*, 1971) and the U.K. (Peffers *et al.*, 1973).

In these last papers the danger of contamination from raw to cooked seafoods and in particular crab meat has been emphasized. Undercooking of whole crabs and claws may also play a part in the survival of the vibrio in cooked crab meat (Peffers *et al.*, 1973).

The organism may be found in the coastal waters of Japan and in the inshore sea creatures in the summer months when the food-poisoning is most common. It may be assumed that sewage from sick persons will also pollute the coastal waters and keep the cycle of man–water–seafood–man in circulation until some other factor, for example, a drop in atmospheric and sea temperature, discourages the organisms from growth in food and survival in water. It is interesting to postulate whether the occasional isolation of *V. parahaemolyticus* around our own southern shores and an occasional outbreak from imported raw and cooked seafood meals could set in motion a similar cycle of events. The answer may depend on the ability of the organisms to adapt to colder conditions or to produce a mutant or variant strain which is able to do so.

Microbiological Standards

Microbiological standards for viable counts and index organisms provide useful guidelines for cleanliness of food manufacture as well as for safety in public health. If the colony count alone were controlled, there would be an assurance that neither spoilage organisms nor intestinal pathogens could reach significant numbers. The number should be related to the level of the particular organisms known to cause spoilage or disease. In order to assess the significance of small numbers of intestinal pathogens in food more information is needed on the approximate dose levels required to initiate disease. Where particular hazards are associated with certain foods then examinations by enrichment methods should be carried out, for salmonellae for example. There should be agreement on the percentage of samples contaminated with salmonellae which ought not to be tolerated on the market. When surveillance studies show an increase in the number of raw food and environmental samples from which a particular serotype of salmonella can be isolated, or deaths of animals from a particular phage-type of *S. typhimurium,* a rise in the incidence of salmonellosis in man due to that serotype or phage-type may be expected. It may be impossible to demonstrate intestinal pathogens in food heavily contaminated with miscellaneous organisms but it is doubtful whether the cells, alive or dead, of vast numbers of bacteria are suitable for consumption. It is not long ago that *Cl. welchii* and *B. cereus* in food were regarded as harmless; how many other organisms of possible significance to health are disregarded even in large numbers?

The significance of index organisms and the relationship of their numbers to intestinal pathogens or simply to the general hygiene of procedures and environment, where hygiene is synonymous with bacterial cleanliness, occupies many hours of discussion. Most probable number techniques, except for water, are notoriously variable from worker to worker and laboratory to laboratory; when numbers are large enough direct plate counts may be used but they are complicated by the need for the identification of colonies.

Enzyme tests are accurate indications of heat treatment and dye tests are rough measures of bacterial content. Sampling procedures must be carefully considered when attempting to assess the safety and keeping quality of foods from the results of microbiological tests.

Prevention

Preventive measures can only be worked out from a knowledge of the habits of the various agents causing food poisoning. Such knowledge is obtained by careful investigation of outbreaks and by laboratory studies carried out on each bacterial agent responsible for food poisoning.

The institution of preventive measures may be related to the economy of industrial processes and catering and also to international trading and politics. The nuisance and loss of work from illness and any fatalities involved may be weighed against the other considerations before efforts are made to correct the faults leading to contamination. Statistical assays of food-poisoning will help to emphasize the areas in which control measures are necessary. Sometimes legislation is required to protect a commodity from faults in production much as the Statutory requirements for the pasteurization of milk, ice-cream mix and bulked liquid egg, the staining and sterilization regulations for condemned meat, and in Scandinavia the orders controlling imports of feedstuffs and production. Industry is usually quick to see the hazards of a process and will work to improve the safety of products. Teaching on food-borne disease should include information on the bacterial agents, the sources, the means of spread and control measures which can be taken in the kitchen, shop and factory, and on the farm where the story begins. The food handler must be convinced of the reality of bacterial contamination so that he learns to avoid it. Education on food hygiene is not only necessary for food handlers, technologists and teachers but for managers, directors, engineers and architects engaged in execution and design of food premises and production. The food microbiologist whether qualified in human or veterinary medicine, science or technology should present the facts to those who handle food in the simplest possible way giving practical demonstrations and illustrations from recent outbreaks.

References

Battey, Y. M., Wallace, R. B., Allan, B. C. & Keeffe, B. M. (1970). Gastro-enteritis in Australia caused by a marine vibrio. *Med. J. Aust.* **1**, 430.

Gilbert, R. J. & Wieneke, A. A. (1973). Staphylococcal food poisoning with special reference to the detection of enterotoxin in food. In *Proceedings of the 8th International Symposium. The Microbiological Safety of Food.* p. 273. London: Academic Press.

Gilbert, R. J., Wieneke, A. A., Lanser, J. & Simkovicova, M. (1972). Serological detection of enterotoxin in foods implicated in staphylococcal food poisoning. *J. Hyg., Camb.* **70**, 755.

Goepfert, J. M., Spira, W. M., Glatz, B. A. & Kim, H. U. (1973). Pathogenicity of *Bacillus cereus*. In *Proceedings of the 8th International Symposium. The Microbiological Safety of Food.* p. 69. London: Academic Press.

Hauschild, A. H. W., Niilo, L. & Dorward, W. J. (1971). The role of enterotoxin in *Clostridium perfringens* type A enteritis. *Can. J. Microbiol.* **17**, 987.

Nickelson, R. & Vanderzant, C. (1971). *Vibrio parahaemolyticus*–A review. *J. Milk Fd Technol.* **34**, 447.

Peffers, A. S. R., Bailey, J., Barrow, G. I. & Hobbs, B. C. (1973). *Vibrio parahaemolyticus* gastroenteritis. *Lancet i,* 143.

Sakazaki, R. (1969). Halophilic vibrio infections. In *Food-borne Infections and Intoxications.* New York: Academic Press.

Simkovicova, M. & Gilbert, R. J. (1971). Serological detection of enterotoxin from food-poisoning strains of *Staphylococcus aureus. J. med. Microbiol.* **4**, 19.

Strong, D. H., Duncan, C. L. & Perna, G. (1971). *Clostridium perfringens* Type A food poisoning. *Infect. Immunol.* **3**, 171.

Sumner, W. A., Moore, J., Bush, A., Nelson, R., Molenda, J. R., Johnson, W., Garber, H. J. & Wentz, B. (1971). *Vibrio parahaemolyticus* gastroenteritis. *Morbidity and Mortality Weekly Rept.* **20**, 256.

Vernon, E. (1970). Food poisoning and *Salmonella* infections in England and Wales, 1968. *Publ. Hlth (Lond.)* **84**, 239.

Williams, L. P., Jr. & Hobbs, B. C. (1973). Enterobacteriaceae infections. In *Diseases Transmitted from Animals to Man.* 6th ed. Eds. Schnurrenberger, McCullock & Huppert. Springfield, Illinois, Charles C. Thomas.

Discussion

Harrigan

We do not yet know the real significance or extent of outbreaks of food poisoning originating in the domestic kitchen due to the purchase of ready cooked meat not maintained under adequate refrigeration up to the time of consumption. I am concerned with the opportunities that exist for cross contamination from raw to cooked meats in the retail establishment. Whilst there are Codes of Practice on Hygiene in the Meat Trades there is ample evidence that many retail establishments take little notice of them and I have met managers of butchers' shops who deny knowledge of them. In many shops one can see raw and cooked meats displayed adjacent to and touching each other, with the same shop assistants serving both without tongs and without intermediate hand washing. Is it time for the Codes of Practice covering the retail trade to be reinforced by more explicit legislation? Should considerably more effort and money be expended to make the general public much more aware of the dangers, so that public pressure on the retail establishments will bring about the necessary improvements?

Bostock

In answer to the plea for legislation to give force to the hygiene Codes of Practice there is food hygiene legislation which requires that food should be protected from contamination. It cannot, however, include the amount of detail given in the Codes. The Food Hygiene Regulations are enforced by the Public Health Inspectors who visit food establishments as much as possible but they have other commitments which limit the time available for this work.

Eklund

In one of your tables you reported that minimum growth temperature of *Cl. perfringens* is $15°$. Do you know of any strain of *Cl. perfringens* which will grow at temperatures lower than $15°$?

Ingram

None of our strains grew at temperatures below $10–15°$. Does Dr Roberts want to fill in any details?

Roberts

In our experiments growth occurred readily in media at $15°$ but not at $10°$. However, in fresh meat which was vacuum packaged, growth has not occurred at $15°$ even after incubation for many weeks and with an initial heavy inoculum.

Wilson

Is the temperature of any great concern in the practical sense since sporulation will not take place in the food?

Ingram

In the case of chilling of meat it is much more significant, for example growth of *Clostridium putrefaciens* in pork.

Woodbine

More interesting is the generation time of *Cl. perfringens* which was shown in the slide as $10–12$ min. This is much lower than the MGT of *E. coli* of about 17 min [Barber (1908) *J. infect. Dis.* 5, 396] and about 20 min for *B. suispestifer (S. choleraesuis)* [Wilson (1922) *J. Bact.* 7, 434]. Does this faster growth operate overall, that is over the whole range of growth temperatures from $15–50°$? If so then, although growth is slower, it could still be faster than one might expect in a food, for example over the time period involved in cooling.

Barnes

The growth rate of *Cl. perfringens* in poultry meat was determined by Mead. A MGT of 10–12 min was obtained in the region of 40° but the organisms grew very slowly indeed at 20°.

Elliott

Yesterday I mentioned that it was difficult to evaluate the effect of processing techniques on naturally occurring salmonellae. Dr Hobbs studied 6 firms, 2 of which were large, 2 medium and 2 small. From each of the small firms she examined only 50 samples. Because the incidence in flocks and herds varies from 0% to 100% one would expect the contamination of the meat to reflect this. If one were to repeat the study of these small firms one might get an entirely different picture.

Hobbs

In reply to Mr Harrigan I agree that it is urgent to pay more attention to the dangers of cross contamination from raw to cooked foods in retail shops and homes. Specific legislation is required in this area although the main objective is to free raw materials such as meat and poultry from salmonellae. Public Health Inspectors need the administrative tools and laboratory co-operation to bring about controls in the critical areas of food contamination.

The significance of the short generation time of *Clostridium welchii* is its rapid growth in slowly cooling bulks of cooked meat and poultry. Maximum growth rates occur between 37° and 47°; food left to cool in warm kitchens, even stored overnight in ovens, and exposed to warming-up procedures may remain at these temperatures for 2–3 or many more hours. Thus rapid accumulation of millions/g of *Cl. welchii* vegetative cells in cooked foods is responsible for the food poisoning which appears to result from a toxic product released during sporulation in the intestine.

The table referred to by Dr Elliott shows the difference in results for salmonella isolations between 2 large firms from which a fair number of samples of sausage were received. I accept his criticism that the good results from the small firms are based on a far smaller number of samples; the survey continues.

Food Poisoning in the U.S.A.

P. S. BRACHMAN, A. TAYLOR, E. J. GANGAROSA, M. H. MERSON AND
W. H. BARKER

Center for Disease Control,
Epidemiology Program,
Atlanta, Georgia, U.S.A.

A specific nationwide surveillance programme for food-borne disease outbreaks
has now been in existence for 6 years. Though reporting has improved somewhat,
this category of diseases remains grossly under-reported; however, the surveillance
data do show trends of disease. In 1971 a total of 320 outbreaks of food-borne
illness were reported from 48 States, Puerto Rico, and the Trust Territories of the
Pacific. The outbreaks involved 13,453 persons. The aetiological agent was
bacterial in 63% of outbreaks and 90% of the individuals involved. Within this
category, staphylococci accounted for 29% of the outbreaks and 38% of the
individuals involved, whereas *Clostridium perfringens* accounted for 16% of the
outbreaks and 29% of the individuals. Of the non-bacterial agents, heavy metals
and other chemicals accounted for 8% of the outbreaks. Viral and parasitic
aetiological agents were related to 3% of the outbreaks. Among agents that were
related to more than 5 outbreaks, *Shigella* accounted for more ill people/outbreak
(an average of 100) than any other agent; *Cl. perfringens* caused symptomatic
disease in an average of 26 people/outbreak; salmonella, 15; and staphylococci, 6.
Bacillus cereus, Cl. botulinum, and heavy metals and other chemicals all had
median attack rates/outbreak in excess of 90%. Vehicles most frequently
associated with food-borne illness were beef, pork (includes ham and salami), fowl
and fish. Beef was most frequently associated with *Cl. perfringens* epidemics, and
pork with staphylococcal epidemics. Approximately 50% of *Cl. perfringens*
outbreaks occurred in restaurants or cafeterias, 29% of staphylococcal and
salmonella outbreaks occurred in restaurants or cafeterias and 38% in homes. Of
particular concern was a multi-State outbreak of enteropathogenic *Escherichia
coli* disease related to imported cheese. Several reported epidemics of *Vibrio
parahaemolyticus* disease were related to contaminated seafood. We are concerned
about the obvious under-reporting of food-borne disease throughout the United
States. The Center for Disease Control has developed a common reporting form,
makes available investigative teams and laboratory support to assist local
authorities, and compiles and distributes regularly a surveillance report—all in the
attempt to improve the investigation and reporting of food-borne disease.

EFFECTIVE CONTROL measures and prevention of any infectious disease
depend on our defining that disease's occurrence by time, place and person.
Surveillance is the active, regular accumulation and analysis of these data.
Surveillance has played an important role in reducing the incidence of such
diseases as smallpox, poliomyelitis, diphtheria and tetanus. Nationwide surveillance of foodborne disease in the United States began 6 years ago in an
attempt to define this entity. Each State health department was requested to

report, utilizing a common reporting form, all outbreaks of food-borne disease to the Center for Disease Control. To stimulate accurate and total reporting, we offered the services of our epidemiologists, statisticians, sanitarians and laboratory personnel, and promised to analyse the data and publish regularly surveillance reports which would be distributed to all participants and other interested persons. A close, productive working relationship was developed with the Food and Drug Administration and the Department of Agriculture.

In this presentation the definition of a food-borne disease outbreak is the development of gastro-intestinal symptoms after ingestion of a pathogenic organism or noxious agent contained in food or water and affecting 2 or more individuals, except in botulism where a single case is classified as an outbreak.

In the 6 years that the programme has been in existence, fewer than 400 outbreaks a year of food-borne disease have been reported; these obviously represent only a small portion of the total number that actually occur. The greatest problem has been to stimulate state and local health departments to report all known or suspected episodes of food-borne disease. There is a lack of reporting from local health units to the state, a lack of local investigations, inadequate laboratory support, and possibly overriding all of these reasons—lack of conviction that food-borne disease is an important public health problem or that anything can be done to prevent outbreaks.

Without doubt, millions of cases of gastro-enteritis occur among all age groups in the United States each year; many of these must result from the ingestion of contaminated foods. However, in 1971, only 320 outbreaks were reported to us. The degree of under-reporting can be only roughly estimated. For example, in 1971, 21,789 isolations of salmonellae from humans were reported to the Salmonella Surveillance Programme (Center for Disease Control, 1971). In the Foodborne Disease Surveillance Programme, 760 individual cases were reported as cases of salmonellosis, only 3·5% of the total salmonella disease picture. With our general knowledge of the epidemiology of salmonellosis, this is obviously an under-estimate.

The fatality rate for salmonellosis is generally considered to be low. It is, however, significant, for in the period 1962–1971 the case fatality rate was 0·41% while in 44 outbreaks investigated in 1971 and involving 1391 individuals it was 0·79%.

Trichinosis provides another example of the degree of under-reporting in our Foodborne Disease Surveillance Programme. In 1971, 4 outbreaks involving 18 patients were reported to this programme, but our trichinosis surveillance programme has data on 15 outbreaks and 115 patients (Center for Disease Control, 1972).

In an outbreak of enteropathogenic *Escherichia coli* related to imported cheese, only 7 (6·5%) of 107 outbreaks that actually occurred were reported to us through the usual channels.

In order to develop an approximate idea of the number of outbreaks that occur annually in the United States, we might look at the data from England and Wales which appears to be more representative of countrywide surveillance. In 1967, 705 general and family outbreaks were reported from England and Wales (Vernon, 1969), whereas 345 outbreaks were reported in the United States in 1968. Considering the population of the involved countries, we can estimate that c. 2800 outbreaks actually occurred in the United States. This would indicate an efficiency of reporting of c. 10%.

A similar analysis of surveillance data from the State of Washington which has an excellent surveillance system, probably the best of any state, suggests the probable occurrence of 3800 outbreaks in the United States in 1971, 12 times the actual number reported. Thus, the current countrywide surveillance programme obtains data from no more than 10% of the actual outbreaks that occur.

It is more likely that a disease of low incidence or with serious consequences is better reported than a disease such as food poisoning, which though of high incidence, is ubiquitous, frequently associated with minimal clinical severity, and involving symptoms not necessarily discussed in "polite society".

In spite of these imperfections, the data collected in our surveillance programme do show trends and can be used to develop generalizations.

In 1971, there were 320 outbreaks involving disease in 13,453 people, reported from 47 states. Three states (Oregon, West Virginia and Wyoming) did not report any outbreaks. The number of outbreaks/state ranged from 1 (11 states) to 31 in California and 57 in Washington. Washington, with one-seventh the population of California, had the greatest number of cases/person, reflecting the sensitivity of their surveillance programme.

The 320 outbreaks include 94 outbreaks (29%) in which the aetiological agent was confirmed in the laboratory, 145 outbreaks (45%) in which the aetiological agent was highly suspect but laboratory confirmation was lacking, and 81 outbreaks (25%) in which the aetiology was unknown. The analyses that follow incorporate data from outbreaks in both the confirmed and highly suspect categories (Table 1).

A total of 201, or 63%, of the outbreaks comprising 12,080, or 90%, of the patients were attributed to bacterial infection. *Staphylococcus aureus* accounted for 92, or 29%, of the outbreaks involving 5115 patients, or 38% of the total patients involved in food-borne outbreaks. *Clostridium perfringens* accounted for 51, or 16%, of the outbreaks and involved 3856, or 29%, of the patients. Salmonella accounted for 30, or 9%, of the outbreaks and involved 760, or 6%, of the patients.

Known non-bacterial agents accounted for 38 outbreaks, 12% of the total, involving 270 patients, or 2% of the total. These included trichinosis with 4 outbreaks, viral hepatitis with 4, and chemical poisoning in 30 outbreaks.

MSF–6

Table 1
Food-borne disease outbreaks United States (1971)

	Outbreaks		Patients	
	Number	%	Number	%
Staphylococcus aureus	92	29	5,115	38
Clostridium perfringens	51	16	3,856	29
Salmonella spp.	30	9	760	6
Other bacterial	28	9	2,349	17
Total bacterial	201	63	12,080	90
Other known	38	12	270	2
Unknown	81	25	1,103	8
Total	320	100	13,453	100

For those agents accounting for more than 5 outbreaks, shigella involved the greatest number of patients/outbreak, with a median of 100 for each of 7 outbreaks. In 89 staphylococcal outbreaks, the median number of patients was 6. The greatest number of people involved in a single outbreak was in a school outbreak in Hawaii; there were 498 cases of streptococcal pharyngitis resulting from consumption of contaminated butter. Nonbacterial outbreaks had a median range of 3–7 cases/outbreak.

In 74 staphylococcal outbreaks, the median attack rate was 72%, in 42 *Cl. perfringens* outbreaks, 51%, and in 28 salmonella outbreaks, 49%. In 10 heavy metal poisoning outbreaks, the overall attack rate was 100%.

The most common vehicles of infection were pork (49) and beef (43). Among staphylococcal outbreaks, 40% were related to contaminated pork (including ham and salami). Thirty-seven per cent of the *Cl. perfringens* outbreaks were related to contaminated beef.

Considering the place where food was known to have become contaminated, food processing establishments were involved in 14% of the outbreaks, food service establishments (restaurants, cafeterias, delicatessens) in 58%, and the home in 28%. Ninety-four per cent of the outbreaks of *Cl. perfringens* infection were due to improper handling in food service establishments; additionally, 60% of the salmonellosis outbreaks and 56% of the staphylococcal disease outbreaks resulted from mishandling in food service establishments, whereas 55% of the chemical intoxications were due to faulty handling in food processing establishments.

Analysis of the site where the contaminated food was ingested reveals that 47% occurred in homes and 33% in restaurants. Among staphylococcal outbreaks, 45% occurred in the home and 36% in restaurants or cafeterias, whereas among *Cl. perfringens* outbreaks, 49% occurred in restaurants and cafeterias and 19% in the home.

Botulism continues to receive more publicity than any other single type of food poisoning within the United States. Commercially processed foods have been implicated in only 15 outbreaks out of 315 since 1940, with only 4 outbreaks being due to commercially canned foods. One outbreak in 1941, with 3 cases including 1 death, implicated canned mushrooms (type E); there were 2 outbreaks in 1963, 1 related to contaminated tuna fish (type E) involving 3 cases with 2 deaths, and the second related to liver paste (type A) with 2 cases and no deaths. In 1971 an outbreak involving 2 cases and 1 death was due to type A contamination of vichysoise soup. Investigations revealed that improper heat processing of the canned product allowed distribution of the pathogen.

During the past 25 years there has been a 75% reduction in the reported number of cases of trichinosis in the United States. In 1971, 115 cases were reported with three deaths (Center for Disease Control, 1972). The decline in clinical trichinosis is reflected in studies of human diaphragms. In one study in which diaphragms from 1931–1943 were examined, 16% harboured trichina larvae (Wright *et al.*, 1943), whereas a more recent study (1966–1970) of autopsy specimens from 8071 people from 48 states revealed that only 4·2% contained trichina larvae (Zimmerman *et al.*, unpub.). Additionally, only 14·3% of the infected diaphragms in the recent study contained viable larvae, whereas 45% of those examined 3 decades earlier harboured living larvae.

Pork products account for 78% of the cases of trichinosis reported in the United States, with the majority of the products involved having been purchased from commercial sources. In 6·9% of the cases, the pork products were prepared and consumed on farms. During recent years, trichinosis acquired from the ingestion of wildlife has increased, with 7·8% of the cases reported during the past 5 years being attributed to the ingestion of bear meat. Twelve of 13 trichinosis cases attributed to bear meat in 1971 were reported from California. A recent survey of 372 black bears in the north-eastern United States revealed that 1·3% were infected, a prevalence rate 13 times greater than that for grain-fed swine and 4 times greater than that for garbage-fed swine in the United States (Harbottle *et al.*, 1971).

A common source outbreak of *E. coli* gastro-enteritis occurred between November 12 and December 8, 1971; at least 387 of the 909 persons exposed in 107 separate outbreaks in 13 states and the District of Columbia experienced acute gastro-enteritis after eating camembert or brie cheese imported from France (Center for Disease Control, 1971). Symptoms included nausea, vomiting, diarrhoea, fever, and headache approximately 24 h after ingestion of cheese. Symptoms persisted for 48 h. *E. coli* serogroup 0 : 124 was isolated from the faeces of ill patients as well as from the cheeses. Both kinds of cheese were produced in the same factory in France and were identical except for their shape.

An unusual outbreak of gastro-enteritis was attributed to candy love beads (Center for Disease Control, 1971). More than 600,000 containers of these items were sold in the United States prior to their recall; only 6 outbreaks involving 13 persons in 3 states were reported. Thirteen of 14 persons who ate the love beads became ill within 10–60 min with symptoms primarily of nausea, vomiting, and abdominal cramps. Some of the patients were markedly lethargic and drowsy; most recovered within several hours. An analysis of the love beads by FDA laboratories revealed over 1000 p/m of cadmium, a toxic heavy metal.

Vibrio parahaemolyticus accounts for over 70% of food poisoning cases in Japan. In Japan, food poisoning due to this organism is seasonal, occurring during the warmer months when the number of vibrios found in coastal waters in Japan is highest. The organism has been isolated from sea water and sea fish. In the United States it has been isolated from coastal waters, oysters, and crabs along the Atlantic Coast, from shrimp along the Gulf Coast, and from marine environments along the Pacific Coast. In August 1971, the first outbreak was described in the United States; it involved 320 of 550 persons who attended a clam bake in Maryland (Center for Disease Control, 1971). The mean incubation period was 15 h, with a range of 8–22 h. Symptoms included diarrhoea (98%), severe abdominal cramps (78%), nausea (76%), vomiting (74%), fever (26%), headache (23%) and chills (10%). The median duration of illness was 2 days, with a range of 1–5 days. Two per cent of the patients were hospitalized; there were no deaths. *V. parahaemolyticus* was isolated from steamed crabs and from faecal specimens from 4 of the ill patients. Subsequently, also in August, another clam bake related outbreak was reported from Maryland. Twenty-five of 75 persons exposed to steamed clams developed symptoms. In both of these outbreaks investigation revealed that the steamed crabs had been placed in a truck beneath baskets of live crabs and driven to the picnics where the outbreaks occurred. This allowed the steamed crabs to become infected from water dripping from the live crabs.

Another reported outbreak in August in Maryland involved 25 of 100 exposed persons who ate contaminated crab salad in a hospital. After an average incubation period of 18 h, the same symptoms were reported with an average duration of 72 h.

A more recent outbreak is the one reported in Hawaii in June of this year, involving *c.* 200 people who attended a *luau*. Thirty-eight of 182 interviewed persons became ill for an overall attack rate of 21%. Symptoms were diarrhoea (97%), abdominal cramps (86%), chills (13%), fever (11%) and vomiting (3%). The median incubation period was 16 h, and 81% of the patients were well within 24 h. Food histories implicated small white crabs that had been caught specifically for the *luau*. Stool specimens from patients were negative for enteric pathogens, but were not examined for *V. parahaemolyticus* and none of the suspect food remained for examination. Additional crabs caught from the same area after investigation of the epidemic were cultured, and *V. parahaemolyticus* was isolated from them. The white crabs came from a particular bay that was grossly polluted with sewage. With knowledge of the epidemiology of this aetiological agent and improved laboratory techniques more cases of *V. parahaemolyticus* gastroenteritis should be reported.

These data are far from complete, but in comparing surveillance data over the past several years one is able to discern trends of food poisoning. We continue to try to improve our overall surveillance, but frankly, we do not see what will improve reporting short of a national calamity or a better reward than a routine surveillance report. Nevertheless, it is possible not only to develop pertinent information concerning trends, but to investigate unusual outbreaks such as those related to *V. parahaemolyticus*, cadmium, or scombroid toxin.

Our ultimate aim must be to control and prevent food-borne disease, some of which can be accomplished by utilization of current knowledge. However, if we are to use more than a shotgun approach to lessening the dangers, surveillance must be improved.

Recommendations

There is an unmet challenge that must be of concern to everyone interested in food-borne diseases. Except for England and Wales with its encompassing system of public health laboratories and the traditions of investigation and reporting of food-borne disease initiated by Sir Graham Wilson and continued by Dr Betty Hobbs and her colleagues, none of the rest of us have developed adequate surveillance programmes for food-borne disease. More complete surveillance data is not only necessary to better define the problem in exacting terms, but is also necessary in order to stimulate better investigations, more adequate reporting, additional research, improved education and, additionally, serve as a guide to modern up-to-date attempts at control and prevention.

Co-operation between countries will allow sharing of mutually important data such as the *Vibrio parahaemolyticus* information from Japan, potentially assist in the control of intercountry related epidemics such as in some outbreaks of salmonellosis, improve cross-fertilization of research ideas, and also serve to stimulate improved domestic surveillance within each country. It is difficult to compare data collected in dissimilar ways using different definitions. We need to develop standards for reporting and investigating outbreaks. We suspect all of us are in agreement, but suggest that we now need to proceed in an expeditious manner.

References

Center for Disease Control (1971). *Morbidity and Mortality Weekly Rep.* **20,** 177.
Center for Disease Control (1971). *Morbidity and Mortality Weekly Rep.* **20,** 356.
Center for Disease Control (1971). *Morbidity and Mortality Weekly Rep.* **20,** 427.
Center for Disease Control (1971). Salmonella Surveillance Annual Summary, 1970.

Center for Disease Control (1972). Trichinosis Surveillance Annual Summary, 1971.

Harbottle, J. E., English, D. K. & Schultz, M. G. (1971). Trichinosis in bears in northeastern United States. *HSMHA Health Rep.* **86,** 473.

Vernon, E. (1969). Food poisoning and *Salmonella* infections in England and Wales, 1967. *Publ. Hlth, Lond.* **83,** 205.

Wright, E. H., Kerr, K. B. & Jacobs, L. (1943). Studies on trichinosis. XV. Summary of findings of *Trichinella spiralis* in a random sampling of the population of the United States. *Public Health Rep. Wash.* **58,** 1293.

Zimmerman, W. J., Steele, J. H. & Kagan, I. G. (1973). Trichiniasis in humans in the United States, 1966–1970. In preparation for submission to *HSMHA Health Reports.*

Discussion

Wilson

What puzzles me is why in England and Wales salmonella cause a high proportion of food poisoning outbreaks compared with staphylococci whereas the converse occurs in the U.S. These are discrepancies which call for thought.

Brachman

The explanation for this discrepancy may be partially attributed to differences in reporting systems in the 2 countries. In the U.S. system, the data for each major aetiological agent of food-borne disease is comprised of all epidemiologically documented, clinically compatible outbreaks which are reported, including those with and without laboratory confirmation. On the other hand, the British data for specific aetiological agents are restricted to those outbreaks for which the presence of the agent has been documented in the laboratory. The majority of staphylococcal outbreaks included in the U.S. summary are not laboratory confirmed, while the converse is true of outbreaks attributed to salmonellosis. Thus by the laboratory documented criteria of the system in Great Britain, a sizable number of U.S. staphylococcal outbreaks would not be included while most U.S. salmonella outbreaks would be included.

Reinius

The present activities of the World Health Organization in relation to surveillance programmes for food-borne disease outbreaks basically take the form of organization of study groups in particular topics. A "Study group on methods for sampling and examination of food and food products for surveillance of food-borne outbreaks" is planned for the summer of 1973.

Smith

Dr Brachman, did you find it odd that type E *Clostridium botulinum* was isolated from mushrooms?

Brachman

I agree that finding type E *Clostridium botulinum* in mushrooms is unexpected. We ourselves were quite suspicious of this report, but have no reason to doubt the laboratory data. As you know type E *Clostridium botulinum* is usually associated with fish or fish products as distinct from types A and B.

Smith

Were the *E. coli* 0124 cases from adults and could you comment on why *V. parahaemolyticus* only appears in Maryland?

Brachman

The cases of *E. coli* 0124 gastro-enteritis were primarily adults. Since the original reports of *Vibrio parahaemolyticus* gastro-enteritis were publicized from

DISCUSSION 151

Maryland in 1971, health departments in several other coastal states have
documented outbreaks with this agent during 1972 (Louisiana, Texas, New
Jersey, Florida, Massachusetts, and Hawaii). The fact that the first outbreaks
were reported from Maryland may be attributed to the special interest of the
State Health department laboratory in instituting proper laboratory procedures
to identify this organism in food-borne disease outbreaks involving seafood.

Ingram

It is unwise to assume that type E *Cl. botulinum* will be confined to marine
products, since it has been isolated from terrestrial soil in various places.

Eklund

Were the mushrooms, involved in the 1941 outbreak, heat processed?

Foster

Apparently yes—records state "Canned mushrooms processed in San
Francisco; mushrooms imported".

Bostock

The statistics presented by Dr Brachman showed a preponderance of
staphylococcal intoxication amongst reported cases of food-borne infection.
Could Dr Brachman offer any explanation for the differences between the U.K.
and U.S. incidence of this type of food poisoning? Is the under-reporting of food
poisoning a factor?

Brachman

The question by Dr Bostock is similar to the one by Sir Graham Wilson.
Under-reporting of certain types of food poisoning could certainly account for
an apparent predominance of other types which are more likely to be reported.
It is quite reasonable to expect that staphylococcal intoxication with relatively
short incubation period and explosive onset might be more frequently reported
to health departments than outbreaks due to other aetiologies which do not
present as dramatically.

Ashton

How representative were the 1971 data on staphylococcal outbreaks since
two large meat packing plants in the U.S. had problems with Genoa Salami?

Brachman

1971 was representative of staphylococcal outbreaks for the last 6 years.

Bryant

Most of the pork-associated staphylococcal intoxication outbreaks reported
in the U.S. over the last 10 years have been attributed to cured hams.

Riemann

The relatively higher reporting of staphylococcal food-poisoning in the U.S.A.
as compared to U.K. might be due to the larger average size of staphylococcal
outbreaks than of salmonella outbreaks in the U.S.

Brachman

No further comment.

Food Poisoning and Salmonella Infections in Australia

R. G. A. SUTTON

*School of Public Health & Tropical Medicine,
University of Sydney,
Sydney, N.S.W. 2006 Australia*

Because food poisoning is not a notifiable disease in most Australian States, official figures on the incidence of this disease are not available. This paper attempts to collect statistics from various sources, and combined with a limited number of results from work carried out on the incidence of food poisoning organisms in Australian food, draws some conclusions on the epidemiology of food poisoning in Australia. Overall the epidemiology is similar to that in Europe, U.K. and North America. The most common causes are *Salmonella* species, *Staphylococcus aureus* and *Clostridium perfringens,* usually associated with consumption of meat or poultry, although seafoods are also of some importance. Most outbreaks appear to follow meals in hospitals, restaurants or at large parties or receptions. School meals play little part, probably due to the fact that in most Australian schools a hot midday meal is not provided. There is a rapid increase in the consumption of "take away" and convenience foods, but to date they have been implicated in only a very small number of food poisoning incidents.

AUSTRALIA is a large country of 3 million square miles with a population of 13 million people, 83% of whom are classified as "urban dwellers". The standard of living is high and access to medical care and education are good. Eating habits generally are similar to those of North America and England. In these regards therefore we might expect that the epidemiology of food poisoning in Australia is similar to that in the U.S.A. and England. In most respects this is true, but some small differences are apparent and these will be mentioned in this paper.

Food poisoning is not a notifiable disease in most Australian states, and reliable records on the yearly incidence of this infection are not available. In preparing this report therefore the assistance of Central Public Health Laboratories in the 4 states of Queensland, Victoria, South Australia and Western Australia, was sought. Based on reports of recent food poisoning outbreaks investigated by these laboratories, together with those from this laboratory, an attempt was made to present a picture of the current epidemiology of food poisoning in Australia. This paper is not meant to represent an official report of any kind.

Incidence of Food Poisoning in Australia

During the past 5 years 48 general outbreaks of food poisoning involving *c.* 2500 cases were investigated in these 5 laboratories. In addition, a total of *c.* 2000 strains of salmonella, isolated from human cases, were submitted each year to 3 salmonella reference laboratories in Adelaide, Perth and Melbourne (see Table 4 below).

In the U.S.A., it is estimated that c. 2,000,000 cases of salmonellosis occur annually, although only 1% of these cases are officially reported (Nat. Acad. of Science, 1969). A similar degree of under-reporting is probably evident in Australia.

When we consider the Australian figures, together with the fact that food poisoning organisms are frequently isolated from foods and food handlers, and that food habits are similar to those in other Western countries, it is reasonable to assume that the relative incidence of food poisoning in Australia is similar to that in the U.S.A. and in England and Wales. Thus, allowing for the degree of under reporting in these countries, it is probable that between 50,000 and 100,000 cases of food poisoning occur annually in Australia.

Types of Foods Responsible for Food Poisoning

Complete epidemiological details were obtained in 33 of the 48 outbreaks of food poisoning investigated. Information is given below on the nature of the food vehicle, location of the outbreak and on the type of organism responsible for the illness.

Meat and poultry are the most common causes of food poisoning in Australia, together being responsible for 21/33 outbreaks (Table 1). These results are similar to those obtained in U.S.A. and England, where meat and poultry are usually responsible for c. 70% of all food poisoning incidents. All the outbreaks of *Clostridium perfringens* food poisoning were due to meat or poultry. Seafoods are also a significant cause of food poisoning, due probably to the large amount of molluscs and crustaceans consumed in this country. With the constant increase in the pollution of our rivers and estuaries, the risk of oysters,

Table 1

Types of food responsible for 33 general outbreaks of food poisoning in Australia for which the food vehicle was established

Food	No. of outbreaks	%
Poultry	11	33
Meat	10	30
Seafoods	6	18
Others	6	18
Total	33	

mussels and other seafoods becoming contaminated with food poisoning
organisms is increasing. Other foods involved included creamed corn spread,
cheese, potato salad, milk and cream cake.

Location of General Outbreaks of Food Poisoning

The location of 33 general outbreaks of food poisoning is given in Table 2.
Receptions and parties catered for by professional caterers were the most
common source of food poisoning incidents. In common with many other
countries, meals served in hospitals are the cause of a significant number of
outbreaks. However, school meals play little part in food poisoning outbreaks in
Australia which is in contrast to the situation in England and some parts of the
U.S.A. In Australia very few schools provide a hot midday meal. School lunch
usually consists of sandwiches and fruit that the pupil brings from home or buys
at a school "tuck-shop". The airline outbreaks both occurred on internal airlines
and were due to coagulase positive *Staph. aureus.*

The information given in Table 2 relates only to general outbreaks. No figures
are available for sporadic cases or small family outbreaks, although many of the
2000 strains of *Salmonella* serotypes identified each year in the reference
laboratories are from these sources.

Table 2

*Location of 33 outbreaks of food poisoning in Australia for which
the aetiology was established*

	No. of outbreaks	%
Parties, receptions, etc.		
(Professional Caterer)	9	27
Hospitals	7	21
Restaurants, hotels	6	18
Private parties	5	15
Canteen	2	6
Schools	2	6
Airline	2	6
Total	33	

Organisms Responsible

Salmonella, Staph. aureus and *Cl. perfringens* are the main causes of general
outbreaks of food poisoning in Australia (Table 3). The figures given in Table 3
are possibly weighted slightly in favour of *Cl. perfringens* food poisoning,
because this laboratory is a reference centre for *Cl. perfringens* and such
outbreaks would be brought to our attention.

Staphylococcal food poisoning was reported from all States and occurs more frequently in the summer months of October to March. Enterotoxin typing has only recently commenced in Australia and useful information on the distribution of enterotoxin types is not yet available. This laboratory has examined the strains of *Staph. aureus* from 3 recent outbreaks and all were producers of enterotoxin type A. The food vehicles in these instances were cooked-peeled prawns, creamed corn sandwich spread, and chicken which was contaminated after cooking.

Table 3
Causative organism in 33 general outbreaks of food poisoning in Australia in which the causative organism was established

	No. of outbreaks	%
Salmonella serotypes	9	27
Staph. aureus	9	27
Cl. perfringens	11	33
Streptococcus spp.	2	6
V. parahaemolyticus	1	3
E. coli	1	3
Total	33	

The incidence of staphylococcal food poisoning is similar to that in the U.S.A. but somewhat higher than that in England (Vernon, 1970; Center for Disease Control, 1970). This may be due to the higher ambient temperatures experienced in Australia, especially in summer, thus allowing more rapid multiplication of the organism than might be expected in England.

Cl. perfringens food poisoning does not appear to occur uniformly throughout all Australian States. Although it has been found to occur frequently in Victoria and to a lesser extent in New South Wales, no outbreaks have been reported in Western Australia, despite a good surveillance scheme operating in this State and repeated efforts by laboratory workers to look for this organism in suspected cases (Mackay-Scollay, pers. comm.). In addition, no reports of *Cl. perfringens* food poisoning have been made from South Australia. This apparent uneven distribution of *Cl. perfringens* food poisoning is confusing. There appears to be no differences in eating habits or methods of food preparation that could explain the situation. It is known that some workers in New South Wales and Victoria have a particular interest in this type of food poisoning and the uneven distribution may be more apparent than real. More information is certainly needed to clarify the situation.

Salmonellosis

If, in addition to large general outbreaks, we consider sporadic cases of food poisoning, salmonellosis is almost certainly the most widespread form of food poisoning in Australia. At least 2000 strains from human cases are identified each year at the 3 main salmonella reference laboratories (Table 4). In addition to human cases, the organism is frequently isolated from meat (Table 5) and poultry, wild and domestic animals (Table 6), reptiles and eggs. From the results of unpublished surveys carried out on frozen chickens and mince meat, it is known that this organism is widely distributed within these foods, 20–30% being contaminated with small numbers of salmonellae. The most common serotype isolated from these foods is *S. typhimurium* (Proudford, pers. comm.).

A review of salmonellosis in Australia up to 1964 is given by Atkinson (1965).

Table 4

Number of salmonella serotypes isolated from human cases and identified in 3 reference laboratories

	Salm. Ref. Lab. Adelaide	Diagnostic Laboratory Melbourne	Public Health Laboratory Perth	Total
1967	508	218	N.A.	
1968	985	254	526	1765
1969	1113	348	618	2079
1970	965	450	664	2079
1971	1009	307	919	2225

These data do not include *S. typhi* and *S. paratyphi* of which only 10–30 isolations are made yearly in Australia.

Table 5

Incidence of salmonella in frozen boneless Australian meat imported into the U.K.*

Type of meat	Years	No. of samples examined	No. positive
Veal	1962–1967	498	31 (6·2%)
Mutton	1963–1967	1072	100 (9·3%)
Horsemeat	1965–1967	401	88 (22%)

* Hobbs & Gilbert (1970).

Table 6

Incidence of salmonellae in farm animals in Australia

Authors	Animal	Time and place of sampling	No. sampled	% positive
Daleel & Frost (1967)	Cattle	At abattoir 2–3 days	2000	11·6
	Calves	At abattoir within 1 day of arrival	300	1·3
Bruce & Hart (1968)	Cattle	At abattoir day of arrival	655	1·5
Chung & Frost (1969)	Pigs	At abattoir after 1 day	1000	8·4
Grau & Brownlie (1965)	Bovine (rumen)	Not known	193	45

The serotypes most frequently isolated from human cases over the past 5 years were *S. typhimurium, S. bovismorbificans, S. anatum, S. chester, S. derby, S. havana, S. meunchen* and *S. newport,* although variations did occur in the isolation rate of particular strains from different states, e.g. 70% of strains of *S. anatum* came from Queensland and Western Australia. Over the past 5 years, this strain has consistently been a major cause of salmonellosis in Queensland. In this regard it is interesting to note that Elder & Simmons (1965) found *S. anatum* to be the most common serotype isolated from meat meal prepared in Queensland. On the other hand *S. bovismorbificans* is isolated frequently in South Australia, New South Wales, Western Australia and Victoria; but rarely in Queensland. There has been a marked increase in the number of *S. bovismorbificans* strains isolated over recent years. The figures for 1969–1971 were, 70, 133 and 206 respectively.

S. typhimurium continues to be the most common strain isolated from human sources in all Australian States, and regularly accounts for 35–45% of all human isolates. In Western Australia however, although it is the most common single serotype isolated, it accounted for only 18–30% of total human isolations over the period 1968–1971. This is primarily due to a high incidence of salmonellosis among aboriginal people in the subtropical rural regions of Western Australia, particularly in the Kimberleys district. Within this district, with a population of only 15,000 people, the incidence of salmonellosis is higher than in the capital, Perth, with a population of 700,000. However, among these people salmonellosis is rarely due to *S. typhimurium,* and this strain accounted for only 48/1641 cases in the Kimberleys during the period 1955–1971 (Table 7). On the other hand, within the metropolitan area of Perth (W.A.), with its large urbanized community, *S. typhimurium* was the cause of 42% of all cases of salmonellosis during this period.

Table 7

Distribution of salmonella groups within 2 selected regions of Western Australia in the period 1955-1971*

Strain	Perth/Freemantle† metropolitan area	Kimberleys area‡
S. typhimurium	546	48
O Groups A–G	650	1055
Other O Groups	86	538

* Iveson *et al.* (1969); Report (1971).

† Perth/Freemantle area; the capital of Western Australia is an urbanized city with a population of *c.* 700,000.

‡ The Kimberleys are a large rural and cattle grazing area of 170,000 square miles in the tropical north. It has a population of *c.* 15,000, including 6000 aborigines.

In the Kimberleys it is noticeable that "O" groups other than A–G accounted for over 30% of the isolates, which is more than 10 times the number of strains of *S. typhimurium* isolated from the area during the same period. Strains commonly found include *S. wandsworth, S. adelaide, S. bahrenfeld, S. emmastad* and *S. hvittingfoss*. This unusual situation arises because within this area the bulk of the cases of salmonellosis occur among aboriginal persons—the main food vehicle responsible for these infections being reptiles (lizards, crocodiles) and occasionally birds which form a common part of the aboriginal diet. MacKay-Scollay (pers. comm.) has frequently isolated *Salmonella* species of "O" groups other than A–G from reptiles in the Western Australian area, but *S. typhimurium* is rarely isolated from these reptiles. Gatti *et al.* (1972) recently isolated *Salmonella* species from 54% of 300 lizards examined in Kinshasa, Zaire. However, they concluded that lizards play no significant role in the epidemiology of human salmonellosis in Kinshasa.

The distribution of serotypes among cases of salmonellosis within these 2 different regions of Western Australia can thus be traced to the distribution of strains within the food vehicles known to cause food poisoning within the particular areas.

Botulism is a rare disease in Australia; only 5 outbreaks of suspected human botulism have been recorded (Table 8). However, botulism among animals is more common, and outbreaks have frequently occurred among cattle, sheep, horses and poultry (domestic and wild). Seddon (1965) reviews the history of animal botulism in Australia. The most common cause of botulism in birds and animals in Australia is *Cl. botulinum,* type C, although cases due to type B have been reported. Type A, B, C and D have been found in soil, but to date no isolations of *Cl. botulinum,* type E, have been reported.

Table 8

Outbreaks of suspected botulism recorded in Australia

Date	Location	Cases	Deaths	Type	Source
1942	Queensland	26	7	B	Canned beetroot
1942	N. Territory	5	1	B	Canned beetroot
1957	Gunnedah N.S.W.	15	–	?	Home bottled beetroot
1963	Victoria	1	1	A	Home preserved cantaloup
1966	Merriwagga N.S.W.	2	–	?	Tinned tuna

Vibrio parahaemolyticus *food poisoning*

Only 1 outbreak of food poisoning due to *V. parahaemolyticus* has been reported in Australia. This occurred at a Queensland holiday island resort and was probably due to frozen crabs, prawns and lobsters which had been allowed to thaw at room temperature in seawater from which *V. parahaemolyticus* had been isolated (Battey *et al.*, 1970; Wallace & Battey, 1971). In this laboratory we are currently investigating the incidence of *V. parahaemolyticus* in oysters and seawater in the Sydney area. The organism can be isolated frequently from oysters although it is usually present in only small numbers—rarely more than 1000/100 g of oyster flesh and usually less than 100/100 g. It has also been isolated from the water in which the oysters were grown. To date only non-haemolytic strains have been isolated in this study.

Escherichia coli Food Poisoning

Only 1 outbreak of food poisoning due to *E. coli* has been reported in Australia (Forsyth & Taplin, 1972). This occurred in a restaurant and was due to a non-motile, lactose negative variant of *E. coli*, which agglutinated with 0111 B/4 antiserum.

Summary

Food poisoning is not a notifiable disease in most Australian States, and even where it is, the infection rate is known to be grossly under-reported. Based on reports from reference laboratories, and comparison with countries of similar living conditions and eating habits, an estimate of 50,000–100,000 cases of food poisoning/year is suggested.

The main causes of food poisoning in Australia are *Salmonella* spp., *S. aureus* and *Cl. perfringens*. Chicken, meat and seafoods are the most common food vehicles and most outbreaks occur following receptions or parties catered for by professional caterers. Hospitals and restaurants are also a major source of food poisoning outbreaks. There is need for a coordinated epidemiological surveillance scheme, so that more detailed information can be obtained on the epidemiology of food poisoning in Australia.

Acknowledgements

Publication of this paper would have been impossible without the assistance given in collecting information about food poisoning incidents occurring in various States, and without the willingness of many colleagues to allow the use of results obtained in their laboratories. In this regard I am especially indebted to Drs J. Forsyth, E. M. Mackay-Scollay, J. Tonge, K. Anderson and W. J. Stevenson.

I am also grateful to the Director-General of Health for permission to publish this paper.

References

Atkinson, N. (1965). Salmonellosis in Australia. In *The World Problem of Salmonellosis*. Ed. E. van Oye. The Hague: W. Junk.
Battey, Y. M., Wallace, R. B., Allan, B. C. & Keeffe, B. M. (1970). Gastroenteritis in Australia caused by a marine Vibrio. *Med. J. Aust.* **1**, 430.
Bruce, J. M. & Hart, B. (1968). Isolation of salmonellae from Northern Territory cattle. *Aust. Vet. J.* **44**, 80.
Chung, G. T. & Frost, A. J. (1969). The occurrence of salmonellae in slaughtered pigs. *Aust. Vet. J.* **45**, 350.
Center for Disease Control (1970). *Food-borne Outbreaks—Annual Summary.* U.S. Dept. Health, Educ. and Welfare. Atlanta, Georgia, U.S.A.
Daleel, E. E. & Frost, A. J. (1967). The isolation of salmonella from cattle at Brisbane abattoirs. *Aust. Vet. J.* **45**, 203.
Elder, J. K. & Simmons, G. C. (1965). Examination of Queensland stock feeds and fertilizers of animal origin for salmonella. *Qld J. Agric. and Animal Sci.*, **22**, 291.
Forsyth, J. R. L. & Taplin, J. (1972). An outbreak of gastro-enteritis among diners at a restaurant in the Dandenongs. *Med. J. Aust. i*, 33.
Gatti, F., Lontie, M., Makulu, A., Bouillon, A., Vandepitte, J., Robinet, R., Maes, L. & Van Oye, E. (1972). Epidemiologie de la Salmonellose a Kinshasa: Role des Lozards (Lacertilia). *Ann. Soc. Belge Méd. trop.* **52**, 127.
Grau, F. H. & Brownlie, L. (1965). Occurrence of salmonellae in bovine rumen. *Aust. vet. J.* **41**, 321.
Hobbs, B. C. & Gilbert, R. J. (1970). Microbiological standards for food: Public health aspects. *Chemy Ind.* p. 215.
Iveson, J. B., Mackay-Scollay, E. M. & Bamford, V. (1969). Salmonella and Arizona in reptiles and man in Western Australia. *J. Hyg., Camb.* 67, 135.
National Academy of Sciences (1969). An Evaluation of the Salmonella Problem, Report by Committee on Salmonella, Division of Biology and Agriculture. National Research Council. Washington, D.C.
Report (1971). Isolations of Salmonella, Arizona, Edwardsiella, Shigella and enteropathogenic Escherichia, Salmonella diagnostic and Reference Laboratory. Perth, W.A.
Seddon, H. R. (1965). *Diseases of Domestic Animals in Australia,* Part 5, Vol. 1, Bacterial Diseases, Revised by H. E. Albiston, 2nd ed., Comm. Dept Hlth Aust.
Vernon, E. (1970). Food poisoning and salmonella infections in England and Wales 1968. *Publ. Hlth., Lond.* 84, 239.
Wallace, R. B. & Battey, Y. M. (1971). *Vibrio parahaemolyticus* in oysters (correspondence). *Med. J. Aust.* **1**, 982.

Discussion

Barrow

If salmonellae are commensal organisms in reptiles, which are the staple diet of aborigines, what is the normal faecal flora of aborigines? Are the salmonellas isolated from them actually causing the disease?

Sutton

To my knowledge the faecal flora of aborigines does not differ significantly from that of Europeans. There may be some small differences, particularly among children, but not of major concern in this context. There is no suggestion that *Salmonella* spp. form part of the normal faecal flora of these people. Indeed the organism does cause disease among them, especially the children, where it is one of many contributory factors to the high infant mortality rate.

Hess

I would like to know the Group and Type of your streptococci related to the 6% of food poisoning outbreaks.

Sutton

Both were type A—the symptoms were pharyngitis and sore throat.

Barrow

The United States and Australia are big countries, one with a large population the other with a small population, each reporting a high incidence of staphylococcal food poisoning in contrast to the low incidence in Britain. I wonder if because of different foods and circumstances, etc., staphylococcal food poisoning in Britain is in fact being under-reported? The organisms may be killed, leaving the toxin and such incidents may not be reported. Perhaps the detection of enterotoxin in foods might increase the number of reported incidents in Britain?

Sutton

This is a possibility but I think it is best answered by the British epidemiologists.

Wilson

There may be something in the under-reporting story for the U.K. but I don't think that is the explanation. It may be that the causes of the outbreaks are not known.

Sutton

I think that the high ambient summer temperature present in Australia and parts of the U.S.A. is a major contributory factor. We know that of all the forms of food poisoning staphylococcal food poisoning has the most pronounced summer "peak" incidence and is rarely found in the winter months particularly in the U.K.

Ingram

Dr Brachman mentioned that 80% of staphylococcal outbreaks came from pork products. It would be interesting to know what proportion of these came from cured pork and especially cooked cured products, since I have the impression that there is a much greater use of such products in the U.S.A. than in the U.K. This would not seem to explain the similarly high proportion of staphylococcal outbreaks in Australia where food habits more resemble those in U.K. and where a further factor may be involved. I have the impression, though not based on adequate systematic data, that salmonellae grow better than staphylococci at temperatures below 20° such as prevail in U.K.

Sutton

In Australia we do not have the situation where pork products are responsible for 80% of staphylococcal outbreaks although hams do play some part.

Foster

The scarcity of staphylococcal poisonings from bakery foods in U.S. may be attributed to widespread practice of refrigerated retail distribution. With respect to the frequency of involvement of cured pork in U.S. staph. poisoning: (1) fermented sausage outbreaks in 1971 from commercial products involved only 3–4 small outbreaks which were insignificant; (2) high frequency in hams is attributed to recontamination and incubation before serving (e.g. at picnics).

Salmonellosis in England and Wales

B. ROWE

Salmonella and Shigella Reference Laboratory,
Central Public Health Laboratory,
Colindale Avenue, London NW9 5HT, England

The Salmonella and Shigella Reference Laboratory, Colindale and its associated regional salmonella centres each year identify *c.* 5000 strains from humans in the British Isles. The genus Salmonella comprises 4 subgenera and from 1966–1971 inclusive the identifications contained 48 strains of subgenus II, 4 of subgenus III (Arizona) and 4 of subgenus IV. The remainder were all subgenus I and *c.* 50% of these belonged to 10 serotypes, whereas the other 50% were from a further 120 serotypes. The food-poisoning reports of the Public Health Laboratory Service in England and Wales show that up to 1967 there was a steady increase each year in the incidents due to Salmonella serotypes other than *Salmonella typhimurium*. Up to that year the total incidents due to *S. typhimurium* exceeded those caused by all other serotypes in any single year. In 1968 for the first time the total incidents due to other serotypes exceeded those due to *S. typhimurium* and this trend has been maintained in 1969. The total number of serotypes involved has not increased significantly and the change is due to the greatly increased prevalence of a few serotypes. *S. panama* has exhibited a sustained rise since 1964, and is now one of the top 5 salmonellae in humans. *S. virchow* and *S.*4,12 : d : − were very prevalent in 1968 and 1969, but the most unexpected rise has been in *S. agona*. This serotype was rare in 1969, but in 1971 almost 700 strains were identified. Other serotypes which have been more prevalent in the last 5 years are *S. enteritidis, S. bredeney, S. heidelberg, S. indiana, S. stanley* and *S. saint-paul*. This latter serotype has increased particularly in 1971. With many of these serotypes the main source of infection is poultry—*S. virchow, S.*4,12 : d : −, *S. bredeney, S. heidelberg, S. indiana, S. stanley* and *S. saint-paul. S. panama's* main reservoir is in pigs. The increase in *S. agona* infections seemed initially due to poultry but now is due partly to infections in pigs and their products.

Introduction

THE SALMONELLA and Shigella Reference Laboratory, Colindale, receives isolates from the whole of the British Isles but, since its main area of responsibility is England and Wales, this review will be restricted to these countries. The review will consider salmonellosis during the years 1967–1971, but will exclude typhoid and paratyphoid.

The Importance of Subgenus Determination

Salmonellas are identified by their antigenic structure and are arranged in the Kauffmann-White scheme according to their somatic "O" and flagellar "H" antigens. Each year new serotypes are added to the scheme which at present contains about 1600 serotypes. According to their biochemical reactions salmonella serotypes may be placed in 1 of 4 subgenera. Subgenus I contains about 1000 serotypes which occur in humans as well as in a wide range of animals both warm and cold blooded. There are about 250 serotypes in subgenus II and these are rare in humans but are usually isolated from cold blooded animals, in particular reptiles and tortoises. The subgenus III salmonellas or Arizonas are rarely isolated from humans but commonly found in reptiles and there are about 300 subgenus III serotypes. Only 30 serotypes are found in subgenus IV and these are similar in their ecology to subgenus II.

The General Picture in England and Wales

The Epidemiological Research Laboratory of the Public Health Laboratory Service publishes information on human salmonellosis in England and Wales (Vernon, 1970). This information is based on reports submitted by the constituent public health laboratories and hospital bacteriologists. In Fig. 1 the information is shown graphically and *Salmonella typhimurium* is considered separately whilst all the other serotypes are summated.

Up to 1967, the incidents due to *S. typhimurium* exceeded those due to all the other serotypes combined, whereas in 1968 and 1969 this situation was

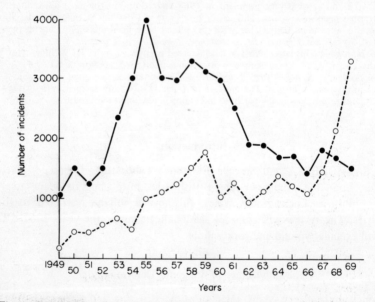

Fig. 1. Reported incidents of salmonella infections in humans in England and Wales. (General outbreaks, family outbreaks and sporadic cases). ●——●, *S. typhimurium*; ○– – –○, other serotypes.

reversed. These other serotypes had shown a steady increase in incidence for many years but in 1968 and 1969 the increase was substantial, as the graph clearly shows. Similar information is not available for the years 1970 onwards and, therefore, for the remainder of this report, the data presented is based on salmonella identifications made at the Salmonella and Shigella Reference Laboratory, Colindale, and the regional salmonella centres. These figures do not accurately reflect incidence since all isolates are not referred for identification. Nevertheless, there is good evidence that the figures reflect trends in the relative frequency of serotypes. The Salmonella and Shigella Reference Laboratory receives only a proportion of the *S. typhimurium* and *S. enteritidis* isolations and consequently the figures do not give an accurate assessment of these 2 serotypes. *S. typhimurium* remains the commonest single serotype infecting humans in England and Wales and will not be considered further. This report will analyse the changes in the epidemiology of the other serotypes, which have been so important in the years 1967 to 1971.

The Importance of Subgenus Determination

Table 1 shows that in the 5 years 1967–1971, 20,678 subgenus I salmonellas were identified from humans, compared with only 52 from the other 3 subgenera.

Although about 1000 serotypes are classified in subgenus I less than 130 are found in humans each year and of these about 100 serotypes have less than 10 isolations each year. Therefore, most of the isolates belong to a few really prevalent serotypes and in any year about 50–60% of all human isolates are accounted for by 10 serotypes.

Human Infection with Subgenus I Salmonellas

Figure 2 shows the marked increase in prevalence of serotypes other than *S. typhimurium* in the years 1968–1971, as presented by the identifications at the Salmonella and Shigella Reference Laboratory. There were 110 different serotypes in 1967 and 133 in 1971 but, since over 90% of these had less than 10 isolates/annum, the most significant factor must have been the greatly increased occurrence of a few prevalent serotypes.

Table 2 shows the most prevalent serotypes for the 10 years 1962–1971. During the first 5 years up to 1966, there was a steady increase in *S. panama* infections and an increase in the incidence of *S. brandenburg* mainly in 1963 to 1965. For both serotypes, pigs were the main source of infection and for *S. brandenburg* pork sausages were probably the main disseminating agents (Bevan Jones *et al.*, 1964).

Table 1

Salmonella isolations from humans in England and Wales

Year	Subgenus I		Subgenus II		Subgenus III		Subgenus IV		All subgenera	
	Isolates	Serotypes	Isolates	Serotypes	Isolates	Serotypes	Isolates	Serotypes	Isolates	Serotypes
1967	2332	105	8	4	1	1	0	0	2341	110
1968	3304	125	5	4	0	0	0	0	3309	129
1969	4914	128	2	2	0	0	3	1	4919	131
1970	4979	129	14	2	1	1	0	0	4994	132
1971	5149	125	17	7	0	0	1	1	5167	133

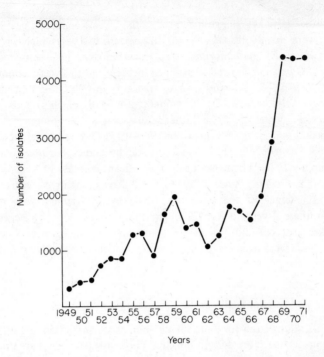

Fig. 2. Salmonellas other than *S. typhimurium*. Isolates from humans in England and Wales identified by the Salmonella and Shigella Reference Laboratory, Colindale and the regional salmonella laboratories.

Table 2

Prevalent salmonella serotypes from humans in England and Wales

Serotype	1962	1963	1964	1965	1966	1967	1968	1969	1970	1971
S. brandenburg	12	106	331	230	106	92	80	65	95	60
S. panama	8	17	59	105	198	225	458	564	608	302
S. virchow	2	11	24	18	1	51	229	361	97	95
S. 4,12 : d : –	–	–	–	–	–	–	170	323	94	60
S. heidelberg	118	43	141	74	60	49	41	281	221	178
S. indiana	1	–	3	30	28	189	98	191	173	162
S. bredeney	131	38	35	39	33	35	52	121	138	172
S. montevideo	12	14	20	16	12	21	203	66	65	74
S. agona	–	–	–	1	–	–	–	2	393	697
S. saint-paul	33	37	36	20	13	17	12	55	71	270
S. enteritidis	106	76	136	218	125	182	335	837	868	644

The most significant changes occurred in the 5 years commencing 1967. Four serotypes were mainly responsible for the pattern in 1968 which was the first year in which *S. typhimurium* was exceeded by the summated "other serotypes". *S. panama* showed a sustained increase and this was accompanied by a marked increase in *S. virchow*. Simultaneously in 1968, the appearance of a new serotype *Salmonella* 4,12 : d:− (monophasic) with a high prevalence greatly assisted in the ascendency of the "other serotypes".

The second part of the 5 years from 1969–1971 featured different serotypes. Since 1968 *S. enteritidis* has increased markedly and *S. panama* continued to increase up to 1970. There has been a significant increase in *S. heidelberg, S. indiana* and *S. bredeney* since 1969. Figure 2 shows that in the years 1970 and 1971 the incidence of the serotypes other than *S. typhimurium* reached a plateau. In these 2 years there was an enormous increase in *S. agona*, previously a very rare serotype, and also in 1971 an important increase in *S. saint-paul*. However, these were balanced by a fall in the incidence of *S. panama, S. virchow* and *S.*4,12 : d:−.

Salmonellas in the Food-Producing Animals

The identifications from food-producing animals are presented in Table 3 under the headings bovine, porcine and poultry. The serotypes named are those which were found to be the most prevalent in humans in the 5 years 1967–1971. The totals are for all serotypes isolated from each particular animal group for each year.

The increase in human infections in 1968 and 1969 due to *S. virchow* and *S.* 4,12 : d:− coincided with their appearance in broiler chickens. In subsequent years both serotypes continued to occur in broiler chickens but with a lower prevalence. It will be seen that *S.* 4,12 : d:− was found in pigs in 1968 and 1969 and this will be considered later.

There was a rise in the human infections with *S. heidelberg, S. indiana* and *S. bredeney* from 1969 onwards and these 3 serotypes also showed an increase in poultry over the same period. Once again broiler chickens accounted for most isolates, although with *S. indiana* some isolates were from ducks. These 3 serotypes were only occasionally isolated from pigs.

There were 11 isolates of *S. saint-paul* from poultry in 1969 but this increased to 74 in 1971. The human pattern showed a similar trend with a rise from 55 in 1969 to 270 in 1971.

The most spectacular feature of human salmonellosis in recent years has been the increase of *S. agona* from 2 in 1969 to 393 in 1970 and 697 in 1971. Apart from a few isolates in bovines, *S. agona* was not identified in food-producing animals prior to 1970. In 1970 isolates were made from pigs and from broiler chickens and in 1971 the isolates from chickens increased.

Table 3

Salmonellas from food-producing animals in England and Wales

Serotype	Bovine 1967	1968	1969	1970	1971	Porcine 1967	1968	1969	1970	1971	Poultry 1967	1968	1969	1970	1971
S. brandenburg	—	—	—	—	—	—	—	3	—	1	—	—	—	—	—
S. panama	—	—	—	—	—	9	11	7	17	14	—	2	2	3	2
S. virchow	—	—	3	2	1	—	2	6	2	—	1	40	6	9	13
S. 4,12 : d : -	—	—	—	6	1	—	15	70	6	2	—	42	106	38	29
S. heidelberg	—	—	1	1	1	—	7	3	10	7	1	10	42	12	11
S. indiana	5	5	1	1	1	1	7	6	1	1	11	10	59	26	82
S. bredeney	2	—	—	—	1	—	5	1	1	1	2	—	8	4	32
S. montevideo	1	—	—	—	1	—	3	8	—	4	—	14	13	12	18
S. agona	—	—	2	4	8	—	—	—	18	8	—	—	—	16	77
S. saint-paul	—	—	2	1	1	—	—	3	1	1	—	—	11	25	74
S. enteritidis	16	2	1	3	—	—	—	—	—	1	2	4	48	32	17
Totals all serotypes	196	162	116	206	103	53	84	294	138	77	58	167	373	253	407

S. montevideo is a frequent isolate from poultry, and it will be seen in Table 2 that the isolates from humans showed a peak in 1968 which coincided with commencement of this prevalence in poultry.

The Epidemiology of Some Important Salmonella Serotypes in England and Wales

The most important reservoir of infection for human salmonellosis in England and Wales is the food-producing animal. The bovine animal continues to be the main source of infection of *S. typhimurium*, but bovines are not the main source of the other serotypes and pigs and more especially poultry must be considered.

There has been a great increase in the production of poultry in the last 5 years and the current output is about 300 million broiler chickens/annum and about 15 million turkeys. This enormous production of broilers is reflected in the nation's consumption of meat which now comprises more poultry and less beef and mutton than previously.

The main change in the pattern of human salmonellosis in England and Wales has been the change in the proportion of human infection due to *S. typhimurium* as compared with those due to other serotypes. With the exception of *S. panama*, it will be shown that broiler chickens are the main source of these other serotypes.

A study of the epidemiology of these serotypes provides valuable information.

S. virchow

There were few human isolates before 1967 but there was an increase in the last quarter of 1967 and most cases were in the Bolton area or Merseyside. The source of infection was broiler chickens and *S. virchow* was isolated at the packing station supplying these chickens. In July 1968 a large outbreak occurred in Liverpool (Semple *et al.*, 1968) and the source of infection was shown to be chickens coming from a packing station in Cheshire. Also, in the summer of 1968 human cases occurred in the Midlands and again chickens from the same packing station were implicated. Brooksbank & Richards (1970) studied the epidemiology of *S. virchow* and found it in the packing station, the broiler rearing units and the hatchery but not in the breeding unit. The broiler units worked on a 3-month rearing cycle and it was suggested that *S. virchow* was introduced at least as early as the preceding cycle and possibly in the autumn of 1967. No definite evidence was presented but the investigators were of the opinion that *S. virchow* had been introduced into the group by infected feedstuff.

In the last quarter of 1968 more cases were reported in the Midlands and an outbreak occurred in a maternity unit (Rowe *et al.*, 1969). It was thought that the infection was introduced into the hospital by a pregnant woman who had consumed chickens which originated in the same packing station. This outbreak was typical of many hospital outbreaks in that the spread of infection was by cross infection, mainly amongst babies.

During 1969 human cases of *S. virchow* infection continued to occur mainly in the North West of England but there were also cases in London and the Home Counties. Most were sporadic cases but some outbreaks were reported (Elwood *et al.*, 1970; Roberts & Marshall, 1970). During 1970 and 1971 there has been a marked drop in human infection due to *S. virchow*

Salmonella *4,12 : d:– (monophasic)*

During the last 5 months of 1968 the Salmonella and Shigella Reference Laboratory, Colindale, received a large number of salmonellas with the structure 4,12 : d : – and it was clear that, although no second phase could be established, this was to be regarded as a specific serotype and a single epidemiological entity. The first human case occurred in July 1968 and in the remaining months of the year 170 human infections occurred mainly in the southern counties of England. The majority of cases were sporadic but outbreaks occurred in several hospitals. *S.* 4,12 : d : – had a peak incidence in 1969 with 323 human isolates and, again, most were from the southern part of England and especially the London area. Human isolates of *S.* 4,12 : d : – in England and Wales dropped to 94 in 1970 and 59 in 1971.

Table 3 shows that most isolates from food-producing animals were from poultry but during 1968 and 1969 many isolates were from pigs. The poultry were mainly broiler chickens but turkeys were involved, particularly in 1968. A high percentage of the poultry isolations came from one large broiler packing station on the south coast.

In June 1968 the serotype was isolated from poultry feedstuffs (type unknown) and in December 1968 it was isolated from meat and bone meal. Numerous isolates were made from fish meal which had been imported from South Africa in 1969. During 1969 the serotype appeared in poultry-offal meal and feather meal and a recurring cycle of infection was set up in the food-producing animals. During 1970, further isolates were made from South African fish pellets but there was none in 1971.

It seems that the most likely source of *S.* 4,12 : d : – was South African fish meal. Cargoes certainly arrived in the United Kingdom in 1969 and it is possible that infected consignments may have arrived in 1968. Turkeys, broiler chickens and pigs became infected and through their food products spread the organism to humans and the main agent was the broiler chicken. It appears that there have been no further imports of fish meal contaminated with this serotype and the incidence in broilers and humans seems past its peak. Nevertheless, many isolates are still made from pigs and products containing pork, predominantly sausages.

S. agona

Prior to 1970, only 3 isolates were received from humans in England and Wales, but there were almost 400 isolates in 1970 and almost 700 in 1971.

S. agona was isolated from a cargo of Peruvian fish meal which arrived in 1970 and soon afterwards isolates were received from poultry and humans. Pigs soon became infected and food containing pork, especially sausages, became a source of human infection. However, broiler chickens have remained the most important source. By 1971, *S. agona* was found in poultry offal meal and feather meal and it is likely that a cycle of infection independent of the original fish meal has been established which may account for the prevalence of *S. agona* in pigs.

S. saint-paul

Between 1962 and 1968 less than 50 isolates were received each year from humans in England and Wales. There was a slight rise in 1969 and 1970 but a sharp increase to 270 in 1971.

S. saint-paul is rarely isolated from cows or pigs but the main animal source is the broiler chicken and isolates from these have almost trebled in 1971. Animal feedstuffs do not seem to be important in the epidemiology of *S. saint-paul* and it is difficult to account for the recent marked increase in the prevalence of this serotype. Most of the isolates received at the reference laboratory were from 1 large packing station and the situation may be similar to that seen with *S. virchow* and *Salmonella* 4,12 : d : −.

S. panama

Prior to 1964, this serotype was an uncommon cause of human salmonellosis in England and Wales but since that year there has been a steady increase in its importance. There were 59 human isolates in 1964 and 608 in 1970, the greatest rate of increase being between 1967 and 1968. Pigs are the main source of infection with pork and bacon, but more especially sausages containing pork, being responsible for the spread to humans.

For many years, *S. panama* was isolated from bones imported from South America, especially from Argentina, and, in addition, isolates have been received from Argentinian horsemeat. It is possible that this imported bone may have been responsible for the introduction of the infection to pigs. However, the infection is probably now maintained in the pig population by a cycle independent of this imported material. Locally-produced animal feed and also the feeding of swill may contribute to the persistence of this serotype.

Conclusions

In the 5 years 1967–1971, the most important aspect of the epidemiology of human salmonellosis in England and Wales has been the upsurge of infections due to serotypes other than *S. typhimurium*. *S. typhimurium* remains the commonest single serotype in humans but the total of the infections with the other serotypes exceeded it in 1968 and in the ensuing years. *S. enteritidis* remains a prevalent serotype and probably second in rank order for humans. Isolates from humans increased in 1969 and 1970 but this was accompanied by an increase in isolates from poultry.

Amongst the remaining serotypes there has been a succession of a few serotypes which became prevalent and for a time accounted for a high proportion of human infections. In 1968 and 1969, *S. virchow* and *S.* 4,12 : d : − were prominent only to fall in incidence by 1970 and be replaced by *S. agona*. In 1971 *S. agona* continued to increase and was joined by *S. saint-paul*. Broiler chickens have been the most important source of all these serotypes.

This succession of serotypes with their peaks of prevalence must be seen superimposed on the less dramatic but nonetheless important pattern of 4 other serotypes. *S. panama* continued to show a rise in its incidence from 1967–1970 but dropped in 1971. *S. heidelberg*, *S. indiana* and *S. bredeney* had less impressive increases over the last 5 years but if considered together are important, and again these 3 serotypes also may be related to broiler chickens.

The Importance of Animal Feedstuffs

There is good evidence that imported animal feedstuffs have played an important part in this epidemiological pattern. The prevalence of *S. panama* in pigs may have been initiated by the importation of the serotype in bones from South America. *S. bredeney* was also frequently isolated from this material in the years 1966–1969.

S. 4,12 : d : − and *S. agona* provide excellent examples of the introduction in imported animal feed with subsequent spread to poultry and to a lesser extent to pigs. A clear chain of infection was established from imported fish meal through broiler chickens and pork products to humans. These 2 serotypes caused a very large number of human infections and it is likely that if the imported fish meal had been subjected to sterilization, the outbreaks would have been prevented or at least reduced in intensity. It is unlikely that sterilization of animal feeds will prevent human salmonellosis entirely but it will certainly reduce it.

Subgenus II Infections

Table 4 shows that the reference laboratory received 46 subgenus II salmonellas from humans in England and Wales in 5 years. The most common serotype was *S. sofia* which is commonly confused with *S. paratyphi* B because of antigenic similarities. The second most frequent serotype was *S. makumira.* All cases were sporadic and no report has been made of a general outbreak in England and Wales due to subgenus II salmonellas.

Table 4

Subgenus II salmonellas from humans and animals in England and Wales (1967-1971)

	Humans	Tortoises	Other animals
S. artis	3	0	0
S. canastel	4	16	0
S. hamburg	1	0	0
S. kraiifontein	1	0	1
S. lethe	1	5	1
S. limbe	1	0	0
S. lindrick	1	0	0
S. makumira	10	0	3
S. sofia	19	7	0
S. uphill	1	14	0
S. 6,8 : a : e,n,x	2	0	0
S. 35 : g,m,s,t : −	2	0	0
Other serotypes	0	9	4
Total	46	51	9

During the same 5 years, there were 60 identifications of subgenus II salmonellas from animals. Tortoises were the most common source of these salmonellas and there is considerable similarity between these isolates and the serotypes found in humans.

There is a large importation of tortoises mainly from North Africa and Jugoslavia and it is not surprising that many of the human infections were in children. When a subgenus II salmonella is isolated, it is essential to enquire about possible contact with a pet tortoise, particularly when the human isolate is from a child.

Subgenus II serotypes are rarely isolated from human or animal foods but Australian kangaroo meat for pet food manufacture was once a common source which seems unimportant at present. Occasional isolates are made from South-west African beef but the problem is small and only 10 isolations were made in 1971.

Subgenus III Salmonellas (Arizona)

Arizona infections in humans in England and Wales are exceedingly rare. In 1966, *Arizona* 26 : 32 : 21 was isolated from a young child with gastro-enteritis and the same serotype was isolated from a sibling who remained symptom-free. This infection was shown to originate in pet terrapins kept in a tank in the home (Plows *et al.*, 1968). In 1967 an adult who had recently returned from France developed diarrhoea and *Arizona* 29 : 33 : 25 was isolated from his stools. In 1971 a man suffered a severe attack of diarrhoea on return from the U.S.A. and *Arizona* 1,3 : 1,2,6 : − was isolated from his stools (Hughes *et al.*, 1971).

Arizonas have been isolated from bones and bone meal of Indian and South American origin (Harvey & Price, 1962) and there have been isolates from imported rabbit meat, Australian kangaroo meat, South African horsemeat, imported egg products and coconut.

In 1968 *Arizona* 7a,7b : 1,7,8 : − was isolated from a batch of turkey poults imported from the U.S.A. The infection was recognized during the quarantine stages and was contained (Timms, 1971). Apart from an isolate of *Arizona* 26 : 29 : 30 from a pig at an abattoir in 1971, no other identifications have been made from food-producing animals.

At the present time the most frequent source of Arizonas in England and Wales is reptiles (usually snakes) in zoological collections and from terrapins.

The Importance of Terrapins as a Cause of Human Salmonellosis in England and Wales

Terrapins (referred to as turtles in the U.S.A.) are small aquatic animals which are imported into England and Wales in very large numbers each year. Most batches come from the southern states of the U.S.A. where they are usually reared in special farms.

Many batches of terrapins are infected with salmonellas and they undoubtedly represent a hazard. Children are very likely to become infected from handling these animals or from the tank water and the keeping of terrapins as pets should be deprecated.

A wide range of salmonella serotypes have been identified from terrapins in England and Wales. Salmonellas of subgenus III (Arizonas) are commonly found as well as subgenus I serotypes. *S. paratyphi* B has been isolated from the water of tanks containing terrapins. Subgenus II serotypes are rarely found in terrapins, and this contrasts with findings in the terrestrial tortoise.

Recent surveys in the U.S.A. suggest that each year 280,000 human cases of salmonellosis may be related to terrapins (Report, 1972). In many states legislation has been passed controlling the sale of these animals and requiring bacteriological examination of batches before sale.

Similar surveys have not been carried out in England and Wales but there is no doubt that terrapins present an unnecessary hazard and their importation should cease.

Summary

The paper reviews the epidemiology of salmonellosis in England and Wales from 1967–1971.

S. typhimurium remained the most prevalent serotype causing human infections. Since 1968 the summated infections due to all the other serotypes have exceeded those due to *S. typhimurium*. This situation was due to a considerable increase in the prevalence of a few serotypes which were previously rare.

The most important serotypes were *S. virchow*, *S.* 4,12 : d : –, *S. agona*, *S. saint-paul*, *S. panama* and *S. enteritidis*. Of lesser importance were *S. heidelberg*, *S. indiana*, *S. bredeney* and *S. montevideo*.

With the exception of *S. panama* the increase in the prevalence of human infections with those serotypes was parallelled by an increase in their isolation from broiler chickens. Imported animal feeding materials, in particular fish meal, were important in the introduction of two serotypes, *S.* 4,12 : d : – and *S. agona*.

Subgenus II salmonellas were rare in humans in England and Wales and most isolates from animals were from tortoises. Arizonas were isolated from four humans and the most frequent sources were reptiles in zoological collections and terrapins.

References

Bevan Jones, H., Farkas, G., Ghosh, A. & Hobbs, B. C. (1964). *Salmonella brandenburg*–an epidemiological study. *Mon. Bull. Minist. Hlth.* **23**, 162.

Brooksbank, N. H. & Richards, D. W. (1970). *Salmonella virchow. State Vet. J.* **25**, 66.

Elwood, W. J., Abbott, J. D., Blackbourn, M. S. & Hughes, M. (1970). An outbreak of *Salmonella virchow* food poisoning amongst hospital staff with experimental evidence on the cooking of chickens. *Med. Offr* **123**, 273.

Harvey, R. W. S. & Price, T. H. (1962). Salmonella serotypes and Arizona paracolons from Indian crushed bone. *Mon. Bull. Minist. Hlth* **21**, 54.

Hughes, M. H., Bartlett, D. I., Baker, M., Dreaper, R. E. & Rowe, B. (1971). Gastro-enteritis due to Salmonella subgenus III (Arizona). A second case diagnosed in Britain. *J. Hyg., Camb.* **69**, 507.

Plows, C. D., Fretwell, G. & Parry, W. H. (1968). An Arizona serotype from a case of gastro-enteritis in Britain. *J. Hyg., Camb.* **66**, 109.

Report (1972). Revision in turtle importation regulations–United States. *Morbidity and Mortality Weekly Rept.* **21**, 447.

Roberts, D. J. & Marshall, J. N. (1970). *Salmonella virchow* in Salford. *Med. Offr* **123**, 278.

Rowe, B., Giles, C. & Brown, G. L. (1969). Outbreak of gastro-enteritis due to *Salmonella virchow* in a maternity hospital. *Br. med. J.* **3**, 561.

Semple, A. B., Turner, G. C. & Lowry, D. M. O. (1968). Outbreak of food-poisoning caused by *Salmonella virchow* in spit-roasted chicken. *Br. med. J.* **4**, 801.

Timms, L. (1971). Arizona infection in turkeys in Great Britain. *J. Med. Lab. Technol.* **28**, 150.

Vernon, E. (1970). Food-poisoning and salmonella infections in England and Wales. *Publ. Hlth, Lond.* **84**, 239.

Discussion

Wilson

In the 1930s *Salmonella enteritidis*, *S. thompson* and *S. typhimurium* were important but *S. thompson* has now disappeared. It was imported originally in Chinese egg products.

Rowe

S. thompson was 1 of the 5 most prevalent serotypes which infected humans in England and Wales during the 1950s. Since 1962 it has remained outside the 10 most prevalent serotypes. The legislation requiring pasteurization of liquid egg became effective in January 1964, but many processors had already introduced pasteurization as early as 1961. This may have helped to reduce the incidence of this organism.

Smith

Have you seen any change in antibiotic resistance patterns over the past years? In particular, have there been any changes in chloramphenicol resistance?

Rowe

I cannot comment on antibiotic resistance patterns since my laboratory does not undertake this work.

Kampelmacher

Antibiotic resistance in salmonellas has been investigated in the Netherlands. During the past 15 years we have examined some 18,000 strains, and resistance to chloramphenicol has always remained low (about 0·3–0·5%). Fluctuation in resistance to tetracyclines and to ampicillin have occurred but in general the pattern has always been similar for strains from human and animal origins.

Brachman

Salmonella surveillance data in the U.S.A. is generally similar to that in England and Wales. Up to 1969 there were *c.* 20,000 human isolates/year but since then there has been an increase to about 21,000–22,000/year, in spite of constant attempts at control. Ten common serotypes account for about 70% of the total annual isolates, and about 150 serotypes occur less than 5 times/year. *S. typhimurium* is the commonest human isolate. Legislation to control a particular problem with "turtles" is being introduced and will necessitate salmonella-free certification before they can be imported or passed across state line.

We commend the W.H.O. for their publicizing salmonella surveillance data from various countries. This should be stimulating to other countries to initiate surveillance and reporting. I should like to ask Dr Rowe to what extent he is successful in attempts to trace a rare salmonella serotype to its source? In the U.S.A. we are only rarely successful.

Rowe

If we identify a rare salmonella serotype we are often successful in our attempts to show that its origin was from abroad. Usually the patient turns out to be of foreign origin or to have returned recently from abroad. However, I have no figures to support my statement.

Ingram

It is notable that with the rapid rise recently in incidents due to new serotypes, introduced probably via feedstuffs as indicated by Dr Rowe, the total number of incidents has remained roughly constant for more than 10 years. To put it mildly, it appears as if a newcomer displaces an equal volume of *S. typhimurium*; it would be interesting to know why. The point seems relevant to Dr Rowe's proposition that to clean up the feeding stuffs would improve the overall picture.

Rowe

My first figure (see text) shows that in the years 1964–1969, the incidents due to *S. typhimurium* remained more or less constant, whereas the incidents due to the other serotypes increased from 1368 in 1964 to over 3000 in 1969. Comparable figures for 1970 and 1971 are not yet available. Nevertheless, my observations lead me to expect a similar ratio to that in 1969. My laboratory returns (see Fig. 2) would support this and show a levelling off of infections due to other serotypes. The critical years were 1967–1969; during these years the tremendous rise in the prevalance of the other serotypes was well out of proportion to the small fall in the incidence of *S. typhimurium*. Prof. Ingram's observation is probably true if we look at the whole 20 years from about 1950. The prevalence of the other serotypes is now at about the same level as was *S. typhimurium* in the 1950s.

Hobbs

Surveillance of foods and feeds may reveal the source of outbreaks of salmonellosis which occur subsequently, for example, an outbreak of sal-monellosis due to *S. typhimurium* U65—a stranger in the U.K.—was traced since it had been isolated from imported poultry examined some months previously. *S. typhimurium* U32 caused 1000 cases (with 12 deaths) from pork products, it was found in pigs. The organism was isolated from the feed but only at the time of the outbreak.

Usually control of salmonella is possible provided the level is not high in foods, feeds and in animals. A build up of a particular serotype overcomes normal measures but possibly only results in fluctuations so that the general level remains constant. In epidemiological investigations, outbreaks of food poisoning due to *Cl. perfringens* were traced back to the food in 80–90% of instances, and staphylococci in 70%. For salmonellosis, 24% of general outbreaks and 1% of family and sporadic incidents are traced back to food.

Rowe

In England and Wales about 30% of the general outbreaks of salmonellosis occur in hospitals. These outbreaks usually occur in wards accommodating young babies or, sometimes, old people. The course of these outbreaks is often protracted and the mechanism is usually cross-infection from an index case. As such these are not food-poisoning and one would not expect to find a common source in a food.

Marine Micro-organisms and Food Poisoning

G. I. BARROW

Public Health Laboratory,
Royal Cornwall Hospital (City),
Truro, Cornwall, England

Food poisoning from marine sources is common throughout the world. Apart from cutaneous or gastro-intestinal hypersensitivity to various seafoods, it is caused either by intoxication or infection mainly from micro-organisms. Intoxications include illness due to consumption of some naturally toxic fish species, but more often from seafoods which contain potent toxins produced by certain marine algae, particularly dinoflagellates. Reference is also made to botulism from seafoods. Infections from marine products may be caused by known bacterial, viral or parasitic pathogens acquired by pollution of the environment, as well as by the marine bacterium *Vibrio parahaemolyticus*. The salient features of botulism, dinoflagellate intoxication and infection with *Vibrio parahaemolyticus* are described, and the role of indicator organisms in assessing the cleanliness and bacterial safety of seafoods is mentioned.

FOOD POISONING from marine sources is almost certainly world-wide in distribution, and perhaps more common than generally appreciated. Apart from individual allergy—cutaneous or enteric—to various seafoods, it includes infections with known bacterial, viral and parasitic pathogens, as well as illness, particularly gastro-enteritis, of unknown aetiology. It also includes poisoning from organic substances and inorganic chemicals, from inherently poisonous fish and from toxins produced by some marine micro-organisms. Adequate investigations of illness thought to be due to marine sources thus involves several disciplines, including clinical medicine, epidemiology, microbiology, virology, marine biology, parasitology, chemistry and toxicology. Their use, however, depends largely on awareness and recognition that seafoods may be implicated in illness. Even when seafoods are incriminated, access to and co-ordination of these services may be difficult unless they are available and organized on a national or regional basis. Many incidents caused by seafoods have probably remained either unrecognized or unidentified because of lack of early epidemiological information as well as satisfactory material for laboratory examination. Seafoods, whether processed or not, may be marketed at considerable distances from their sources, often in different countries or even continents, and relevant information is essential for adequate investigations and surveillance.

Food poisoning from plants such as toadstools (*Amanita* species) or deadly nightshade *(Atropa belladonna)* are well known, but poisoning from animals, though probably recognized locally, is perhaps less well known. Animal tissues may become poisonous by bacterial or enzyme action, but some are naturally toxic even when consumed fresh. Many animal toxins are heat resistant but water soluble, so that thorough washing and discarding of cooking water are wise precautions. Table 1 modified from Scott (1969) illustrates the extent and geographical distributions of outbreaks from animal foods in various parts of the world. It also indicates clearly the importance of fish and shellfish as causes of food poisoning. Millions of tons of seafish and shellfish, however, are consumed annually throughout the world, and it is hardly surprising that food poisoning should sometimes occur. This may be due either to intoxication or to infection, mainly from certain micro-organisms which live and breed in the sea. The main features of these illnesses are given in Table 2. Other organisms which may be acquired through pollution of the marine environment or during the subsequent processing and handling of seafoods are also mentioned in this review.

Table 1

Illustrative outbreaks of food poisoning from animal sources*

Place	Vector	Cases	Deaths
Hawaii	Fish	433	7
California	Mussels, clams	479	33
Fanning Island	Fish	95	0
Saipan	Eel	57	2
Minamata, Japan	Fish	83	10
Indo-Pacific	Fish	40,000	?
Canada	Oyster, clam	94	?
Alaska	Little Neck clams	25	0
Wake Island	Mullet	29	0

* Modified from Scott (1969).

Parasitic infestations, which are widespread, especially in the Far East, are not considered, although many depend on the consumption of raw fish which act as intermediate hosts for one of the phases in the complex parasite life cycles. Most species, such as *Paragonimus westermani*, *Diphyllobothrium latum*, *Clonorchis sinensis* and *Gnothostoma spinergum* are acquired mainly from freshwater fish, crustacea or molluscs. *Anisakis* species which may cause visceral eosinophilic granulomata involve seafish, and in Europe are associated with the consumption of raw or pickled herring (Thiel *et al.*, 1960; Ashby *et al.*, 1964).

Intoxications from Seafoods

Natural toxicity

Illness, varying from gastro-enteritis to neurological disturbances, sometimes fatal, may be due to natural toxicity of actual fish tissues. This occurs mostly in

Table 2

Salient features of food poisoning from seafoods

Type of food poisoning	Food vehicle	Incubation period	Nature of illness	Nature of toxin	Treatment	Epidemiology
Fish toxicity	Tropical fish—viscera, liver, ovaries. Natural: e.g. Puffer fish Secondary: e.g. Scombroid fish	Rapid onset A few minutes to a few hours	Nausea, vomiting Numbness, paraesthesia of face and limbs Muscular weakness incoordination, paralysis Usually short duration Often fatal	Varies with species: Tetrodotoxin-like Water soluble Small molecule Rapid absorption Excreted unchanged in urine Not antigenic	Emetic, gastric lavage Symptomatic with artificial respiration Stimulation of urinary output No antidote available	Seasonal, limited in extent Depends on local eating habits Mainly tropics, Pacific, and Far East
Botulism Type E	Raw, fermented or inadequately preserved foods eaten without sufficient final heating	Usually 12–36 h May vary from a few hours to several days	Nausea, vomiting Ocular, pharyngeal and respiratory paralysis Temperature sub-normal Duration several weeks Often fatal	Specific exotoxins blocking synthesis and/or release of acetyl choline at nerve endings Large molecules Antigenic	Gastric lavage with alkali Symptomatic, with artificial respiration Specific antitoxin given early	Limited in extent Depends mainly on local eating habits Sporadic incidents Northern hemisphere
Dinoflagellate intoxication	Shellfish, raw or cooked, especially mussels, oysters and clams	Rapid onset A few minutes to a few hours	Numbness, paraesthesia of face and limbs Muscular weakness, incoordination, paralysis Diarrhoea and vomiting *not* usual Duration 1–3 days Sometimes fatal	Varies with species: Saxitoxin-like Water soluble, heat and acid stable Small molecule Rapid absorption Interferes with action potentials. Excreted unchanged in urine Not antigenic	Emetic, gastric lavage Symptomatic with artificial respiration Stimulation of urinary output No antidote available	Sporadic outbreaks, mainly in tropics but can occur in temperate zones if meteorological conditions suitable
V. parahaemo-lyticus infection	Fish, shellfish and processed seafoods	Usually 12–24 h May vary from 2–48 h	Acute gastroenteritis with pyrexia and dehydration Occasionally dysentery-like Duration: a few days Rarely fatal	Marine bacterium ? Cholera-like enterotoxin causing electrolyte and fluid loss into gut lumen	Supportive with fluid replacement Tetracycline if severe	Probably worldwide Occurs often in countries where raw or semi-preserved seafoods eaten. Direct transfer from man to man doubtful

the Pacific, the Far East and the tropics from fish dishes such as octopus, squid, eels, sea urchins and sea anemones, as well as from seasonally toxic fish eggs. The nature of some of these toxins is still undetermined, but many are probably similar to that of the deadly puffer fish *(Arothron hispidus)* and related Tetraodontidae species which are still considered as delicacies in parts of the Far East. Their water soluble toxins, which are seasonally present mainly in the liver and ovaries, have been characterized chemically and are referred to as tetrodotoxin, TTX or "fugu poison". Such toxins are absorbed rapidly from the gastro-intestinal tract and act quickly on the neuromuscular system, probably by interference with action potentials (Evans, 1969). These toxins cannot be neutralized and treatment depends on artificial respiration and increasing urinary output.

Secondary toxicity

Ciguatera, another kind of neuromuscular poisoning, is caused by the consumption of various fish species ecologically associated with coral reefs. The nature of the toxins involved, which are heat-stable but not water soluble, are unknown, but they are probably formed initially by marine algae and pass through a food chain from small herbivorous fish to the larger carnivores. Secondary toxicity may also occur in scombroid fish, which include tuna and mackerel, due to the formation of saurine-like toxins from histidine by bacterial action (Scott, 1969). Unless commercially canned or frozen, scombroid fish should therefore be eaten fresh. In the tropics, other rare kinds of poisoning from fish, including hallucinatory intoxication from mullett, also occur.

Whether primary or secondary, the epidemiology of poisoning from fish is essentially limited in character and depends mainly on local eating habits.

Botulism

Although not a true marine micro-organism in the strict sense, *Clostridium botulinum* is mentioned because seafoods have often been responsible for human botulism. The 6 known types of this anaerobe each produce antigenically distinct exotoxins, some of which may be activated and potentiated by trypsin during digestion (Sakaguchi & Sakaguchi, 1967). They are water soluble, heat-sensitive, acid stable and highly neurotoxic, probably acting by preventing the synthesis and/or release of acetyl choline at nerve endings (Heyningen & Arseculeratne, 1964). Although botulism in man due to types A, B and more recently F, have occasionally been reported from fish products, type E has previously been particularly associated with seafoods—so much so that it was thought at one time to be a marine organism. More recent studies, however, have shown that it is terrestrial in origin and carried to the sea where it is harboured by coastal fish (Riemann, 1969). This probably reflects the ability of the spores to survive in marine environments, as the organism has been isolated in sea areas remote from human habitation. In Britain, *Cl. botulinum* types B and E may be expected to occur in fish on sale, but probably never more than 1 or 2 spores/fish (Shewan, 1970).

Outbreaks of botulism occur sporadically in man, but animal and aquatic life may be affected on a large scale, particularly by types C and D. Human botulism is almost invariably caused by food which has been stored after inadequate heating or preservation, and is eaten either cold or with minimal final heating. Clinically, muscular inco-ordination followed by paralysis, often preceded by gastro-intestinal upset, ensue after an incubation period of 12–36 h. Epidemiologically, the incidence of botulism caused by different types of toxin varies from one country to another and depends largely on local eating habits. Type E botulism seems to occur mainly in countries in the northern hemisphere, where uncooked or preserved seafoods are traditionally eaten (Table 3). In Japan, for example, seafoods have been responsible for nearly all cases of botulism. Type E strains of *Cl. botulinum* have also been associated with smoked, vacuum-packed as well as pickled and salted fish. In Russia, however, although home-processed fish accounted for more than half of the cases of botulism between 1958 and 1964, the majority were caused by type A. The mortality rate, including some cases due to type E, was 23% compared with about 5% in the 15% of incidents due to type B poisoning, mostly from home preserved ham (Sakaguchi, 1969).

Table 3

*Incidence of Type E Botulism in different countries**

	Place	Year	Outbreaks	Cases	Deaths
Japan	Hokkaido	1951–64	31	225	42
	Tohoku area	1953–63	19	82	35
United States	Alaska	1950–62	8	22	8
	Mainland	1932–63	8	37	15
Canada		1944–63	13	36	22
Poland		1962–63	5	23	5
U.S.S.R.		1958–64	13	25	?
Denmark		1951–64	4	9	2
Sweden		1960–62	3	5	1
Norway		1963	1	5	1

* Modified from Sakaguchi (1969).

In Britain, the last cases of botulism occurred in 1955 and were due to type A toxin. Two persons were affected after eating fish pickled 4 weeks previously in Mauritius (Mackay-Scollay, 1958).

MSF–7*

Botulinum exotoxins are antigenic, and antitoxin, if given early in illness, is the only specific treatment. Because of its high mortality rate the dangers of botulism have played a considerable part in the determination of methods in food technology.

Dinoflagellate toxicity

Outbreaks of neurotoxic or paralytic food poisoning from seafoods—almost exclusively shellfish, however—are more commonly attributable to phyto-plankton. This is due to the ingestion of certain unicellular algae, particularly dinoflagellates, by fish and shellfish. Dinoflagellates together with other photosynthesizing micro-organisms, constitute the grass of the sea and almost all marine life ultimately depends on them for food either directly or indirectly. They multiply normally about once every 24 h by binary fission, but cysts, formed by sexual fusion, remain viable for long periods in the seabed until conditions are suitable for germination. In summer and autumn, with warmth and long hours of daylight, these micro-organisms multiply rapidly. Under certain conditions of wind and tide and in the presence of suitable concentra-tions of nutrients, population explosions occur which result in the accumulation of millions of dinoflagellates/l at the sea surface. These extensive "blooms" vary in colour from milky-white to red, and are often luminescent. The so-called red tides are well known in warm and tropical waters, but blooms do sometimes occur in temperate seas, and they precede marine and human morbidity by a few days to a few weeks (Rounsefell & Nelson, 1966).

During normal metabolism, some dinoflagellates, notably *Gonyaulax catanella*, *Gonyaulax tamarensis* var. *excavata*, *Gymnodinium aureoleum*, *Gymnodinium breve*, *Pyrodinium phoneus* and possibly some *Prorocentrum* species produce potent heat-stable alkaloid toxins (Seaton, 1972). These unicellular organisms are very much larger than bacteria, ranging from 20–60 μm in size, and the toxins they produce affect only vertebrate animals. When such dinoflagellates are ingested by fish, not only can considerable marine mortality ensue, but other animals including birds, which feed on fish containing the toxin, may themselves be poisoned. Man is not usually at risk because dead fish are not harvested. Invertebrate fish, however, particularly filter feeding bivalve molluscs, including different species of oysters, mussels, cockles, scallops and clams throughout the world are particularly dangerous to man because they are able to accumulate such toxins in their tissues without harm to themselves. Such apparently healthy shellfish can thus cause illness by "transvection" of dinoflagellate toxins. The toxins are concentrated in the digestive glands, gills or siphons, and are so powerful that a single shellfish may contain a fatal dose for man. Burning sensations affecting the mouth and face with progressive paraesthesia and numbness throughout the body are followed by muscular inco-ordination, paralysis and sometimes death. In contrast to infective bacterial food poisoning, abdominal pain, diarrhoea and vomiting are unusual. Full details of this and other kinds of fish poisoning are given in the extensive reviews of Halstead (1965, 1967) and Kao (1966).

The specific toxins produced by dinoflagellates vary with the species concerned, but their general properties and actions are similar. They are closely related to the toxin produced by the butter clam *Saxodonus giganteus*. This toxin has been extracted, defined chemically and is referred to as saxitoxin or SXT. It is water soluble, heat resistant and stable under acid conditions. It is rapidly absorbed through the gut and acts on the neuromuscular system, probably like tetrodotoxin by interference with action potentials (Evans, 1969). In contrast to botulinum toxins, both saxitoxin and tetrodotoxin are small molecules which are therefore not antigenic. Antitoxins or effective antidotes are not available and acquired resistance to such poisoning does not occur. The toxins are excreted largely unchanged in the urine, and treatment is therefore symptomatic with stimulation of urinary output.

Outbreaks of neurotoxic shellfish poisoning have often been reported from the Pacific and Atlantic coasts of North America, the Gulf of Mexico, Japan and Europe, particularly Scandinavia and Portugal and they undoubtedly occur also in other parts of the world (Rounsefell & Nelson, 1966). Epidemiologically they depend on meteorological conditions, and outbreaks are therefore sporadic and again geographically limited in extent. Although dinoflagellate blooms may occur at sea, those which affect man must occur in the shallow waters inhabited by shellfish. In Britain, outbreaks are uncommon and very few have been reported. The first outbreak for some considerable time occurred in the summer of 1968 on the north-east coast of England (McCollum *et al.*, 1968). It was associated mainly with mussels *(Mytilus edulis)* and at least 78 people were known to have been affected. The clinical picture was typical, though most of the cases were mild, probably because the mussels were cooked before consumption and significant amounts of the heat stable toxin discarded with the cooking water. The dinoflagellate involved was *Gonyaulax tamarensis* var. *excavata* and the brown bloom affected other marine life, particularly non-commercial bivalves as well as sea birds (Clark, 1968). As a result of this outbreak, annual surveys of toxin levels in molluscan shellfish have been carried out in Britain (Ingham *et al.*, 1968; Conn & Farrand, 1970). In areas such as Florida, California and some parts of Canada where toxic blooms are known to occur frequently, dinoflagellate species and numbers as well as the presence of toxin in shellfish are monitored. Toxin levels are assayed by their lethal effect on mice (Report, 1970), but it is important to identify the "bloom" organism quickly as toxin extraction methods vary with different species. When critical toxin levels are found in any area, the harvesting and sale of shellfish are prohibited until safe. Not all shellfish, however, pass through commercial channels, and although dinoflagellate blooms may not persist long, their damage is already done.

Infections from Seafoods

As well as illness caused by bacteria which inhabit the sea or which pollute coastal waters, sea products, like other foods, may be contaminated with a variety of micro-organisms during processing: the poorer the hygiene, the greater the risks. Apart from accepted food poisoning organisms such as salmonellae, staphylococci and *Cl. perfringens,* many other organisms may cause illness, either individually or collectively, especially if present in large numbers and particularly in travellers (Pilsworth, 1952; Rowe *et al.,* 1970). Numerous low temperature and salt tolerant organisms are often found in seafoods, but their possible role in disease is problematic.

In Britain, little infection has been attributed to marine foods other than shellfish. Though uncommon now, the association of typhoid with the consumption of raw shellfish, especially oysters and mussels, is well-known. In coastal waters heavily polluted with sewage, other fish eaten raw or imperfectly cooked may also cause typhoid, though paratyphoid is unusual. In the Phillipines and the Far East, cholera outbreaks have been attributed on epidemiological grounds to oysters, shrimps and sun-dried fish. Indeed, it has even been suggested that fish may act as reservoirs of infection and thus contribute to maintaining cholera endemicity, although the evidence is doubtful (Pollitzer, 1959). With the spread of cholera into Europe, however, the human sewage–fish–man chain might assume more importance.

In Britain edible shellfish include oysters *(Ostrea edulis* and *Crassostrea angulata),* mussels *(Mytilus edulis),* cockles *(Cardium edule),* scallops *(Pecten maximus),* clams *(Mercenaria mercenaria),* whelks *(Buccinum undatum)* and periwinkles *(Littorina littorea).* Except for the scavenging whelks and peri-winkles, they feed by filtering microscopic particles, including bacteria, from the large amount of water passed through their gills, subsequently conveying this material by ciliary action to the gut. In sewage polluted waters such as estuaries, shellfish may thus concentrate large numbers of organisms, including those pathogenic to man. Cockles, whelks and periwinkles are cooked commercially before sale and the risk of human infection, bacterial or viral, from them because of sewage pollution is therefore minimal. Some shellfish, particularly oysters and occasionally mussels, are normally sold alive and eaten uncooked, and are thus potentially hazardous. This danger is avoided by allowing them to eliminate ingested sewage organisms in their mucous pseudo-faeces by feeding in clean or purified water (Dodgson, 1928, 1936). In Britain, oysters are usually purified in large tanks capable of holding several thousand shellfish in which the water is sterilized by ultra-violet irradiation, recirculated by pump and aerated by cascade (Wood, 1969). After purification for at least 36 h, such shellfish should be virtually free from sewage organisms. The efficiency of cleansing is judged by standard bacteriological tests using coliform organisms as indicators of faecal pollution, shellfish containing more than 5 *Escherichia coli* I/ml of tissue being regarded with suspicion (Clegg & Sherwood, 1947; Knott, 1951; Sherwood & Thomson, 1953). Such purification plants, or alternative cleansing methods in other countries, coupled with grading tests have undoubtedly ensured supplies of bacteriologically cleaner shellfish, and enterobacteria associated with gastro-enteritis are usually absent. These procedures do not, however, guarantee freedom from marine bacteria, from enteroviruses, from biological toxins, or from chemical substances present in the marine environment, all of which can be concentrated physiologically by shellfish without harm to themselves.

Outbreaks of viral hepatitis have been attributed on epidemiological grounds to raw shellfish from sewage polluted waters (Roos, 1956; Mason & McLean, 1964). It would be surprising if other enteroviruses, including that of epidemic vomiting disease (Clarke *et al.*, 1972), were not also involved. Uncooked shellfish, by repute, often cause "indigestion", and gastro-enteritis, usually of unknown aetiology, is probably the commonest kind of illness following their consumption. In 1 such outbreak, some patients also subsequently developed hepatitis (Dismukes *et al.*, 1969). Recent outbreaks of gastro-enteritis in Britain were traced to oysters from 1 producer in Cornwall. This producer was compared with a nearby producer of trouble-free oysters (Barrow & Miller, 1969). The suspect plant apparently operated efficiently yielding oysters free from coliform organism, and thus of satisfactory bacterial quality. No known causal agent was identified, but many halophilic organisms were consistently found in cleansed oysters from the suspect plant, but not in similar oysters from the satisfactory plant. Inadequate circulation was thought to have limited ultra-violet irradiation of the water, allowing a build-up of these organisms in the cleansing tanks and this was confirmed by subsequent monitoring of the plant after improving the rate of circulation. These organisms comprised a variety of marine vibrios, mainly *V. alginolyticus* and *V. anguillarum,* but the presence of *V. parahaemolyticus* has since been confirmed (Barrow & Miller, 1972). This marine vibrio was first identified in Japan by Fujino and his colleagues in 1951 as the cause of acute gastro-enteritis which affected 272 persons who had eaten *shirasu*—a Japanese delicacy prepared from boiled and semi-dried sardines. It is now recognized as the commonest cause of food poisoning in summer, due largely to the national custom of eating raw fish and other sea products (Sakazaki, 1969). Extensive work has since been done on this organism in Japan, but as many of the earlier papers were in Japanese it is only in recent years that its significance has been appreciated elsewhere. After taxonomic studies, it was finally placed in the genus *Vibrio* by Sakazaki and his colleagues (1963).

Vibrio parahaemolyticus

This organism was at first thought to be limited to Japan and the Far East, but it has since been isolated from fish (Nakanishi *et al.,* 1968), shellfish (Fishbein, Mehlman & Pitcher, 1970), crabmeat (Baross & Liston, 1970) and other marine sources (Vanderzant, Nickelson & Parker, 1970) in various countries, and it is almost certainly widely distributed throughout the world, especially in warm coastal waters. Apart from the Far East and Pacific, it has been recognized as the cause of outbreaks of food poisoning in Australia (Battey *et al.,* 1970) and the U.S.A. (Sumner *et al.,* 1971). In Britain, it has been isolated from a fatal case of food poisoning that followed consumption of cockles, in which enterotoxin-producing staphylococci were also found (Barrow, 1972, unpublished). It has also recently been identified as the cause of food poisoning among airline passengers and crew returning from Bangkok to London. The food vehicle was crabmeat in *hors d'oeuvres* dressed with crab claws (Peffers *et al.,* 1973).

Clinically, illness caused by *V. parahaemolyticus* varies from acute cholera-like gastro-enteritis to mild diarrhoea, or it may occasionally be dysentery-like with blood and mucus in the stool. The average incubation period is between 12 and 20 h, but ranges from 2–48 h, presumably depending partly on the size of the infecting dose, the acidity of the stomach and the nature of the food. Treatment is supportive with fluid replacement if necessary until the bacteriological diagnosis is established. If severe, tetracycline, neomycin or streptomycin may be used; it is resistant to ampicillin. The infective dose is considered to be large, but the generation time of the organism is short—about 10 min—and it can thus multiply more quickly, as well as at lower temperatures, than most other pathogens (Smith, 1971). As in cholera, *V. parahaemolyticus* is excreted in large numbers during illness, but the vibrios diminish rapidly with clinical recovery. It is therefore important to examine specimens as early as possible during illness. If this is not possible or if delay in transit is likely, transport media should be used to prevent complete swamping of the vibrios by other organisms before inoculation of selective media.

Adequate clinical and epidemiological information is also important to ensure that suitable cultural techniques are employed. Fortunately, *V. parahaemolyticus* can be isolated by the same media and methods used for cholera organisms, though salt-colistin or glucose-salt-teepol broth may be used for enrichment culture as well as alkaline peptone water Enrichment culture is particularly important for its isolation from foods, though other marine organisms can be troublesome. *V. parahaemolyticus* forms large characteristic green or blue-green colonies on thiosulphate citrate bile-salt sucrose (TCBS) or bromothymol-blue salt teepol (BTBST) media in contrast to the yellow sucrose fermenting colonies of *V. cholerae* and *V. alginolyticus* (Sakazaki, 1969). Whatever media used, any Gram-negative organisms which are motile, oxidase and catalase positive, and sensitive to methylene blue (Yan, 1969) and vibriostatic agent 0/129 (2,4-diamino-6,7-di-iso-propylpteridine phosphate) should be regarded as vibrios and further identified. If this is done, infection with non-cholera vibrios, including *V. parahaemolyticus*, may well be recognized more often. Broth subcultures (4–6 h) of such organisms are inoculated into triple sugar iron agar (TSI) slopes and sulphide-indole-motility (SIM) medium. Further differential tests are performed on isolates which show motility, are indole positive and H_2S negative and give an acid (yellow) butt and alkaline (red) slant without gas production in TSI medium. The minimum characteristics essential for the identification of *Vibrio parahaemolyticus* and its differentiation from some other organisms are shown in Tables 4 and 5.

Table 4

Minimum additional characteristics for identification of V. parahaemolyticus

Motile	+
Good growth in 1% trypticase (BBL) peptone water overnight shake cultures at 37°C	
(1) without added NaCl	–
(2) with 6% NaCl	+
(3) with 8% NaCl	+
(4) with 10% NaCl	–
Good growth in 1% trypticase (BBL) peptone water with 2% NaCl at 42–43° overnight	+
Acid without gas from glucose anaerobically in Hugh & Leifson's oxidation-fermentation test	+
Voges-Proskauer test in semi-solid medium	–
Indole production	+
H_2S production	–
Lysine decarboxylase	+
Arginine dihydrolase	–

Except where stated, these tests are performed at 37° with media containing 3% NaCl.

All strains of *Vibrio parahaemolyticus* have identical H antigens, but the majority of strains can currently be differentiated into 50–60 different sero-types by agglutination tests with 11 O and 52 K antisera (Sakazaki, Iwanami & Tamura, 1968). Certain serotypes are able to produce β lysis of human blood in an agar medium—referred to as the Kanagawa test (Sakazaki *et al.*, 1968). The disease is an infection in that the organisms multiply rapidly in the gut, possibly producing a cholera-like enterotoxin (Craig, 1972). Unlike cholera vibrios or salmonellas, however, spread from 1 person to another does not seem to be important. On the other hand, many of the strains isolated from food have failed to induce illness in human feeding tests, whereas strains isolated from patients have produced the disease in volunteers. This correlates well with

Table 5

Differentiation of V. parahaemolyticus *from other organisms*

Organism	Colonial appearance on TCBS	TSI slant/butt	Lysine-decarbo-xylase	Voges-Proskauer	Growth in 8% NaCl
V. parahaemolyticus	Large, dark green centre	Alk/Acid	+	–	+
V. cholerae	Medium yellow	Acid/Acid	+	d* or +	–
V. alginolyticus	Large yellow	Acid/Acid	+	+	+
V. anguillarum	No growth	Acid/Acid	–	+	–
Other marine vibrios	Large green	d/d	d	–	–
Aeromonas	Small yellow or no growth	Acid/Acid (gas d)	–	d	–
Pseudomonas sp.	Small, pale green or colourless	Alk/Alk	–	–	–
Plesimonas	No growth	Alk/Acid	+	–	–
Proteus sp.	Small, yellow black or green	Alk/Acid (d)/(gas d)	–	(mirabilis) (d)	–

* d = variable.

the Kanagawa test, as the majority of strains isolated from food sources are unable to lyse human blood. The reasons for this are not clear, but multiple serotypes are often present in seafoods and sometimes in patients (Zen-Yoji *et al.*, 1970).

The role of carriers in the spread of infection is not known and more epidemiological work, including serological typing, is needed as the mode and vehicle of transmission may vary in different countries. Seafoods are frequently involved, but other foods may become cross-contaminated. In Calcutta, for example, infection with *V. parahaemolyticus* is known to occur, but seafoods do not form part of the staple diet (Chatterjee *et al.*, 1970). Although the tolerance of different serotypes varies, the organism is fully susceptible to heat and disinfection, and its viability is greatly reduced at low temperatures. Adequate cooking of food shortly before consumption is therefore the best safeguard against infection.

Future Control

Hygienic handling and plant sanitation can yield good bacterial standards in seafoods. Despite this, it must be recognized that some marine products, even when fresh, will contain certain micro-organisms which can cause illness and may increase greatly in number during processing or storage. With a vast international trade in frozen and processed seafoods, particularly crabmeat, shrimps, prawns and shellfish, however, together with changes in packaging and treatment, medical and food microbiologists must be aware of potential hazards, especially from imported products. Even with better control of marine environmental pollution, a different approach may be needed to the usual methods for assessing cleanliness and safety in seafood products. In addition to coliform and enteric organisms as indicators, marine bacteria should perhaps also be used, as they may be present in large numbers. Indeed, marine micro-organisms may assume more importance in the future because of warm effluents from nuclear power plants and the growth of fish farming in many countries. Safety, however, ultimately depends not so much on microbiological standards as on education in hygiene and local eating habits.

Acknowledgement

I am grateful to Dr G. T. Boalch of The Marine Biological Association, Plymouth, for his advice on dinoflagellates and intoxications from fish.

References

Ashby, B. S., Appleton, P. J. & Dawson, I. (1964). Eosinophilic granuloma of gastrointestinal tract caused by the herring parasite—*Eustoma rotundatum*. *Br. med. J.* **1,** 1141.

Baross, J. & Liston, J. (1970). Occurrence of *Vibrio parahaemolyticus* and related haemolytic vibrios in marine environments of Washington state. *Appl. Microbiol.* **20,** 179.

Barrow, G. I. (1972). Unpublished.

Barrow, G. I. & Miller, D. C. (1969). Marine bacteria in oysters purified for human consumption. *Lancet* **2,** 421.

Barrow, G. I. & Miller, D. C. (1972). *Vibrio parahaemolyticus*: a potential pathogen from marine sources in Britain. *Lancet* **1,** 485.

Battey, Y. M., Wallace, R. B., Allan, B. C. & Keeffe, B. M. (1970). Gastro-enteritis in Australia caused by a marine vibrio. *Med. J. Aust.* **1,** 430.

Chatterjee, B. D., Neogy, K. N. & Gorbach, S. L. (1970). Study of *Vibrio parahaemolyticus* from cases of diarrhoea in Calcutta. *Indian J. med. Res.* **58,** 234.

Clark, R. B. (1968). Biological causes and effects of paralytic shellfish poisoning. *Lancet* **2,** 770.

Clarke, S. K. R., Cook, G. T., Egglestone, S. I., Hall, T. S., Miller, D. L., Reed, S. E., Rubenstein, D., Smith, A. J. & Tyrell, D. A. J. (1972). A virus from epidemic vomiting disease. *Br. med. J.* **3,** 86.

Clegg, L. F. L. & Sherwood, H. P. (1947). The bacteriological examination of molluscan shellfish. *J. Hyg., Camb.* **45,** 504.

Conn, N. K. & Farrand, R. J. (1970). Shellfish toxicity in Scotland, 1969. *Scott. med. J.* **15,** 179.

Craig, J. P. (1972). The enterotoxic enteropathies. In *Microbial Pathogenicity in Man and Animals.* London: Cambridge University Press.

Dismukes, W. E., Bisno, A. L., Katz, S. & Johnson, R. F. (1969). An outbreak of gastro-enteritis and infectious hepatitis attributed to raw clams. *Am. J. Epidem.* **89,** 555.

Dodgson, R. W. (1928). Report on mussel purification. *Fish. Invest., Lond.* Ser. II. 10, No. 1. London: H.M.S.O.

Dodgson, R. W. (1936). Shellfish and the public's health. *Br. med. J.* **2**, 169.

Evans, M. H. (1969). Mechanism of saxitoxin and tetrodotoxin poisoning. *Br. med. Bull.* **25**, 263.

Fishbein, M., Melhman, I. J. & Pitcher, J. (1970). Isolation of *Vibrio parahaemolyticus* from the processed meat of Chesapeake Bay blue crabs. *Appl. Microbiol.* **20**, 176.

Fujino, T., Okuno, Y., Nakada, D., Aoyama, A., Fukai, K., Murai, K. & Ueho, T. (1951). On the bacteriological examination of *shirasu* food poisoning. *Jap. J. ass. Infect. Dis.* **25**, 11. (In Japanese.)

Halstead, B. W. (1965). *Poisonous and Venomous Marine Animals of the World,* Vol. 1: Invertebrates. Washington, D.C.: U.S. Govt. Printing Office.

Halstead, B. W. (1967). *Poisonous and Venomous Marine Animals of the World,* Vol. 2: Vertebrates. Washington, D.C.: U.S. Govt. Printing Office.

Heyningen, W. E. van & Arseculeratne, S. N. (1964). Exotoxins. *Ann. Rev. Microbiol.* **18**, 195.

Ingham, H. R., Mason, J. & Wood, P. C. (1968). Distribution of toxin in molluscan shellfish following the occurrence of mussel toxicity in north-east England. *Nature (Lond.)* **220**, 25.

Kao, C. Y. (1966). Tetrodotoxin, saxitoxin and their significance in the study of excitation phenomena. *Pharmacol. Rev.* **18**, 997.

Knott, F. A. (1951). *Memorandum on the Principles and Standards employed by the Worshipful Company of Fishmongers in the Bacteriological Control of Shellfish in the London markets.* London: Fishmongers Company.

McCollum, J. P. K., Pearson, R. C. M., Ingham, H. R., Wood, P. C. & Dewar, H. A. (1968). An epidemic of mussel poisoning in north-east England. *Lancet* **2**, 767.

MacKay-Scollay, E. M. (1958). Two cases of botulism. *J. Path. Bact.* **75**, 482.

Mason, J. O. & McLean, W. R. (1964). Infectious hepatitis traced to the consumption of raw oysters. An epidemiologic study. *Am. J. Hyg.* **75**, 90.

Nakanishi, H., Leistner, L., Hechelmann, H. & Baumgart, J. (1968). Weitre Untersuchungen uber das Vorkommen von *Vibrio parahaemolyticus* und *Vibrio alginolyticus* bei Seefischen in Deutschland. *Archs Lebensmittelhyg.* **19**, 49.

Peffers, A. S. R., Bailey, J., Barrow, G. I. & Hobbs, B. C. (1973). *Vibrio parahaemolyticus* gastro-enteritis in international air travel. *Lancet i,* 143.

Pilsworth, R. (1952). Bacteriological studies on cooked shellfish. *Mon. Bull. Minist. Hlth* **11**, 128.

Pollitzer, R. (1959). *Cholera.* Monograph series, No. 43. Geneva: W.H.O.

Report (1970). Bioassay for shellfish toxins. In *Recommended Procedures for the Examination of Sea Water and Shellfish.* A.P.H.A.: New York.

Riemann, H. (1969). Botulism-types A, B and F. In *Food-borne Infections and Intoxications.* Ed. H. Riemann, p. 291. Academic Press: New York.

Roos, B. (1956). Hepatitis epidemic transmitted by oysters. *Svensk. Lakartidn.* **53**, 989.

Rounsefell, G. A. & Nelson, W. R. (1966). *Red Tide Research Summarized to 1964 including an Annotated Bibliography.* U.S. Fish and Wildlife Service: Special Scientific Report—Fisheries No. 535.

Rowe, B., Taylor, J. & Bettelheim, K. A. (1970). An investigation of travellers' diarrhoea. *Lancet* **1**, 1.

Sakaguchi, G. (1969). Botulism-Type E. In *Food-borne Infections and Intoxications.* Ed. H. Riemann, p. 329. Academic Press: New York.

Sakaguchi, G. & Sakaguchi, S. (1967). Some observations on activation of *Cl. botulinum* Type E toxin by trypsin. In *Botulism 1966.* Eds. M. Ingram & T. A. Roberts), p. 266. London: Chapman and Hall.

Sakazaki, R. (1969). Halophilic vibrio infections. In *Food-borne Infections and Intoxications.* Ed. H. Riemann, p. 115. Academic Press: New York.

Sakazaki, R., Iwanami, S. & Fukumi, H. (1963). Studies on the enteropathogenic, facultatively halophilic bacteria, *Vibrio parahaemolyticus.* I. *Jap. J. med. Sci. Biol.* **16**, 161.

Sakazaki, R., Iwanami, S. & Tamura, K. (1968). Studies on the enteropathogenic, facultatively halophilic bacteria, *Vibrio parahaemolyticus*. II. *Jap. J. med. Sci. Biol.* **21**, 313.

Sakazaki, R., Tamura, K., Kato, T., Obara, Y., Yamai, S. & Hobo, K. (1968). Studies on the enteropathogenic, facultatively halophilic bacteria, *Vibrio parahaemolyticus*. III. *Jap. J. med. Sci. Biol.* **21**, 325.

Scott, H. G. (1969). Parasitic infections. In *Food-borne Infections and Intoxications*. Ed. H. Riemann, p. 578. Academic Press: New York.

Seaton, D. D. (1972). *Dinoflagellates associated with mass mortalities of marine organisms and/or capable of toxin production with particular reference to the North Sea and adjacent sea areas.* Dept. of Agriculture and Fisheries for Scotland: IR72-2. Unpublished Typescript.

Sherwood, H. P. & Thomson, S. (1953). Bacteriological examination of shellfish as a basis for sanitary control. *Mon. Bull. Minist. Hlth* **12**, 103.

Shewan, J. M. (1970). Bacteriological standards for fish and fishery products. *Chemy Ind.*, p. 193.

Smith, M. R. (1971). Two newly described bacterial enteric pathogens: *Vibrio parahaemolyticus* and *Yersinia enterocolitica*. *Clin. Med.* **78**, 22.

Sumner, W. A., Moore, J., Bush, A., Nelson, R., Molenda, J. R., Johnson, W., Garber, H. J. & Wentz, B. (1971). *Vibrio parahaemolyticus* gastro-enteritis–Maryland. *Morbid. Mortal.* **20**, 356.

Thiel, P. H. van., Kuipers, F. C. & Roskam, R. Th. (1960). A nematode parasitic to herring causing acute abdominal syndromes in man. *Trop. geogr. Med.* **12**, 97.

Vanderzant, C., Nickelson, R. & Parker, J. C. (1970). Isolation of *Vibrio parahaemolyticus* from Gulf Coast shrimp. *J. Milk Fd Technol.* **33**, 161.

Wood, P. C. (1969). *The Production of Clean Shellfish*. Laboratory Leaflet No. 20. Burnham-on-Crouch. Ministry of Agriculture, Fisheries and Food.

Yan, W. K. (1969). Methylene blue sensitivity of Vibrio organisms. *J. Med. Lab. Technol.* **26**, 90.

Zen-Yoji, H., Sakai, S., Kudoh, Y., Itoh, T. & Terryama, T. (1970). Antigenic schema and epidemiology of *Vibrio parahaemolyticus*. *Hlth Lab. Sci.* **7**, 100.

Discussion

Beuchat
Please comment on potential toxin-producing algae in freshwater fish.
Barrow
Some species of freshwater algae are known to produce toxins which may affect vertebrates including fish. This is known to occur, for example, in freshwater lakes in Scandinavia and Israel.
Ingram
In freshwater fish culture, a recognized hazard is the flagellate *Prymnesium panvulum* which kills the fish. I don't know whether it is dangerous to man.
Barrow
Not so far as I know.
Hobbs
What is the heat sensitivity of *V. parahaemolyticus* cooked in crabmeat (in claws and bodies of crabs)? Can the *Vibrio* survive the cooking? There is some indication that this may be so. Thus contamination of cooked crab may be cross contamination from raw to cooked or survival through cooking followed by growth in the cooked food.
Barrow
I think the answer to this question depends on the actual details of the cooking method used for crabs, e.g. boiling, steaming, the duration, etc., but I

certainly agree that survival in the body is possible. The claws, however, are probably exposed to maximal heat and I would have thought that any *V. parahaemolyticus* or other organisms in the claw meat would probably be killed. Nevertheless either survival or cross contamination from uncooked marine products may occur.

Ellermann

As far as I realized from your paper no case of viral hepatitis had been seen due to consumption of oysters. Do you have any import of oysters in your country?

Barrow

As far as I know, no outbreaks of viral hepatitis following consumption of uncooked shellfish have been reported in Britain. However, I do think that sporadic cases of hepatitis can be attributed to oysters.

Linderholm

One outbreak of virus hepatitis caused by oysters has been reported in Sweden during the fifties. There were several cases in different places in Sweden all originating from oysters from the same bay on the Swedish west coast. One case of hepatitis was found in a family living on the bay shore with the toilet outlet going into the bay where the oysters were held for some days before they were sold to different places in the country. It was considered that the virus accumulated in the oysters during the holding time because the toilet outlet was highly diluted.

Barrow

Yes, I knew about this report. It is very interesting and there is no doubt whatever that the oysters were responsible.

Mocquot

Concerning the cooking of crabs in relation to the kill of micro-organisms, one must remember that crabs are at their best from the organoleptic point of view when cooked in salted water (at least the salt content of sea water). This raises the boiling point.

Barrow

This is a very good point and I will look into it.

Hobbs

Oysters from oyster beds in Cornwall are known to have caused recognizable gastro-enteritis disease in man. There were about 10 outbreaks and symptoms were also observed in human volunteers. However, viruses were not found in the oysters either in the U.K. or by Dr Cliver at Wisconsin. Could Dr Larkin comment on the involvement of shellfish as vectors in transmission of viruses?

Larkin

No one has isolated the virus of infectious hepatitis. However, a number of epidemiological studies have definitely shown raw shellfish consumption to be responsible for a number of outbreaks of food-borne hepatitis. I should prefer to reserve any further comments to my paper this afternoon.

Salmonellae in Poultry in Great Britain

J. A. LEE

Epidemiological Research Laboratory,
Central Public Health Laboratory,
Colindale
London NW9 5HT England

In recent years there has been an increase in human incidence of certain serotypes which have been isolated from poultry and feed ingredients. Investigation of incidents due to some of these serotypes have showed that they originated from domestic poultry and feeding stuff was a likely source of infection. Responsibility for control of the poultry chain from farm to consumer is divided between public health and veterinary authorities. The processing plant forms the dividing line. Infected birds introduce contamination into the processing plants, which are not a primary source of infection. Surveys have shown that substantial (over 20%) of processed poultry is contaminated with salmonellae. Recognized dangers in kitchens are cross contamination and long warm storage of cooked birds. Veterinary investigations have pointed to 2 important controllable factors— contaminated feeding stuffs and transmission via eggs. To reduce human salmonellosis from poultry, control of infection in flocks is required. To achieve this, there is need for collaboration between the animal feed and poultry industry and veterinary and public health authorities.

Introduction

THE INTENSIVE farming of poultry introduced because of the economic necessity of producing poultry at a reasonable price has led to an increase of infectious diseases in poultry. Some of these diseases have been serious veterinary problems and have caused great economic losses to the industry. The occurrence of salmonellae in poultry has been important because of its public health implications and as such has been a cause of concern both to public health authorities and to the industry.

Poultry as a Vehicle of Salmonellae and a Cause of Human Salmonellosis

Infections caused by *Salmonella pullorum* and *S. gallinarum* are no longer of major economic importance to the poultry industry. In recent years, however, there has been an increase in the number of isolations of other salmonella serotypes in poultry. Mortality in chicks from salmonella infections, excluding *S. typhimurium* infections, has been relatively low, but the serotypes found have caused increasing numbers of human incidents and poultry has been shown to be the vehicle of infection in human outbreaks.

The reported number of salmonella incidents and cases has increased over the past 5 years and in 1970 the number was the highest since 1961 (Reports, 1961–1970). Figure 1 compares the number of incidents due to *S. typhimurium* and those due to serotypes other than *S. typhimurium*—the "other salmonellae". There was a fall in the incidence of both groups in 1962. Between 1962 and 1966 *S. typhimurium* incidents declined steadily while those due to other serotypes remained constant except for a rise in 1964. With the exception of 1968, *S. typhimurium* incidents rose slightly after 1966, but the increase in incidents due to the other serotypes was much greater and more than tripled between 1966 and 1970.

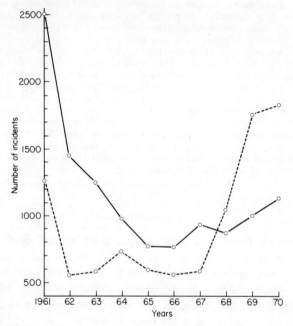

Fig. 1. Comparison of trends in human incidents of food poisoning of *S. typhimurium* with other salmonellae 1961–1970. o——o, *S. typhimurium;* o– – –o, other salmonellae.

The increase in the total number of incidents since 1966 is largely due to an increase of the "other salmonellae", which have now become of greater importance than *S. typhimurium*. In 1961 these serotypes accounted for 34% of all incidents, but in 1970 for 63%. Amongst these certain serotypes in particular have markedly increased in prevalence during this time and have been largely responsible for this changing trend. These serotypes include *S. enteritidis, S. virchow, S. heidelberg, S. 4,12 : d:–, S. infantis, S. agona, S. indiana* and *S. saint-paul.* They have been commonly isolated from processed poultry, poultry processing plants, poultry on farms and animal feed ingredients including poultry offal meal. In addition these serotypes have been implicated in human outbreaks due to poultry.

Table 1 shows the number of salmonella food poisoning outbreaks in which the vehicle of infection was established and those due to poultry from 1966–1968 (Vernon, 1967–1970) and 1969–1970 (Vernon, unpublished). Only in a small proportion of all outbreaks is it possible to implicate a vehicle of infection. A substantial proportion of such outbreaks has been shown to be due to poultry. The proportion of such outbreaks caused by poultry increased sharply after 1965. During the 5 year period 1966–1970 poultry caused 42% of such outbreaks. In the preceding 5 years 1961–1965 it was responsible for only 13% (Reports, 1962–1964; Vernon, 1965–1966). It is noteworthy that the increase in incidents caused by the "other salmonellae"—as shown in Fig. 1— coincides with the increase in the proportion of outbreaks caused by poultry.

Table 1

Outbreaks of salmonella food poisoning in which the vehicle was identified and those due to poultry 1966–1970

	Years				
	1966	1967	1968	1969	1970
All outbreaks	25 (100)	18 (100)	29 (100)	32 (100)	38 (100)
Due to poultry	8 (32)	4 (22)	15 (52)	15 (47)	17 (45)

Percentages are in brackets.

Table 2 shows the type of poultry found to be the vehicle of infection in outbreaks in 1968, 1969 and 1970. Chicken was the commonest vehicle, but turkey caused a number of outbreaks as well. There were no recorded outbreaks due to duck.

Table 2

Outbreaks due to poultry 1968–1970 in which the vehicle was further specified

	Years		
	1968	1969	1970
Chicken	11	9	11
Turkey	4	6	6

Epidemiological Investigations of Human Incidents due to Poultry

S. saint-paul

During the years 1957–1961 this serotype caused a large number of human incidents, some of which were studied (Galbraith, Archer & Tee, 1961). In 1959 the Central Veterinary Laboratory reported a number of isolations of *S. saint-paul* from outbreaks in cattle and poultry in the southern half of England and Wales and human incidents occurred in the same area in 1958 and 1959. An outbreak of chicken enteritis was thought to have originated from contaminated poultry meal on the farm. *S. saint-paul* was not isolated from this meal, although it contained other salmonella serotypes. The chief constituent of the meal was meat imported from America, where *S. saint-paul* was widespread.

This American meat meal was heavily contaminated with salmonellae; it had been imported by 1 large firm and used in their compound feeding stuffs since 1958, but only at their factories in the south of England. Thus, the use of this material corresponded geographically and in time to the occurrence of human and animal *S. saint-paul* infection in the south of England.

Galbraith *et al.* (1962) described an outbreak of *S. saint-paul* infection due to the consumption of cooked meats which were contaminated by the intestinal contents of infected ducks during preparation at a butcher's shop. Subsequently, *S. saint-paul* was isolated at the farm which supplied ducks to the shop. The feeding stuffs used on the farm were investigated and meat meal from America was found to be heavily contaminated with salmonellae, but *S. saint-paul* was not isolated. The implication of the same material in a previous incident due to *S. saint-paul* suggested that it was the source of infection in the ducks.

Galbraith (1961) stated that on epidemiological grounds *S. saint-paul* infections originated from imported feeding stuffs.

S. infantis

Newell *et al.* (1959) observed that certain farms in Northern Ireland used a feeding stuff known to be contaminated with *S. infantis* and *S. orion*. Hens became infected with *S. infantis* and produced eggs containing the organism whilst *S. orion* was selected by pigs. Subsequently in 1958 the first isolations of *S. orion* and *S. infantis* were made in Northern Ireland from human cases after it had been confirmed that these types were imported in feeding stuffs and had caused animal infections.

S. virchow

Semple *et al.* (1968) described a food poisoning outbreak in north-west England due to *S. virchow* associated with chickens. The source was traced to a packing station and *S. virchow* was isolated from chicken carcasses. Pennington *et al.* (1968) made further investigations of this outbreak and found *S. virchow* in eviscerated carcasses sampled at the packing station. There was no evidence of gross contamination within the plant and the standard of hygiene was adequate. Rearing farms which supplied the packing station were studied and *S. virchow* was found in hut droppings and cloacal swabs in 9 out of 14 of these farms and in samples of dust at the hatchery. The serotype was not found in the breeding flocks which supplied the hatchery. In a similar outbreak in the same area (Lee *et al.*, unpublished, 1967) due to *S. virchow*, this organism was found in 10% of cloacal swabs of chickens arriving at the packing station from 6 different farms.

Table 3 shows the number of human incidents and number of isolations of *S. virchow* from processed poultry and feed ingredients from 1966–1968 (Vernon, 1967, 1969, 1970). A great increase in incidents occurred in 1968, the majority in the north-west. It is noteworthy that in 1967, the year preceding the increase in human incidents, *S. virchow* was isolated on 7 occasions from feed ingredients. It is possible that *S. virchow* was introduced by a feeding stuff into 1 or more chicken breeding flocks in the north-west.

Table 3

Isolations of S. virchow *from humans, poultry and feed ingredients 1966–1968*

Year	Human incidents	Poultry (food)	Feed ingredients
1966	2	–	1
1967	31	1	7
1968	253	34	3

These investigations suggested that certain salmonella outbreaks involving contaminated chickens could be traced back to poultry packing stations, broiler rearing farms and hatcheries supplying the rearing farms. All are links in the poultry chain of infection which is shown diagrammatically in Fig. 2.

Responsibility for Investigation and Control of Salmonellae in Poultry

This responsibility is divided between 2 ministries. The Ministry of Agriculture, Food and Fisheries (M.A.F.F.) is responsible for the health of birds and hygienic standards on farms and in hatcheries which supply processing plants. The Department of Health and Social Security (D.H.S.S.) is responsible for the practice of hygiene in processing plants and in shops and retailers. Local Authority Public Health Departments undertake the inspection and supervision of premises which process or retail poultry for human consumption and with the

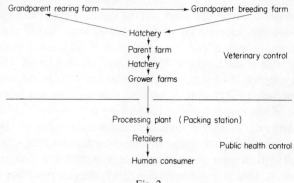

Fig. 2.

laboratories of the Public Health Laboratory Service (P.H.L.S.) investigate human outbreaks of food poisoning and take whatever action is considered necessary to control or prevent outbreaks. There is contact between veterinary and public health authorities in the investigation of outbreaks shown to have originated from farms, but there has been no planned joint investigation of the origin and transmission of salmonellae in poultry flocks.

Public Health Investigations

Hitherto, public health authorities who are responsible for salmonellae in processed poultry for human consumption have investigated and attempted to control only that section of the chain under their authority.

Crabb & Walker (1971) examined nearly 700 chicken samples from 10 different suppliers in 1968 and 1969 and found that 33% of chicken portions and 24% of whole bird samples were contaminated with salmonellae. Contamination rates of samples from individual suppliers varied from 0–64%, with only 1 supplier with a contamination rate of less than 10%. Gould & Rhodes (1969) reported that 43% of 151 sets of giblets from different packing stations were contaminated.

Work has recently been carried out on the survival and multiplication of salmonellae in frozen chickens under varying conditions of thawing, cooking and storage (Roberts, 1972). Recommendations were made for these practices which, if followed, would no doubt reduce the risk of clinical infection. The 2 main hazards in preparation were found to be the danger of cross contamination from raw to cooked birds by hands, surfaces, equipment and utensils; and long warm storage of cooked birds. An average of 35% of raw chickens from different packing stations examined in this study contained salmonellae. It was felt that while a high proportion of raw chickens were contaminated with salmonellae, the danger of salmonella food poisoning could not be ruled out, and that efforts should be directed towards the provision of salmonella-free flocks so that dressed poultry could be free from salmonellae before retail sale.

In the outbreak due to *S. virchow* in 1967 (Lee *et al.*, unpublished) symptomless excreters of *S. virchow* were found amongst food handlers in 3 chicken retail establishments and swabs from working equipment in 1 yielded *S. virchow*. This indicated that when salmonellae were introduced in poultry further contamination hazards arose from food handlers and equipment.

Dixon & Pooley (1961) studied the occurrence of salmonellae in the environment of a poultry processing plant. They concluded that dissemination of salmonellae probably occurred during the evisceration process and also during the chilling of carcasses in water tanks. The chlorination of the chilling tanks was considered to be relatively ineffective because of the high content of organic matter in the water. Timoney *et al.* (1970) showed that the predominant serotype isolated from the processing plant environment changed when a new broiler or hen crop was processed, and that plant equipment did not remain contaminated although some stages of processing contributed to the further spread of salmonellae amongst carcasses. He concluded that processing plants were not important primary sources of salmonellae, but that organisms were introduced with birds from certain infected flocks.

Stewart (1965) indicated the degree of bacterial spread that could occur in a chicken processing plant. By injecting *Serratia* into a chicken prior to evisceration and following subsequent birds down the line, he showed that the next 150 birds were contaminated. Hermans (1969) also found that during the slaughtering process salmonellae were transferred from 1 chicken to another and concluded that the only way to prevent contamination of processed chickens was to ensure that birds brought into the slaughterhouse were not carrying salmonellae.

Veterinary Investigations

Gordon & Tucker (1965) designed an experiment using *S. menston* and showed that by feeding a ration containing *c.* 1 organism/g of food, infection could be established in the adult fowl and that the eggs laid by these fowls, the chicks hatched from their eggs and the eggs laid by the progeny on maturity were all infected with *S. menston*.

Crabb & Walker (1971) referred to a paper given by McCoy at the British Veterinary Poultry Association's 1969 winter meeting which indicated that egg transmission of salmonellae occurred with *S. enteritidis, S. indiana, S. menston, S. panama, S. thompson* and *S. typhimurium*.

These investigations along with those of outbreaks described point to 2 important controllable factors in salmonella epizootiology—introduction by feeding stuffs and transmission via eggs.

Present and Future Control

It has not been possible to reduce the high rate of contamination of processed poultry or to prevent human infection by traditional public health hygiene measures. Poultry flocks constitute a large reservoir of salmonellae of public health importance and there is clearly a need for a national programme aimed at the control of this reservoir. This has not come about hitherto mainly because of the divided ministerial responsibility for the chain of infection and the lack of contact between the 2 ministries and the poultry industry. There has also been and perhaps still is the belief that contamination of poultry carcasses can be prevented in the processing plant by such means as chlorination of spin chiller water.

The M.A.F.F. are chiefly concerned with disease causing mortality and clinical illness in livestock. Under the present poultry health scheme, membership of which is not compulsory, the survivors of infected batches of poultry, implicated in outbreaks of salmonella infection, may not be used for breeding purposes. However, most salmonellae of public health importance are carried asymptomatically by flocks. It is understood that the poultry health scheme is being revised and that the control of salmonellae other than *S. pullorum* and *S. gallinarum* is receiving detailed attention.

Certain companies of the poultry industry have themselves monitored and investigated salmonella infections and introduced certain control measures, but, because of the lack of contact with public health authorities, it has not been possible to assess the effect of the measures on contamination rates of processed poultry and on the incidence of human salmonellosis.

Public health authorities have been able to trace human outbreaks to flocks and hatcheries, but they have not had the authority to investigate fully the reservoir of infection in poultry, nor to recommend measures for its control.

Control measures are required to prevent the introduction of salmonellae into flocks and the spread and transmission of salmonellae that are already in flocks. For the control of the introduction of salmonellae it is necessary to reduce the contamination of feeding stuffs, to ensure the salmonella-free status of breeding flocks and to avoid contact with other sources of salmonellae such as wild birds and rodents. To control the spread and transmission of salmonellae already in flocks it is necessary to detect infected birds so that they are not used for breeding purposes and to prevent egg transmission and the environmental contamination of hen houses and hatcheries. With the exception of measures to reduce the contamination of feeding stuffs which are not under their control, these measures are to a certain extent being applied by the poultry industry. There is need, however, for further work to determine the best methods for the detection of infection in flocks and to clarify the role of ovarian infection in egg transmission. There is also a need to assess the efficacy of the measures already being applied.

There are now good signs that the lack of contact between the 2 ministries and the poultry industry may be coming to an end. Recent consultations between members of the Public Health Laboratory Service and the industry have led to an understanding of each others position and an exchange of information. It is agreed that feeding stuffs are an important method of introduction of salmonellae into poultry, and that action is needed to reduce contamination of feed ingredients of animal origin.

Developments following on these consultations have raised hopes that there may soon be a co-ordinated effort between the 2 ministries, the poultry industry and the animal feed industry aimed at a better understanding and improved control of salmonellae in poultry in Britain.

References

Crabb, W. E. & Walker, M. (1971). The control of salmonella in broiler chickens. In *Hygiene and Food Production,* p. 119. Ed. A. Fox. Edinburgh and London: Churchill Livingstone.

Dixon, J. M. S. & Pooley, F. E. (1961). Salmonellae in a poultry-processing plant. *Mon. Bull. Minist. Hlth* **20**, 30.

Galbraith, N. S. (1961). Studies of human salmonellosis in relation to infection in animals. *Vet. Rec.* **73**, 1296.

Galbraith, N. S., Archer, J. F. & Tee, G. H. (1961). *Salmonella saint-paul* infection in England and Wales in 1959. *J. Hyg., Camb.* **59**, 133.

Galbraith, N. S., Mawson, K. N., Mason, G. E. & Stone, D. M. (1962). An outbreak of human salmonellosis due to *Salmonella saint-paul* associated with infection in poultry. *Mon. Bull. Minist. Hlth* **21**, 209.

Gordon, R. F. & Tucker, J. F. (1965). The epizootiology of *Salmonella menston* infection of fowls and the effect of feeding poultry food artificially infected with Salmonella. *Br. Poult. Sci.* **6**, 251.

Gould, L. M. & Rhodes, F. (1969). Salmonellae in chicken giblets. *Med. Offr.* **122**, 31.

Hermans, K. H. (1969). The prevention of salmonella contamination in poultry processing plants. Proceedings of the Fifth World Symposium of the Association of Veterinary Food Hygienists, Opatija, 1969.

Newell, K. W., McClarin, R., Murdock, C. R., MacDonald, W. N. & Hutchinson, H. L. (1959). Salmonellosis in Northern Ireland with special reference to pigs and Salmonella-contaminated pig meal. *J. Hyg., Camb.,* **57**, 92.

Pennington, J. H., Brooksbank, N. H., Poole, P. M. & Seymour, F. (1968). *Salmonella virchow* in a chicken packing station and associated rearing units. *Br. med. J.* **4**, 804.

Reports (1961–1970). On the State of the Public Health. The Annual Reports of the Chief Medical Officer of the Department of Health and Social Security for the years 1961–1970. H.M.S.O.: London.

Reports (1962–1964). Food poisoning in England and Wales 1961–1963. *Mon. Bull. Minist. Hlth* **21–23**.

Roberts, D. (1972). Observations on procedures for thawing and spit roasting frozen dressed chickens, and post-cooking care and storage: with particular reference to food poisoning bacteria. *J. Hyg., Camb.* **70,** 565.

Semple, A. B., Turner, G. C. & Lowry, D. M. O. (1968). Outbreak of food poisoning caused by *Salmonella virchow* in spit-roasted chicken. *Br. med. J.* **4,** 801.

Stewart, D. J. (1965). The occurrence of enteropathogenic *E. coli* and salmonellae in processed broilers. *Proc. First International Congr. of Fd Sci. Technol.* **2,** 485.

Timoney, J., Kelly, W. R., Hannan, J. & Reeves, D. (1970). A study of salmonella contamination in some Dublin poultry processing plants. *Vet. Rec.* **87,** 158.

Vernon, E. (1965–1967). Food poisoning in England and Wales 1964–1966. *Mon. Bull. Minist. Hlth* **24–26.**

Vernon, E. (1969–1970). Food poisoning and salmonella infections in England and Wales 1967–1968. *Publ. Hlth, Lond.* **83–84.**

Discussion

Barnes

I should like to support the control of the feedstuffs for the elimination of salmonellae in order to lower the incidence of salmonellae in the live birds. Once in the processing plant it is very difficult to prevent the spread of salmonellae even with good hygienic conditions. Cross contamination of carcasses can occur right at the beginning of the line, in the scald tanks, pluckers, etc., even before evisceration or chilling.

Legislation for the control of feedstuffs is required so that conscientious processors will not be penalized at the expense of others.

Foster

If feedingstuffs are the primary source of animal infection why so frequently does one find one (or more) serotype in the feed and an entirely different one in the infected animals?

Lee

This is almost certainly because different serotypes behave differently in different environments—in this case the feedingstuff (treated animal by-products) and the animal's intestine. Some serotypes, such as *S. typhimurium*, may multiply easily in the animals' intestines whilst others, for example *S. senftenberg*, appear to survive easily the heat treatment of processed feedstuffs. We also know that animals select diet serotypes from feeds and so act as a biological filter. Thirdly, methods of isolation are important in determining the presence of certain serotypes in feedstuffs.

Kampelmacher

The question raised by Prof Foster is an old one. (1) We control minimal amounts of the total feed input. (2) What happens in the enrichment medium? Perhaps there is competition of different types, or different sensitivities of strains relative to that of the total flora. (3) Most important: how many colonies are picked from positive plates? Usually not more than 1 to 3 I would say. These considerations may be of great significance when evaluating the situation. When we investigated whole sacks of fish meal many years ago we found an answer to some of the questions which have just been raised.

Linton

In reply to the question why do we find certain *Salmonella* serotypes in animal feed yet isolate other serotype(s) from the animal, the isolation of serotypes from food depends to a great degree on the methodology used. One method of enrichment culture (or pre-enrichment) may favour 1 serotype preferentially to others. Harvey has shown that if he suppressed certain serotypes by antisera he obtained a much higher percentage of, say, *S. typhimurium*. On the other hand different animal hosts act as biological filters of different serotypes. Newell *et al.* demonstrated this and we have confirmed it in pigs. Thus the host may select, from the same feed, one serotype whilst the laboratory may select another.

Skovgaard

Cross contamination during the slaughtering operations plays an important role in the spreading of salmonellosis from poultry to man. Yesterday we learned that the spread of salmonellae in pigs in the lairages plays an important role in the spreading of salmonellosis but the same could be the case for poultry. We can obtain salmonellae in pure cultures from the bone marrow. The presence of salmonellae here is due to circulation in the blood stream prior to the killing. This is "improved" to some extent by transport stress.

Session 2

Epidemiology of Food-Borne Infection in Man and Animals

Part 2: Animals
Chairman: Professor B. Weitz

The Occurrence of Salmonellosis in Poultry in Denmark 1935-1971 and the Eradication Programme Established

H. E. MARTHEDAL

Royal Veterinary & Agricultural University
Copenhagen, Denmark

The report comprises a short survey of the occurrence of salmonellosis, channels of infection and measures of combat applied in Denmark. In the period 1946–1962 a considerable number of salmonella types occurred in chickens, but from 1962 onwards *S. typhimurium* has gradually become dominant; a new closely related serotype, Salmonella O form (1.4.5.12), has since appeared and today these two types together are responsible for most outbreaks of salmonellosis among hens, ducks and geese in Denmark. In pigeons salmonellosis occurs frequently; about 15% of the pigeons sent to the Poultry Research Laboratory are infected with a special subtype of *S. typhimurium,* often designated the IV-variant ("the pigeon type"). The most important measures in the fight against salmonellosis are control of: 1, breeding centres, →2, hatcheries, →3, the feed. With regard to the first point, the main emphasis is laid on the breeding centres because when the disease is controlled at the central stage the beneficial effect is seen at other stages of production. Also the current control of the hatcheries has proved valuable and the same applies to control of feed, in particular of meat- and bone-meal. Since the SPF-programme commenced in 1963 the preventive control of salmonellosis in poultry in Denmark has given favourable results.

SALMONELLOSIS in poultry plays a most important part in both human and veterinary medicine. For producers of poultry the disease may result in considerable losses, not only owing to the deaths caused by it, but also because of losses in connection with condemnations at the processing plant.

In man the occurrence of salmonella bacteria in foodstuffs involves danger of food poisoning, which under adverse circumstances may involve a large number of persons.

Occurrence

Hens

The first outbreaks of salmonellosis* in Denmark were diagnosed in 1934. Right up to 1944 only 2 salmonella types—*Salmonella typhimurium* and *S. enteritidis*—had been recognized as the cause of salmonellosis in poultry. From 1944 to 1968, 15 types were added: *S. nilosse, S. newport, S. chester, S. muenchen, S. chicago, S. oranienburg, S. montevideo, S. infantis, S. schwarzengrund, S. vejle,*

---

* *Salmonella pullorum* and *S. gallinarum* not included.

211

S. newington, S. amager, S. kentucky, S. indiana and S.O-form 1,4,5,12 and a small number which were not identified. The types are mentioned in the order in which they were isolated (Table 1). However some types were isolated only once or a few times, for example *S. newington, S. amager* and *S. montevideo,* and they are, therefore, of no importance so far. From Table 1 it appears that until 1959 infections with *S. typhimurium* in hens were on the whole of small extent. In 1949 and 1956 small increases were registered; a far more considerable rise from 1961 to 1966 was due to the fact that some big hatcheries and breeding centres were infected with *S. typhimurium,* which resulted in dissemination via chickens.

In the period 1946–1961 *S. typhimurium* was not the dominating type, but since 1963 it has become predominant. In 1968 and in the following years *Salmonella* O-form 1,4,5,12 (without H-antigens) was isolated. Apart from the lack of H-antigens this serotype resembles *S. typhimurium* and its fermentation pattern corresponds to type 17 (the so-called "duck-type"). Today about 70–80% of all outbreaks of salmonellosis in hens are caused by these 2 types (Fig. 1).

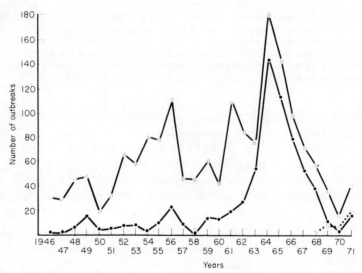

Fig. 1. Occurrence of salmonellosis in chickens 1946–1971. ○——○, All salmonella types; ●——● *S. typhimurium*; ○········○, O form: 1.4.5.12.

Web-footed birds

Figure 2 shows the occurrence of salmonellosis in ducklings and goslings in the period 1946–1971. The majority of outbreaks were caused by *S. typhimurium* (about 87%), 8% by *S. enteritidis* and 5% by other salmonella types *(S. niloese,*

Table 1

Survey of the occurrence of salmonella types in chickens

	S. typhi murium	S. enteritidis	S. niloese	S. newport	S. chester	S. muenchen	S. chicago	S. oranienburg	S. montevideo	S. infantis	S. schwarzengrund	S. vejle	S. newington	S. amager	S. kentucky	S. indiana	O-form 1,4,5,12	S. not identified	Total
1934	1																		1
1935	2																		2
1936	5																		5
1937	1																		1
1938	1																		1
1939	2																		2
1940	3																		3
1941																			
1942																			
1943	1																		1
19	1		13																14
1945	6	1	8																15
1946	1		29																30
1947	1		28																29
1948	6		41																47
1949	16	3	23	6	1														49
1950	4	1	12	1															18
1951	5		20		5	3													33
1952	8	2	23		19	1	12	1	1										67
1953	9	1	18		21		8		1										58
1954	4	4	29	1	33	4	6												81
1955	11	3	32	1	21		6	1		1	2								78
1956	24	1	19	4	45		6			1	8	5	1						114
1957	10		11	2	6		13			2	2	1						1	48
1958	2		21		8		10				2							3	46
1959	19		20		1		17											4	61
1960	14		5	2	3		6			7			1		1			4	43
1961	23		6		5	1	51	2		6		1						5	100
1962	32		3				27			22					1				85
1963	54		2	1		1	7	4		6						1			76
1964	145		1				15	1		11			1		6	1		1	182
1965	115						21			10								2	148
1966	80					1	8			4					4			2	99
1967	54						13			3					2			1	73
1968	39						11			4					1	1	3		59
1969	13						9			5							11		38
1970	4						4			1							6		15
1971	19									1					2		20		42

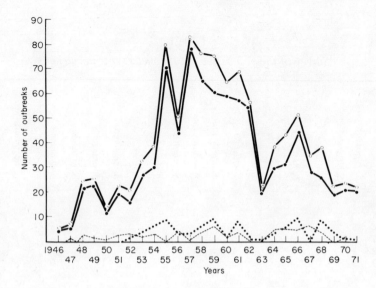

Fig. 2. Occurrence of salmonellosis in ducklings and goslings 1946–1971. ○──○, All salmonella types; ●──●, *S. typhimurium*; . . . ., *S. enteridis*; - - - -, other salmonella types.

S. chester, S. infantis, S. newport, S. muenchen, S. vejle and *S. indiana).* Serotypes in the last category have generally caused solitary outbreaks only, a factor that should be borne in mind in the assessment of cases under observation.

Turkeys

Stocks of turkeys in Denmark are limited in numbers and salmonellosis in this fowl has so far been of relatively small importance (Table 2).

Pigeons

Outbreaks of salmonellosis in these birds are frequent, as is shown in Fig. 3. For 26 years (1946–1971) infection in pigeons sent to The Poultry Research Laboratory has been ascertained in about 15% of the birds received. The special IV-variant ("the pigeon type") of *S. typhimurium* has always been involved; this type has seldom been isolated from domesticated poultry and it is, therefore, unimportant in this connection.

"The pigeon type", of which there are a number of biochemical subtypes, must be assumed to have a strong affinity for pigeons and closely related birds.

Table 2

Survey of the occurrence of salmonella types in turkeys

	S. typhimurium	S. chester	S. enteritidis	S. niloese	S. newport	S. infantis	S. vejle	S. chicago	S. montevideo	S. muenchen	S. indiana	S. senftenberg	S. not identified	Total
1946														0
1947														0
1948	1													1
1949	1													1
1950														0
1951														0
1952		1												1
1953			1											1
1954	4			2										6
1955	5			1	1									7
1956	8			1		5	1							15
1957	2	1												3
1958	1	1						2						4
1959	3			1				1	1					6
1960	2					1							1	4
1961	2					1		2						5
1962	4					1				1				6
1963	4													4
1964	3												3	6
1965	2		1								1		1	5
1966	7		1			2						1	1	12
1967	3				1	3					1			8
1968	2							2						4
1969							5							5
1970	5													5
1971	3					1					1			5

Cage birds and wild birds

In canaries, parrots, bullfinches, crossbills, waxwings and others, salmonellosis occurs relatively frequently. In the period 1946–1971, 80 outbreaks were diagnosed. Mainly they were caused by *S. typhimurium* and the majority were of phage types rarely found in poultry in Denmark.

Transmission of Infection

Salmonella infections in poultry may be considered under the following headings: (1) infection via breeding stock; (2) infection via hatcheries; (3)

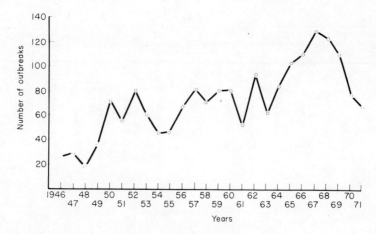

Fig. 3. Occurrence of salmonellosis in pigeons 1946–1971.

infection via feed; (4) infection via wild birds; (5) infection via mice and rats.

The importance of these potential channels of infection in Denmark will be dealt with separately.

Infection via breeding stock

Around 1960 there were 10 breeding centres in Denmark within the broiler industry. By May 1972 the number was reduced to 2, 1 of which covers over 90% of the total production. As a natural consequence of the reduction in the number of breeding centres, 1 of the remaining centres, which is a fusion of 3, is of considerable size. As the state of health of the breeding flocks exerts great influence on the other stages of production, it is absolutely necessary that the centres are subjected to current veterinary control. In order that the control can become effective, the structure of the breeding centre must be of suitable design and disposition. Figure 4 shows in schematic form the structure of broiler production; it will be noted that a breeding centre must be a closed entity, which means that no material should enter from outside, neither in the form of animals nor of eggs. If importation of animals is desirable, this is allowed only from a breeding centre where the state of health is known or from one of the two breeding stations (Fig. 5) which belong to the Danish Poultry Breeding Committee and which have since 1960 been subject to control by the Poultry Research Laboratory.

There are 2 breeding stations in Denmark, named Skrillings and Strynø; they are directed and run by the Danish Poultry Breeding Committee together with the Research Institute for Animal Husbandry, Department of Poultry Research. The main purpose of these breeding stations is to select through intensive

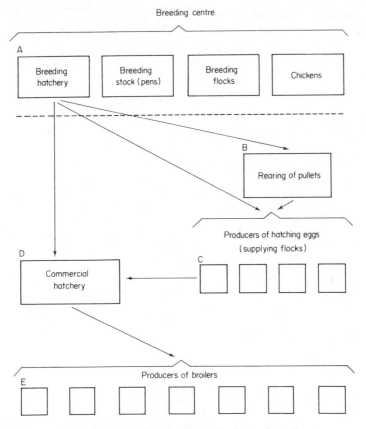

Fig. 4. Structure of the broiler-production—schematic.

breeding work strains or lines with the desired properties—better growth, lower feed consumption/kg growth and higher yield of eggs. All new imports of breeding material come to the breeding stations and after satisfactory examination in quarantine they are released for further production. The hatching eggs from the imported animals are sent to the breeding stations, where their value in the breeding work is investigated before their offspring can be bought by the breeding centres. By means of such procedures the possibility of controlling all diseases including salmonellosis in animals is good. As mentioned above it is our experience that it is important to keep the breeding material completely free from disease.

Figure 6 portrays the cycle of salmonella infection from parent through eggs to chicks and older birds. Figures 7–9 show lesions in liver, spleen and heart caused by *S. typhimurium* infection in chicken.

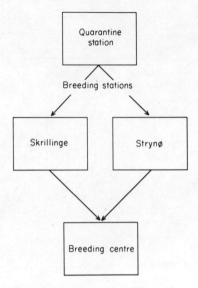

Fig. 5. Quarantine and breeding stations.

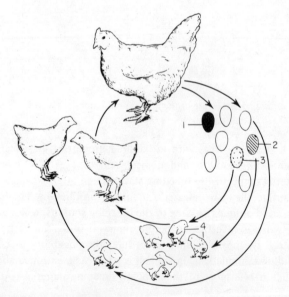

Fig. 6. The cycle of *S. typhimurium* infection—schematic. 1, Badly infected egg which produces no chicken (addled); 2, moderately infected egg giving rise to infected chick (4) which dies; 3, lightly infected egg from which an infected chick (4) is hatched, surviving and growing into a bird which may continue to excrete salmonellae.

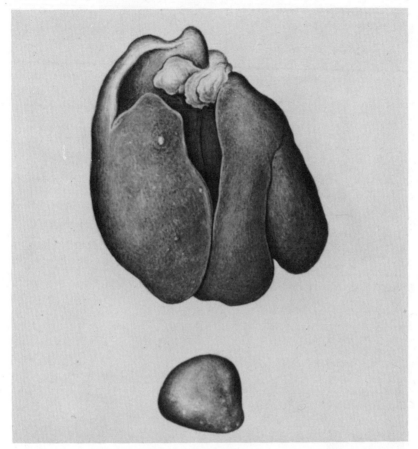

Fig. 7. Liver and spleen with a few necrotic foci from an 8-week old chicken.

Hatcheries

The appearance of fresh salmonella types only partly explains the rise in salmonellosis (Fig. 1). A highly contributory and essential factor must be sought in the building and development of poultry production in the Fifties. In this period a tremendous development took place within the poultry sector, which involved not only hatching of chickens but also the hatching of eggs from ducks, geese and turkeys which was undertaken by many hatcheries. Such a mixture proved to have many disadvantages as the majority of all salmonella outbreaks caused by *S. typhimurium* could, almost without exception, be traced back to these hatcheries. On the 7th December 1964 an Order was issued to stop this

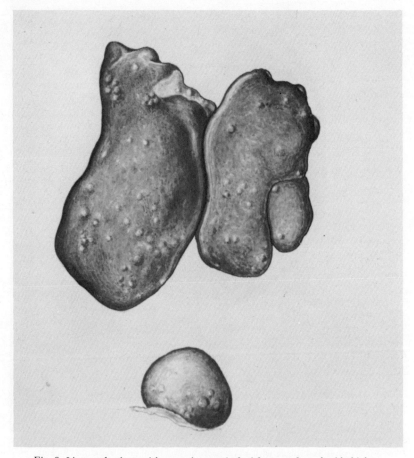

Fig. 8. Liver and spleen with several necrotic foci from an 8-week old chicken.

practice; among other things it was stated that "geese, ducks and turkeys must at no time be found on a property on which there is a hatchery, or from which hatching eggs or poultry are supplied to a hatchery".

If hatching eggs from 1 supplier have been infected with salmonellae, transmission will occur during hatching to chickens originating from other suppliers, and later after hatching—and specially during the first weeks—there will be ample opportunity for the spread of infection (Fig. 5).

Infection via feed

As mentioned earlier, new salmonella types were isolated in the period 1944–1964, a considerable number which must, without doubt, be attributed to imported feedstuffs, in particular meat and bone-meal (Müller, 1952, 1954).

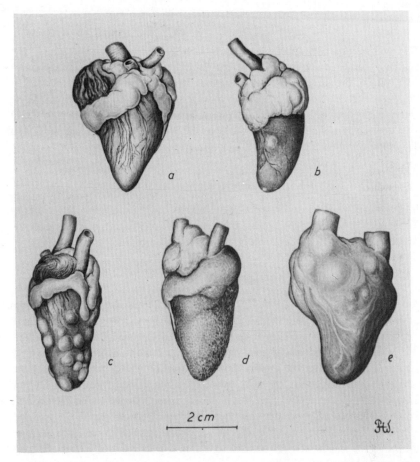

Fig. 9. (a) The normal heart; (b) with a few necrotic foci; (c) with several necrotic foci; (d) fibro-purulent pericarditis; (e) fibro-purulent pericarditis and necrotic foci in the myocardium.

Samples of these feedstuffs were examined in the State Veterinary Serum Laboratory and a very large number of salmonella types were isolated, several of which had not formerly been found in domestic animals in Denmark. Through feedstuffs these new types were spread to breeding poultry and hence via hatcheries to broiler flocks. It must be assumed that various serotypes have specific affinities for the species of poultry involved as there are types which have no tendency to become established in poultry, for example, *S. dublin* and *S. enteritidis* var. *danysz*. According to the government circular of January 1954, imported meat and bone-meal must be reheated, so that the possibilities of introducing new salmonella types through this feed vehicle should now be reduced.

Infection via wild birds and others

This will occur under special circumstances only; for example, when birds that regularly forage on a dump come to a farm and infect the feed. If poultry involved in the production of hatching eggs are involved, transmission of infection will be possible. We have seen a concrete example of this in 1965 where a breeding centre appeared to be infected by birds from a nearby dump. The salmonella isolated from the stock of the breeding centre—a rare subtype of *S. typhimurium*—was also found in seagulls shot on the dump.

Infection via rats and mice

In Denmark we have no evidence of infections originating from rats and mice. There is an effective fight against these noxious animals and it is not expected that they will play a part as carriers of salmonella.

Eradication Programme

As a control measure in the work of preventing infectious diseases, importation of birds—first and foremost domesticated poultry—has been forbidden since 1924–1925. When the National Poultry Breeding Committee finds it desirable, the Ministry of Agriculture may under certain conditions give permission to import; in such a case the imported animals must be placed in quarantine. Since 1960 the quarantine time has been 5–6 months, during which period the animals are examined for infectious diseases, including salmonella infections. In 1963 the Poultry Research Laboratory began an SPF programme, which is primarily aimed at the broiler sector (Marthedal, 1965). Under this programme there are measures for the preventive control of salmonellosis; gradually these measures have been made more stringent, mainly as a consequence of the concentration of animals which has taken place during recent years. The main weight is laid on control of the breeding centres, based on the consideration that if diseases are controlled at the central stages, the effect will be spread to the other stages of production (Fig. 4).

The most important measures which are carried through in collaboration with the breeding centres and on voluntary basis are described below.

General control measures

Until 1956 the control of salmonella infections in poultry was the whole and in practice limited to directions given by the Veterinary Department and the Poultry Research Laboratory. However, in the period 1952–1956 there was a rise in the number of cases of food poisoning in man and these infections must in some cases have originated from poultry products (eggs and meat). Thus an order about salmonella infections in domestic animals was issued in 1956; as far as poultry was concerned, this order aimed at first and foremost obtaining a better separation between hens on one side and ducks, geese and turkeys on the other. Epizootiological investigations made in the period 1946–1955 showed that some hatcheries in which all categories of poultry (hens, ducks, geese and turkeys) were hatched were permanently infected with *S. typhimurium* (and sometimes also with *S. enteritidis*) and that many cases of food poisoning could be traced back to such entities.

The separation of the 4 categories of poultry was not completed at that time but the hatching rooms for hens eggs were to be separated from those used for hatching eggs from ducks, geese and turkeys.

Only after a revision of the 1963 order was it possible to carry through a complete separation between the categories of poultry. In the Order of 29th May 1963 about Salmonella Infections in Domestic Animals, etc. it stated:

"Ducks, geese and turkeys are to be totally separated from hens supplying hatching eggs, i.e. breeding must be on separate farms, and for the purpose of avoiding possible spread of infection, in particular of *S. typhimurium*, the supplying flock should not at any time during growth be near web-footed birds and turkeys".

In our experience the central source of dissemination of infection with salmonellae is the hatcheries and, as the breeding centres have the greatest influence on overall production, it was decided through close co-operation between the Poultry Research Laboratory and the Poultry Breeding Committee in connection with the SPF programme started in 1963 (Marthedal, 1965) to include salmonellosis in the preventive programme.

The programme for the prevention of salmonellosis mainly comprises the following provisions:

Special control measures for salmonellosis

(1) *Breeding centres.* The breeding stock of the centres shall be blood tested for specific *S. typhimurium* reaction, all animals being examined at the age of 5–5½ months, when they have just started laying and before the eggs are taken for hatching. If the *S. typhimurium* reaction is positive, the reactors are sent for bacteriological control.

(2) A representative selection from each hatching (2 x 10–15 chickens) should be forwarded at 8–10 day intervals, beginning 1 week after hatching, to the Poultry Research Laboratory for bacteriological examination.

(3) From the age of 1 month a representative selection of animals is sent in for laboratory (including bacteriological) examination.

(4) Shortly before the breeding stock is sent to the processing plant blood samples are again taken from all animals and examined for *S. typhimurium* infection.

(5) Grandparent animals are mostly subjected to the same examinations described under points 1 – 4.

Systematic laboratory examinations

(1) *Blood examinations.* For serology the O-antigen (*S. typhimurium* strains, alcohol treated) and H-antigen (the i-phase, formalin treated) are used. Dilutions of 1 : 20, 1 : 40, 1 : 80, and so on are used; the H-reaction is read after 4 h and the O-reaction after 20 h incubation at 37°.

(2) *Dust samples* are taken from the incubator. In the laboratory 15 ml of selenite broth are added, the tubes are incubated for 20 h and the broths inoculated on to brilliant green agar.

(3) *Bacteriological examinations.* A representative number of reactors to the agglutination test are forwarded to the laboratory for bacteriological study. In no case will the results of the agglutination tests be accepted as evidence of infection without confirmation by bacteriological examination of reactors. The liver is plated directly on to brilliant green agar. Samples from the liver, gall-bladder, spleen, duodenum, caecum, rectum, ovary and kidneys are incubated at 37–38° in selenite broth. After 20 h the cultures are inoculated on to brilliant green agar.

(4) *Identification.* The isolated cultures undergo both serological and biochemical examinations. Slide agglutination may be supplemented with tube agglutination. In the biochemical examinations the most important sugars mentioned by Kauffmann (1966) are included. By such combination of serological and biochemical examinations it is possible to perform a sure and quick determination of serotype. In the case of *S. typhimurium,* several subtypes of which may occur, it is very important to obtain a quick and exact determination of subtype to assist in tracing the source of infection (Marthedal, 1960).

(5) *Epizootiological examinations.* An essential part of the fight against salmonellosis in poultry is to trace potential sources of infection. Such investigations will be made in collaboration with the Veterinary Department. They will take place if salmonellosis is diagnosed in poultry (broilers or replacement birds) received in the Poultry Research Laboratory. In such cases the following information is required: (a) the origin of the chickens (hatchery); (b) the suppliers of the hatching eggs.

If a number of flocks from different suppliers appear to be sources of infection, the following procedure is observed: (a) blood samples are immediately taken from suspect flocks for examination for *S. typhimurium* (about

25%). Reactors are sent in for bacteriological examination; (b) hatching eggs from suspect flocks are hatched separately, and at the end of hatching dust and shell samples are taken for examination for salmonella; (c) a representative number of chickens from suspect flocks is sent to the Poultry Research Laboratory for bacteriological examination on the same lines as mentioned under section B, point 2.

By this procedure it has been possible to trace the source of infection quickly and thereby to effect further control.

Discussion

There is no doubt that the introduction of an SPF programme in 1963 cleared the way for a far more effective programme for control of all infectious poultry diseases, including salmonellosis (Marthedal et al., 1969). Before that time the broiler producers faced great pathological problems, first chronic respiratory disease and second salmonellosis, especially infections with *S. typhimurium*. In 1964–1966 control measures against salmonellosis intensified. The results obtained after 4 years of systematic combat were presented at the XIV World's Poultry Congress, Madrid (Hansen & Marthedal, 1970). The results were favourable and the measures extended. In 1972 there was a small rise in the number of cases, originating mainly from 2 sources of infection which have now been eliminated.

Even if the preventive measures described under "Eradication programme" are carried through and extended, there will always be risks from other channels of infection. In this connection feed constitutes a potential risk (Morris et al., 1969). The Order issued in Denmark relating to re-sterilization of certain imported feedstuffs applied to meat-, bone- and blood-meals, that is meal of animal origin, but vegetable feedstuffs have not been subjected to the same demands, even though it is known from several investigations that they too may contain salmonella bacteria (Timoney et al., 1970) although to a lower extent.

To reduce dissemination of infection by feed, some countries (for example, Sweden) have decided that all producers of hatching eggs who voluntarily agree to a preventive salmonella programme shall use controlled feed. Experiments in Sweden and other countries have shown that during the pelleting process there is a considerable reduction of salmonellae. The Poultry Research Laboratory recommends that the breeding centres should use only pelleted feed for the breeding stock.

References

Hansen, H. C. & Marthedal, H. E. (1970). *Salmonella typhi-murium* in poultry (four years of systematic investigations in breeding centres). *XIV World's Poultry Congress, Madrid*, Vol. III, 489.

Kauffmann, F. (1966). *The Bacteriology of Enterobacteriaceae.* Copenhagen, Munksgaard.
Marthedal, H. E. (1960). *Salmonella gallinarum, Salmonella pullorum* og *Salmonella typhi-murium.* Thesis. A/S Carl F. Mortensen, Copenhagen.
Marthedal, H. E. (1965). Hovedlinier ved den fremtidige forebyggende bekaempelse af fjerkraesygdomme. *Erhvervsfjerkrae* **4,** 173.
Marthedal, H. E., Velling, G. & Badstue, P. B. (1969). Control of avian mycoplasmosis in Denmark. *Off. int. Epiz.* **72,** 387.
Morris, G. K., McMurray, B. L., Galton, M. M. & Wells, J. G. (1969). A study of the dissemination of salmonellosis in a commercial broiler chicken operation. *Am. J. vet. Res.* **30,** 1413.
Müller, J. (1952). Bacteriological examination of imported meat- and bone-meal and the like. *Nord. VetMed.* **4,** 290.
Müller, J. (1954). Salmonellainfektioner i danske kvaegbesaetninger. *VII Nordiske Veterinaermøde, Oslo,* p. 525.
Timoney, J., Kelly, W. R., Hannan, J. & Reeves, D. (1970). A study of salmonella contamination in some Dublin poultry processing plants. *Vet. Rec.* **87,** 158.

Discussion

Abrahamsson

I have been working a great deal with *Salmonella* in chicken and I am amazed how few salmonellae you have found. I would like to know if the numbers given in Table 1 refer to a big outbreak or just to single birds?

Marthedal

They were from both big and small outbreaks and were not just a few birds.

Abrahamsson

Does your report cover the results only from your laboratory or from the whole of Denmark?

Marthedal

This is from all Denmark.

Abrahamsson

I have examined much bone- and meat-meal produced in Denmark and have found that often the salmonellae in the bone-meal can be related to the feathermeal added to the meal. We have found *Salmonella* types in the feathermeal which you have reported not to occur in Denmark during recent years.

Tomkin

Is the programme you describe voluntary or is it a mandatory programme supported by legislation?

Marthedal

The programme is on a voluntary basis.

Weitz

How is it enforced?

Marthedal

By good relations between the testing laboratories and the farmers.

Linton

Is the Salmonella O-form (1,4,5,12) serotype mentioned in your paper the same as the unnamed type found in the U.K. (i.e. 0,4;d,–) or a different serotype devoid of flagella antigen?

Marthedal

Salmonella O-form 1,4,5,12 is without H-antigens so I think it is not identical with the serotype you mentioned.

Tomkin

On what basis are you able to convince the farmer that he should participate in the programme—economic factors or concern for public health?

Marthedal

The programme is set up first of all from a public health viewpoint but it is also an economic question.

Kampelmacher

Salmonella control in the EEC is on a voluntary basis. One problem is in the dissemination of salmonellae through sewage, farm effluents, etc. which are used as fertilizers.

Hess

Salmonellae at an average level of $10^5/l$ were found in 45 of 48 samples (93·8%) of fresh sludge obtained from 43 Swiss sewage plants. In fermented sludge we found salmonellae at an average level of $10^3/l$ in 114 of 128 samples (89·1%). We propose therefore that sludge should be pasteurized or irradiated prior to use as a fertilizer.

Food Infection Communicated from Animal to Man

H. J. SINELL

Institut fur Lebensmittelhygiene,
Berlin,
Germany

Salmonellosis is still the most common among the true enteric food infections. In Germany there is a steady increase of *Salmonella dublin* amongst cattle and *Salmonella typhimurium* dominates now, as before, amongst calves. Most of the salmonella food infections in man are caused by egg products and poultry. *S. typhimurium* and *S. enteritidis* are the prevailing isolates in these cases. The contamination rate in poultry at present amounts to 30%. In this connection the results and recommendations of the Round Table Conference on Poultry Hygiene and Inspection held at Hanita, Israel on 3–7 April, 1972, are to be stressed. Milk borne infections have taken a dominating role among food infections. *Streptococcus agalactiae* infections of the cow's udder have more significance with regard to human health than is usually presumed. Based on statistical data, Hahn *et al.* (1972) established a strong correlation between the consumption of raw milk and the frequency of *Streptococcus* group B infections in human beings. The struggle against *Streptococcus agalactiae* infection of the bovine mammary gland must be seen in the future not only under economical but also under public health aspects. Staphylococcal infections in the living animal may cause entero-intoxications. The enterotoxins C and D seem to occur more frequently among mastitis staphylococci than among other staphylococcal populations. The results of phage typing as well as epidemiological studies indicate that most of the enterotoxigenic staphylococci contaminating food originate in humans, which means that the animal constitutes a minor risk for the consumer in this respect.

A CONSIDERABLE number of agents causing infectious diseases of animals may be the cause of corresponding diseases in man. The transfer of such agents *per os* via food generally represents only 1 of several possible ways of infection. Food infections in the narrow sense, i.e. enteric diseases, constitute a very small proportion of these infections. Based upon reviews in the literature, Table 1 summarizes some of the infections which are known or at least suspected to be transmitted by the oral route (Bryan, 1969; Joint FAO/WHO Expert Committees, 1967, 1970; Kaplan *et al.,* 1962; Olitzki, 1972).

Group 1, covers the typical enteric infections and toxi-infections in which the common pathway of infection is via the digestive tract.

Group 2, comprises diseases usually summarized under the heading of "milk-borne infections". Since the infective organisms, having passed the mammary gland, are excreted with the milk, the ingestion of raw or insufficiently pasteurized milk may be the cause of disease. Within this group other paths of infection are possible and are certainly more common in most instances. On the

229

Table 1

Organisms causing food infections

Group 1	*Salmonella* Enteropathogenic *Escherichia coli* *Vibrio parahaemolyticus*
Group 2	*Brucella* *Mycobacterium tuberculosis* β-haemolytic *Streptococcus* *Streptococcus agalactiae* *Staphylococcus aureus* *Coxiella* Tick-borne encephalitis virus Foot and mouth disease virus Vaccinia virus
Group 3	*Bacillus anthracis* *Pasteurella tularensis*
Group 4	*Listeria* *Erysipelothrix* *Leptospira* *Pasteurella pseudotuberculosis*
Group 5	Parasites

other hand, the infections listed in groups 1, 3 and 4 may also be communicated by milk, for example, due to faecal or other excretal contamination.

Group 3, although alimentary infection by organisms of this group is possible, it is not common.

Group 4, these organisms have not yet been established as causes of food infection in the narrow sense. Transmission by the oral route has been proved only in a very few cases.

Group 5, among the parasitic infections, taeniasis and trichinosis must be mentioned in particular, as well as the disease, recently described, which follows ingestion of living nematode larvae *(Anisakis marina)* in the muscle of fish (van Thiel *et al.,* 1960; van Thiel, 1966; Healy & Gleason, 1969; Priebe, 1971). These are exclusively alimentary infections. Helminthosis, however, is not included in this text since the subject requires special treatment. Particular attention should also be paid to toxoplasmosis in view of the recent information concerning the life cycle of the parasite (Frenkel *et al.,* 1970; Hutchison *et al.,* 1970; Janitschke, 1971; Boch & Kuhn, 1972).

Several extremely important food infections and intoxications have not been listed in Table 1. This catalogue, therefore, is not a complete list of the microbial food infections and intoxications. Out of the various categories 3 examples will be presented, thus narrowing the scope of this paper.

Salmonellosis, now as ever, is the "classical" food infection. Most economically advanced countries attempt to collect data on the frequency of this disease. The proportion of cases not reported is repeatedly estimated in the literature as 99% (Steele, 1969). Indeed, the statistics on the actual spread of salmonellosis are imperfect, although careful investigations have been carried out in several countries, especially England and the United States. Based on reports from many regions over the past 10 years, however, a shift in the frequency of the vectors causing salmonellosis may be observed. The primary cause of food infections has shifted from the large slaughter animal infected intravitally to poultry and egg products.

The persistent increase in salmonella isolations and the continual appearance of new types led to the establishment of a Central Institution for Salmonella Research in the Federal Republic of Germany in 1961. This Institution is engaged exclusively with the registration of isolations from the living animal, its excreta, foods of animal origin, abattoirs and slaughter-houses as well as food factories. Between 1937 and 1944 a similar Institution existed at the Institut für Lebensmittelhygiene der Friedrich-Wilhelms-Universität in Berlin.

Within the first 5 years of the establishment of the present Central Institution, at the Federal Health Institute, more than 25,000 strains of salmonellae have been identified. The maximum number was reached in 1970, when 10,399 strains were registered and evaluated (Pietzsch & Bulling, 1970, 1971). About one-third of these strains was isolated during bacteriological examination within the official meat inspection scheme, and two-thirds came from foods, faeces, feeds, and also samples of sewage and soil. Diagnostic examinations of animals from zoological gardens and wild animals are included as well as successive bacteriological controls in food factories and slaughter-houses. The number of isolations from foods and from cases of food poisoning is not as spectacular as the imposing results demonstrated, for example, by Weissman & Carpenter (1969). In 5 abattoirs in Georgia, U.S.A., these authors found that 56% of the pig carcasses and 74% of the beef carcasses were infected or contaminated with salmonellae. Nevertheless within the statistics of sources of food infections, the meat of slaughter animals in the U.S.A. is given a subordinate position. Reference is made primarily to eggs and egg products (Steele & Galton, 1966; Foster, 1969). During the period between 1934 and 1964, some 25,000 salmonella strains of animal origin were isolated, two-thirds came from domestic fowl (Steele & Galton, 1966). Sixty-four per cent of the salmonella outbreaks and 88% of individual cases were caused by eggs, egg products, and poultry as reported by the Communicable Diseases Centre, Atlanta, for the period between 1962 and 1967 (Foster, 1969). A similar development is taking place in other countries (Takacs & Nagy, 1969) and apparently also in the Federal Republic of Germany. Compared with 1963, the number of salmonella isolations from egg products increased five-fold in 1970

(Pietzsch & Bulling, 1971). Approximately 90% of these isolations are distributed among the types *S. thompson, enteriditis, typhimurium, reading, infantis* and *braenderup*. Unfortunately, in Germany the statistics of sources responsible for food infections are somewhat incomplete. Nevertheless, there is obviously an upward trend in the infections caused by poultry, poultry products, eggs and egg products in which *S. typhimurium* and *S. enteritidis* predominate. To a certain degree this could be correlated with the increase in the total number of salmonella cases in man in the Federal Republic of Germany. In 1970, 12,410 cases were reported, which is double the number reported in the previous year. In the German Democratic Republic too, the number of isolations has increased steadily (Nass, 1970), particularly from children. The cause of this may include the generally enhanced susceptibility of the young and the increase in the consumption of food of animal origin.

Salmonellosis in slaughter poultry does not show any reduction (Patterson, 1972). The rate of infection or contamination in deep frozen and fresh commercial poultry is about 30% in Germany (Pietzsch & Bulling, 1971; Sinell, unpublished data) which is in marked contrast to information published only a few years ago (Sadler & Corstvet, 1965) in the U.S. The statements and recommendations of a Round Table Conference on Poultry Hygiene and Inspection held at Hanita, Israel in April, 1972, are of particular significance in this context. At this conference several measures were recommended including the following: continuous health control of the broiler stock, with examination of the feeds and the environment. In addition to the hygienic measures at abattoir level, a permanent flow of information between those responsible for *ante-* and *post-mortem* examination and the production line (hatcheries and rearing farms) is an urgent need. Thus it is essential that veterinary certificates are provided prior to slaughter, as a health passport.

Among the large slaughter animals and in particular among cattle (but not calves) there is steady increase in *S. dublin*. There is a noteworthy reduction in salmonella isolations from horses in Germany (Bulling & Pietzsch, 1968; Pietzsch & Bulling, 1971), but this is in contrast to the data published in other countries (Steele & Galton, 1966). The demand that each horse be examined bacteriologically before it is declared fit for consumption cannot be justified (Kuhlmann, 1964).

Among the other infections communicated from animal to man, the "milk-borne" infections take a primary position. In many countries tuberculosis and brucellosis are eradicated or at least considerably reduced. Correspondingly the rate of tubercular infection in man has also decreased markedly. But there remains another unsolved problem—mastitis of cattle.

Several organisms causing mastitis can be transmitted to man and vice versa. It is a well established fact that group A streptococci can be communicated from man to the cow's udder where they infect the milk. There is abundant literature available describing several outbreaks of scarlet fever and septic sore throat following the consumption of raw or insufficiently pasteurized milk (see Bryan, 1969). The best known example is the great epidemic in Chicago, which affected more than 10,000 people in 1912. In this field an additional infection will be described well known in man but as yet not generally recognized as a food infection.

It is known that Lancefield group B streptococci may provoke infections in man. There were, however, a considerable number of reports in the literature referring to 4160 individual cases where *Streptococcus agalactiae* was isolated from various materials originating from man (reviewed by Müller, 1967 and Hahn *et al.*, 1970). Most of these infections are connected with the puerperium, with or without septic symptoms, and the urinary and sexual tract. Isolations have also been made in cases of endocarditis, meningitis and last but not least in many lethal infections of new-born infants. Group B streptococci have been found also as concomitants in the normal puerperium of healthy women as well as in the nasopharyngeal cavity. Their frequency amounts to between 0·2 and 25·2% of the total streptococci isolated from humans. They should not be regarded as harmless saprophytes (Erbslöh & Grün, 1949; Kahler & Aicher, 1952; Kexel, 1965; Kexel & Schonbohm, 1965; Müller, 1967; McKnight *et al.*, 1969; Steinitz *et al.*, 1971). The significance of these germs in livestock requires no further interpretation. The infection of the bovine mammary gland still represents one of the major problems in veterinary medicine. In developed countries this infection is probably responsible for the heaviest financial losses in animal production. It has been extremely doubtful for a long time whether there is any connection between the disease of the animal and of man (Seelemann, 1963). Most authors do not think that transmission from animal directly to man is likely. This assumption is supported by the fact that the biochemical properties of the human strains are different from those isolated from the animal. Within the human strains the polysaccharide antigens Lancefield Ia to III prevail, whereas in the bovine mastitis strains protein antigens RX can be detected more often. Furthermore, there are differences in the susceptibility to bacitracin (Hahn *et al.*, 1970). Group B streptococcal infections in man and in cattle, therefore, have been regarded over many years as independent infections (Butter, 1966; Butter & de Moor, 1967). Butter wrote in 1966 ". . . it is obvious that in the Netherlands at present cattle do not constitute a considerable source of infection of group B streptococci for man".

On the other hand transmissions of the infection from cattle to man and vice versa were described many years ago (Erbslöh & Grün, 1949; Butter, 1966; Livoni, 1965). The association with the consumption of raw milk was not absolutely clear in all these cases. An important new public health aspect was revealed by the investigations of Hahn *et al.* (1972). In the Streptokokken-Zentrale, Kiel, 250 strains of group B streptococci were selected from a group of diseased persons. Those involved were requested to fill in a questionnaire. The answers revealed an extremely high proportion, 27·7%, who had consumed raw milk, either temporarily or continuously. The percentage of consumers of the so-called "Vorzugsmilch" (a type of grade A milk) amounted to 19·7%. But it should not have exceeded 0·5% on the basis of a balanced distribution as required by statistical laws, in fact 1/40 of the actual value. Incidentally this proportion correlates well with the percentage of commercial raw milk positive for group B streptococci. A general survey of 3465 samples of Vorzugsmilch proved 706 or 20·4% to be infected with *Streptococcus agalactiae*. This shows that the consumption of raw milk and group B streptococcal infections are connected to a degree which is more than accidental. In this respect it should be mentioned that *Strep. agalactiae* has been isolated also from several other animal species, for instance from pigs (Kohler & Mochmann, 1958) and also even in epidemic infections of fish (Milojovic & Teofanovic, 1963; Robinson & Meyer, 1966).

Although there are biochemical differences in the behaviour of group B streptococci found in cattle and man, the extensive overlap does not permit separation between different species or variants. It is presumed that there is a considerable adaptation of the various ecovariants to a new environment. The loss of ability to synthesize β-galactosidase supports this assumption. Overall it seems that the occurrence of streptococcal infections in the bovine mammary gland must be regarded not only as an economic problem, but in future must be given greater emphasis as a public health problem. Milk containing *Strep. agalactiae* must be excluded from human consumption under all circumstances. The measures to be taken by the official control authorities must be intensified correspondingly. Furthermore, it has been recommended that a regular and special test for group B streptococci be introduced into the examination of pregnant women (Hahn *et al.*, 1972).

One of the most frequently recorded causes of food poisoning is the entero-intoxication from *Staphylococcus aureus*. The food is contaminated in nearly all instances by human carriers. But strains originating from the animal may also produce enterotoxins. Do these strains represent a hazard to human health? This question must be answered in the affirmative so far as raw milk is concerned. The hygienic significance of those staphylococci forming enterotoxin which infect the udder has been investigated repeatedly since the first reports of Barber (1914) and this still remains a problem. The proportion of staphylococci positive for enterotoxin from individual milkings varies according to different authors, between 2% and 67·5% (see Untermann, 1971). Most of the data indicate about 10% or less. The distribution of the enterotoxins in staphylococcal populations in the bovine udder obviously shows a different pattern to those of the staphylococci isolated from cases of food poisoning. Among 157 strains from acute bovine mastitis, Olson *et al.* (1970) found that 11 produced toxin type C, 11 type D and in 1 instance both C and D toxins were detected. Toxins A and B were never found, although in cases of food poisoning type A toxin is predominant. In the results from tests on the collection of strains in our Institute there was confirmation of this distribution of toxin types, although among the mastitis strains type A enterotoxin was detected occasionally (Untermann, 1971; Untermann & Lupke, 1971).

The origin of the milk strains, however, showed considerable variation. The strains mentioned above, particularly those from milk and cheese, were by no means all "typical" animal strains belonging to the phage groups II and IV. Our findings showed that merely 18·4% of 233 cheese strains belonged to the phage groups II and IV. By far the predominant proportion of these strains, especially those which formed enterotoxins, were included in groups I and III (Sinell & Mentz, 1971; Untermann, 1971). This indicated that the strains originated mainly from human contamination.

Little is known about the occurrence of enterotoxic "animal" staphylococci in meat. It is also doubtful whether such strains have caused cases of food poisoning. We have tested slaughter pigs, several raw materials at various stages of production, as well as ready-to-eat products for the presence of staphylococci producing enterotoxin (Siems et al., 1971; Sinell et al., 1970). An exclusive animal-specific staphylococcal flora was found both inside and on the surface of the heads of pigs even after scalding. Among 74 strains tested for enterotoxins, 10 (13·5%) were positive, 4 produced enterotoxin A and 6 enterotoxin C (Untermann, 1971).

We followed the development of the staphyloccal population in the course of the production of pig's head in aspic, from the initial material to the finished product. The proportion of "animal" staphylococci decreased continually during the various stages of production, so that the finished product contained only those strains which, according to their phage type pattern, appeared to come from man (Siems et al., 1971). Similar results were obtained from fresh minced meat; the strains of staphylococci isolated from this material also pointed more towards human than animal contamination. Out of 177 strains merely 21 belonged to phage group II. A proportion of group II strains was also found in collections of human isolates (Pulverer, 1963; Sinell et al., 1970).

These findings reaffirm the conclusion that the major hazard of staphylococcal enterotoxins in foods of animal origin arises not from staphylococci in animals but from the carriage of the organisms by man; it is these staphylococci which contaminate food.

References

Barber, M. A. (1914). Milk poisoning due to a type of *Staphylococcus albus* occurring in the udder of a healthy cow. Philipp. J. Sci., Sect. B, Trop. Med. **9**, 515.

Boch, J. & Kühn, D. (1972). Zur Epidemiologie der Toxoplasmose. *Fortschr. VetMed.,* *Beih. Zentbl. VetMed.* **H.17,** 207.

Bryan, F. L. (1969). Infections due to miscellaneous microorgansims. In *Food Infections and Intoxications,* p. 224. Ed. H. Riemann. New York and London: Academic Press.

Bulling, E. & Pietzsch, O. (1968). Ergebnisse und Schlußfolgerungen aus fünfjährigen Salmonellose-Untersuchungen (1961–1965). *Zentbl. VetMed., Reihe B* **15,** 913.

Butter, M. N. W. (1966). Streptokokken van groep A en van groep B bij nuljarigen. Een epidemiologisch onderzoek. Proefschrift, Amsterdam.

Butter, M. N. W. & de Moor, C. E. (1967). *Streptococcus agalactiae* as a cause of mengingitis in the newborn, and of bacteraemia in adults. *Antonie van Leeuwenhoek.* **33,** 439.

Erbslöh, F. & Grün, L. (1949). Galtstreptokokken der serologischen Gruppe B als Erreger der Endokarditis lenta. *Dt. med. Rdsch.* **3,** 508.

Foster, E. M. (1969). The problem of salmonellae in foods. *Fd Technol.* **23,** 1178.

Frenkel, J. K., Dubey, I. P. & Miller. N. L. (1970). *Toxoplasma gondii* in cats: Fecal stages identified as coccidian oocysts. *Science, N.Y.* **167,** 893.

Hahn, G., Heeschen, W. & Tolle, A. (1970). Streptococcus–Eine Studie zur Struktur Biochemie, Kultur und Klassifizierung. *Kieler milchw. Forsch. Ber.* **22,** 335.

Hahn, G., Heeschen, W., Reichmuth, I. & Tolle, A. (1972). Wechselbeziehungen zwischen Infektionen mit Streptokokken der serologischen Gruppe B bei Mensch und Rind. *Fortschr. VetMed., Beih. Zentbl. VetMed.* **H.17,** 184.

Healy, G. R. & Gleason, N. N. (1969). Parasitic infections. In *Food Infections and Intoxications,* p. 175. Ed. H. Riemann. New York and London: Academic Press.

Hutchison, W. M., Dunachie, J. F., Slim, J. Chr. & Work, K. (1970). Coccidian-like nature of *Toxoplasma gondii. Br. med. J.* **1,** 142.

Janitschke, K. (1971). Die Bedeutung von Tieren als Infektionsquelle des Menschen. *Dt. med. Wschr.* **96,** 1.

Kahler, J. & Aicher, J. (1952). Der Galtstreptokokkus als Erreger einer ulzero-polypösen Endokarditis beim Menschen. *Z. allg. Path.* **88,** 312.

Kaplan, M. M., Abdussalam, M. & Bijlenga, G. (1962). Diseases transmitted through milk. In *Milk hygiene.* Geneva: W.H.O.

Kexel, G. (1965). Uber das Vorkommen der B-Streptokokken beim Menschen. *Z. Hyg. InfektKrankh.* **151,** 336.

Kexel, G. & Schönbohm, S. (1965). *Streptococcus agalactiae* als Erreger von Säuglings-meningitiden. *Dt. med. Wschr.* **90,** 258.

Köhler, W. & Mochmann, H. (1958). Streptokokkenfunde der Tonsillen und Antistrepto-lysintiter bei Schlachtschweinen. *Zentbl. Bakt. Abt. I.* **173,** 58.

Kuhlmann, W. (1964). Uber das Vorkommen von Salmonellen bei geschlachteten Pferden. *Mh. VetMed.* **19,** 790.

Livoni, P. (1965). Gruppe B-Streptokokinfektionen hos mennesker og hos dyr. *Ugaskr. Laeg.* **127,** 1353.

McKnight, J. F., Ellis, J., Jensen, K. A. & Franz, B. (1969). Group B streptococci in neonatal deaths. *Appl. Microbiol.* **17,** 926.

Milojovic, Z. & Teofanovic, M. (1963). Parallel investigations of CAMP tests with sheep, calf and cattle blood. *Acta vet. Beograd.* **13,** 51.

Müller, G. (1967). Die Typisierung der Streptokokken der Gruppe B. *Arch. exp. VetMed.* **21,** 43.

Nass, W. (1970). Ergebnisse bakteriologischer Untersuchungen über Salmonellosen (1958–1967). *Z. ges. Hyg.* **16,** 106.

Olitzki, A. (1972). *Enteric Fevers.* Basel/New York: S. Karger.

Olson, J. C., Jr., Casman, E. P., Baer, E. F. & Stone, J. E. (1970). Enterotoxigenicity of *Staphylococcus aureus* cultures isolated from acute cases of bovine mastitis. *Appl. Microbiol.* **20,** 605.

Patterson, J. T. (1972). Salmonellae in processed poultry. *Ktavim, Rec. agric. Res.* **20,** 1.

Pietzsch, O. & Bulling, E. (1968). Verbreitung der Salmonella-Infektionen bei Tieren, tierischen Lebens- und Futtermitteln in der Bundesrepublik Deutschland einschließlich Berlin (West). *Bd. gesd. bl.* **11,** 233.

Pietzsch, O. & Bulling, E. (1970). Verbreitung der Salmonella-Infektionen bei Tieren, tierischen Lebens- und Futtermitteln in der Bundesrepublik Deutschland einschließlich Berlin (West) Jahresbericht 1968. *Bd. gesd. bl.* **13**, 5.

Pietzsch, O. & Bulling, E. (1971). Verbreitung der Salmonella-Infektionen bei Tieren, tierischen Lebens- und Futtermitteln in der Bundesrepublik Deutschland einschließlich Berlin (West). *Bd. gesd. bl.* **14**, 369.

Priebe, K. (1971). Die lebensmittelhygienische Bedeutung des Nematodenlarvenbefalls bei Seefischen. *Archs Lebensmittelhyg.* **22**, 193.

Pulverer, G. (1963). Pathogene Staphylokokken bei Mensch und Tier sowie im Freiland. *Med. Habil. schr. Köln.*

Robinson, J. A. & Meyer, F. P. (1966). Streptococcal fish pathogen. *J. Bact.* **92**, 512.

Sadler, W. W. & Corstvet R. E. (1965). Second survey of market poultry for salmonella infection. *Appl. Microbiol.* **13**, 348.

Seelemann, M. (1963). Zur Frage der Pathogenität des Sc. agalactiae (Galtstreptokokkus) und der Gesundheitsschädlichkeit der Milch von mastitiskranken Kühen. für den Menschen. *Mh. Tierheilk.* **15**, 199.

Siems, H., Kusch, D., Sinell, H.-J. & Untermann, F. (1971). Vorkommen und Eigenschaften von Staphylokokken in verschiedenen Produktionsstufen bei der Fleischverarbeitung. *Fleischwirtschaft* **51**, 1529.

Sinell, H.-J., Kusch, D. & Untermann F. (1970). Zum Vorkommen von koagulasepositiven Staphylokokken in Hackfleisch. *Zentbl. VetMed., B* **17**, 429.

Sinell, H.-J. & Mentz, I. (1971). Enterotoxinbildende Staphylokokken in Käseproben. *Alimenta* **S.22.**

Steele, J. H. (1969). Epidemiology of salmonellosis. *J. Am. Oil Chem. Soc.* **46**, 219.

Steele, J. H. & Galton, M. M. (1966). Salmonellosis. *J. Am. Vet. med. Ass.* **149**, 1079.

Steinitz, H., Schuchmann, L. & Wegner, G. (1971). Leberzirrhose, Meningitis und Endokarditis ulceropolyposa bei einer Neugeborenensepsis durch B-Streptokokken *(Strept. agalactiae). Arch. Kinderheildk.* **183**, 382.

Takács, J. & Nagy, G. Y. (1969). The incidence of salmonellae in meat and meat products and in animal materials in the years 1947–1967 in Hungary. *Zentbl. Bakt. Abt. I.* **218**, 305.

van Thiel, P. H. (1966). The final hosts of the herringworm *Anisakis marina. Trop. geogr. Med.* **18**, 310.

van Thiel, P. H., Kuipers, F. C. & Roskam, R. Th. (1960). A nematode parasitic to herring causing acute abdominal syndromes in man. *Trop. geogr. Med.* **12**, 97.

Untermann, F. (1971). Untersuchungen zum kulturellen und serologischen Nachweis enterotoxinbildender Staphylokokken sowie über die Verbreitung dieser Keime. *Vet. med. Habil. schr., Berlin (West).*

Untermann, F. & Lupke, H. (1971). Bedeutung von Mastitis-Staphylokokken als Ursache von Lebensmittelvergiftungen. *Dt. ges. Milch., Berlin, Okt.* (Vortrag).

Weissman, M. A. & Carpenter, J. A. (1969). Incidence of salmonellae in meat and meat products. *Appl. Microbiol.* **17**, 899.

Joint F.A.O./W.H.O. Expert Committee on Zoonoses (1967). Third Report, W.H.O. Techn. Rep. Ser. No. 378. Rome: FAO.

Joint F.A.O./W.H.O. Expert Committee on Milk Hygiene (1970). Third Report, W.H.O. Techn. Rep. Ser. No. 453. Rome: FAO.

Discussion

Weitz

How do you establish that *Str. agalactiae* even when isolated from man causes disease in man? What symptoms are produced? *Str. agalactiae* can come from many sources (e.g. walls, bedding, etc.) associated with infected animals.

Barrow

Would you elaborate on the statistical data suggesting human disease from Group B streptococci? What kind of clinical illness occurred and what people were involved, i.e. adults or children?

Sinell

There is much information available on this point in the medical literature. Summaries on this subject are given by Müller (1967), Hahn, Heeschen & Tolle (1970) and Hahn, Heeschen & Reichmuth (1972); individual cases are described, for instance, by Erbslöh & Grün (1949) Kahler and Aicher (1952), Kexel & Schönbohm (1965), McKnight *et al.* (1969) and many others (the references are given in the paper above).

Bergdoll

How many phage group II strains of staphylococci were found in cheese? We have found all group II strains examined to produce enterotoxin, for the most part an unidentified toxin. Only 1 case of food poisoning with which I am familiar was caused by a phage group II staphylococcus. Only phage group II strains are involved in "scalded skin disease" in babies.

Sinell

Of 233 strains of staphylococci isolated from cheese 18·4% belonged to phage groups II and IV.

Kampelmacher

Prof. Sinell spoke about the Conference on Poultry Hygiene held in Israel this spring. If members of this audience are interested in the report of this Conference it can be obtained free of charge by writing to Dr M. van Schothorst, Secretary-Treasurer World Association of Veterinary Food Hygienists, P.O. Box 1, Bilthoven, The Netherlands.

Hess

Prof Sinell states that "Vorzugsmilch" containing group B streptococci may be a source of infection for man. Raw milk with group B streptococci is by no means a "Vorzugsmilch".

Sinell

That is correct in theory. Unfortunately practice has shown that a considerable proportion of "Vorzugsmilch" contains viable group B strepto-cocci, which is in contradiction of the legal regulations (see data in literature cited in paper).

The Origin of Pathogens in the Live Animal in Relation to Food as Part of the Zoonoses Chain

N. Skovgaard

Institute of Hygiene and Microbiology,
Copenhagen,
Denmark

There is an increasing number of reports on occupational diseases amongst food-handlers and cases of human diseases where food is suspected to be the causative agent. The importance of the live animal as a reservoir for such pathogens is illustrated by examples which also show the influence of the host-parasite interrelationship. It is advocated that the food hygienist should make more effort to understand the importance of food in the epidemiological pattern. Foods as possible vectors of diseases like toxoplasmosis and yersiniosis are discussed. Figures are given of the incidence of staphylococci and β-hemolytic streptococci in pigs which explains their constant presence in raw pork products. The route of contamination is indicated in 2 diagrams.

THE IMPORTANCE of food as the cause of infection and disease in man is under constant re-evaluation as epidemiological data and information on the origin of micro-organisms and their mode of transmission to man are accumulated.

There is an increasing number of reports in the medical literature on occupational diseases amongst food handlers and workers engaged in food producing factories, such as slaughterhouses and dairies. Reports are also increasing of infections in man where the causative agents are tentatively believed to originate in animals. Epidemiological studies in the animal kingdom are needed to clarify these areas. These should reveal whether or not the organism in question is present in the animal species and whether it can be isolated as part of the normal flora of that animal. Such investigations should also show whether the organism can produce clinical disease in the animal. Should the latter be the case there will often be a greater risk for infection in man. This is because the number of bacteria to which man is likely to be exposed will be much greater when propagation has taken place in infected tissues of the animal than when the organisms constitute part of the normal flora.

Animal Food as a Vector for Pathogens Present in the Live Animal

The question of whether an animal and the food derived from it are to be incriminated as the source of disease in man is closely linked with the problem of a possible host specificity for the organism in question or the host-parasite interrelationship.

A few examples will illustrate this. *Salmonella typhi* is specific to man. This is also to a large extent the case with *S. paratyphi* A and B. Therefore these organisms originate in man, although food might serve as a vehicle for transmission and propagation. Group A streptococci are seldom isolated from foods other than milk and some specific food items, like peeled, frozen shrimps. Man is the predominant reservoir and thus the most likely source.

It is now well-established that the origin of salmonellosis in man is the animal kingdom. The demonstration of this organism in a food generally indicates its presence in the animal from which the food was derived. Although many possible pathways for contamination exist, the animal–food–man chain of infection is the most common. Some streptococci, such as groups L and E, are predominantly pathogenic to animals and have little or no virulence to man. The origin of such organisms in food is consequently likely to be the live animal and the importance of their presence in food must be evaluated in the light of their virulence to man.

Other micro-organisms, e.g. staphylococci, show equal pathogenicity to man and animals and may be present with high frequency in both. Typing by biochemical criteria, bacteriophage and other means must be applied to determine whether the source is man or animal. Whether food is incriminated in a zoonosis depends to a large extent on the host preference of the micro-organism as noted above. Is the micro-organism in question predominant in man or in the animal, is it to be found in both, or is it only seldom present in one of them? This, combined with the virulence of the organism, to a certain extent determines whether a food-borne chain of infection is man–food–man or animal–food–man.

Whether food is implicated in the chain of infection will further depend on the ability of the organism to survive and to propagate in the food. Some micro-organisms, such as salmonellae and Lancefield group L streptococci, propagate readily in foods, while others, such as group A streptococci, are unable to do so. Whether man or animal or both serve as foci for the natural dissemination of the causative agent, with food as the vector, cannot be estimated unless the origin of the organism is known. Here the food hygienist is usually following the rather narrow traditional programmes of microbiological investigation. He is looking for spoilage organisms and so-called indicator organisms, concentrating mainly on indicators of faecal contamination. The value of such investigations should not be underestimated, but they reveal or indicate the presence of only very few of the many micro-organisms which may be transmitted through food.

One of the main foci for pathogenic micro-organisms in both man and animal is the pharyngeal and tonsillar microflora. The absence of faecal indicators in a given food is no proof of the absence of pathogenic bacteria from such sources, as their presence in food is unrelated to faecal contamination. For such reasons the food hygienist should extend his activities from these rather passive, often unproductive and personally uninspiring areas to cover the epidemiological aspects of the problem.

The importance of an integrated meat inspection service with ante-mortem inspection at farm level and a feed-back system of information from the post-mortem inspection to the farmer is rapidly becoming accepted. Apart from the laboratory investigation of milk, feed-back systems are not working in the field of food microbiology and hygiene. In his search for specific pathogenic organisms the food hygienist is greatly hampered by the natural saprophytic microflora predominant in raw food. In clinical specimens it is often much easier to detect specific organisms.

For this reason it is natural in food epidemiology to concentrate on systematic investigation of the presumed natural food of the pathogenic organisms. I am not advocating that the food hygienist should be responsible for basic research in the field of animal microbiological ecology. But I am emphasizing that due emphasis should be placed on the proper understanding of the pathways by which pathogenic micro-organisms enter foods. When speaking of zoonoses the food hygienist should consider himself one of the diagnostic links in the epidemiological chain.

The Importance of Foods of Animal Origin in the Line of Transmission

The magnitude of the problem of food as a vector for zoonoses will be outlined by analysing some examples of the transmission of pathogens from the animal as natural host to man with food serving as the vector.

In recent surveys a high proportion of a large number of Danish pigs were found to carry staphylococci and β-haemolytic streptococci on the skin of the live animal. The carrier rate of staphylococci is not less than 50% and is probably much higher when an adequate technique is used. β-haemolytic streptococci may be isolated from 10–15% of the skin samples from the live animals.

Micrococci are part of the normal resident skin flora of pigs and staphylococci seem to establish themselves in this microflora with a rather high frequency. The scalding, singeing and subsequent back-scraping will not remove all staphylococci. These procedures will only open the sweat and sebaceous glands where the bacteria propagate on the live animal. Thus staphylococci will be present in low numbers in the normal microflora on the finally dressed and chilled carcass. The β-haemolytic streptococci are part of the transient skin-flora and do not normally establish themselves as part of the resident flora.

In the live animal the main reservoir of staphylococci and β-haemolytic streptococci will be the pharyngeal and tonsillar flora. The carrier rate of staphylococci in these organs was not less than 70%. It is characteristic that when staphylococci are present in the tonsillar tissue, their numbers are in the order of millions/g of tissue. β-haemolytic streptococci were likewise present in more than 50% of samples from the tonsillar tissue of pigs.

Swabs from bacon sides have shown that staphylococci can be isolated from nearly 90% of the sides. They are also present in low numbers (5–10/ml) in the corresponding brine.

The plucks removal has further shown to be a very important procedure in spreading specific pathogenic organisms; the tonsillar or pharyngeal tissue is cut with the same knife as is used for loosening these organs. This means that staphylococci as well as β-haemolytic streptococci will be present in raw pork products in low numbers simply because they are part of the normal flora of the live animal and the dressed carcass. This problem is important in the human staphylococci carrier problem, the meat itself being by far the greatest source of staphylococci.

A recent survey in a factory where there were high staphylococcal counts in sliced bacon showed that several workers in the cutting room were suffering from wounds, which were covered with adhesive material, from which staphylococci were isolated. Phage typing of staphylococci isolated from the workers in the cutting room and from the sliced bacon indicated that the strains in the bacon originated from the meat itself and not from the workers, since the bacon strains were non-typable and considered to be of animal origin while the strains isolated from the workers were readily typable and none of these were found in the final product.

In considering the risk of the animal staphylococci causing food poisoning it is noteworthy that probably not less than 2/3 of the animal strains produced enterotoxin.

Food is considered to play an important role in dissemination of such frequent human diseases as toxoplasmosis and yersiniosis. In some countries 30–50% of the adult population gives a positive *Toxoplasma* sero-reaction, but the infection is frequently not diagnosed because of weak clinical symptoms.

Toxoplasmosis is transmitted to man: (1) transplacentally; (2) through consumption of raw and partially cooked meat; and (3) from exposure to the faeces of cats (Center for Disease Control, 1971). In the U.S.A. the percentage of animals with a positive antibody reaction varies from 14–33. In Denmark and Germany a positive reaction has been reported in 20 and 65% of pigs. The corresponding rates in sheep in Denmark and Germany were 61 and 89% respectively.

Most cases of acquisitive toxoplasmosis are believed to be of alimentary origin and the main source of infection seems to be meat, either meat containing infectious cysts or surface contaminated meat. A recent report of an epidemic of toxoplasmosis amongst medical students in New York City who had eaten undercooked hamburgers indicated that the risk is not merely theoretical. We have still a long way to go before food hygienists are able to supply information on this point.

Human yersiniosis is another disease in which food probably plays an important role. Man is infected via the intestinal tract. The disease is caused by a small Gram-negative rod, *Y. enterocolitica.* formerly listed among the *Pasteurella* bacteria but now grouped taxonomically with *Y. pestis* and *Y. pseudo-tuberculosis.* It is interesting to note that certain *Yersinia* types show cross-reactions with *Brucella suis* antigens. The organism causes uncharacteristic cases of abdominal infections with mild fever. Arthritis is a common complication. The disease is most frequent in children. It is tentatively assumed that this disease is one of the most common bacterial intestinal infections in Denmark, perhaps the most common; 1–2% of the normal population is estimated to have suffered from it in Denmark (Larsen, 1972) and similar figures are reported from Sweden. In the Netherlands the organism has been shown to be present in pigs. Some workers attach even more importance to the *Yersinia* organisms than to salmonellae and this may well be justified in countries where salmonellosis is effectively controlled.

It is well-established that *Listeria* may be present in animals for slaughter but the importance of this in relation to food as a cause of human listeriosis has yet to be adequately investigated. The same applies to streptococcal diseases in human beings resulting from infections with Lancefield groups R, S, T and U streptococci. These bacteria are known to be present in domestic animals, at least in pigs.

The above-mentioned organisms (staphylococci, toxoplasma and yersenia etc.) are listed to underline the importance of the animal as a reservoir of food-borne zoonoses. Organisms pathogenic to man and present in animal foods are most likely to originate in the animal from which the food is ultimately prepared. It is well-established that the reservoir of salmonellae in food is the animal carrier but this is equally true for many other organisms.

Surveys of the possible lines of transmission are given in Figs 1 and 2. Where

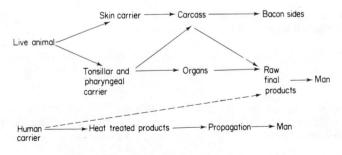

Fig. 1. Line of transmission of staphylococci.

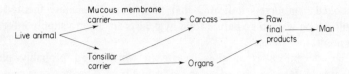

Fig. 2. Line of transmission of pathogens other than staphylococci from live animals to man.

raw products are concerned the indicated pathways from animal to man are of great importance. In the case of heat treated food the recontamination from both man and animal reservoirs must be considered.

Recommendations

(1) Improvement and development of methods for detection of pathogenic micro-organisms in foods.

(2) Zoonotic and epidemiological studies on the importance of food in diseases of man.

References

Center for Disease Control (1971). Parasitic diseases–Toxoplasmosis. *Veterinary Public Health Notes,* Georgia. December 1971 p. 3.

Larsen, J. H. (1972). Human Yersiniose. Resultat af serologisk undersøgelse for infektion med *Yersinia enterocolitica. Ugeskr. Laeg.* **134,** 431.

Discussion

Foster

How does beef become infested with Toxoplasma (if it does) and do the Dutch suffer from this problem owing to their taste for "raw meatballs"?

Skovgaard

I cannot give any answer to the first question. It is possible that the organism might survive the heat treatment in the interior of meatballs.

Weitz

Does the contamination of slaughtered meat come from the infection of cattle or might it arise from human or animal vectors?

Skovgaard

Contamination from cats is certainly the most important vector of the infectious cysts. This contamination may happen in the private home as well as in shops.

Riemann

I believe that the cat and maybe the dog are the only known sources of infective forms of toxoplasmosis in humans. In the case where meat is the vehicle the most likely source of contamination is probably the cat.

Olson

Would you please repeat your statement relative to human infection by

Yersinia in Denmark?

Skovgaard

1-2% of the population are supposed to have suffered from this infection. These estimates are based on positive serum reactions.

Lewis

What is your opinion of the value of routine tests for *Staphylococcus aureus* as an indicator of nasopharyngeal or epithelial contamination of foods, particularly those of animal origin?

Skovgaard

It would be taking things too far to use it as an indicator of contamination from muciparous sources.

Mechanism and Prevention of Salmonella Infections in Animals

W. EDEL, M. VAN SCHOTHORST, P. A. M. GUINEE AND E. H. KAMPELMACHER

Rijks Instituut Voor de Volksgezondheid,
Bilthoven,
The Netherlands

Salmonella infection in animals is only a part of the whole salmonella problem, which is still increasing despite the greatly increased preventive measures taken during the last decennia. For the year 1971, in The Netherlands, the registered isolations from man were nearly 10,000/annum, a record number; it is estimated that the real morbidity is at least 20 times as high. There are several reasons for this rise in incidence. Of paramount significance are: (i) imported contaminated feed components and certain foods; and (ii) the increasing environmental pollution. The incidence of salmonella infection in man, slaughter animals and the environment is given. *Salmonella typhimurium* is the most common serotype (>50% of all isolates) and 50% of strains are of the fermentative phage type II 505. It has been demonstrated that there is a close connection between infections in man, animal and the environment. The infection is maintained by vicious circles, which are very difficult to break. The main preventive measures recommended to control the salmonella situation are, in order of effectiveness: (i) adequate pelletisation of mixed feeds; (ii) disinfection of effluents from slaughter houses and processing plants, as well as those from the populations of cities and villages, by efficient purification plants; (iii) sanitation in animal quarters before slaughter; (iv) adequate and well monitored slaughterhouse meat processing and butcher shop sanitation; (v) educational activities for and monitoring of catering establishments and food handling in households.

SALMONELLOSIS is an important public health problem in the Netherlands. In recent years the number of cases in man has increased although several preventive measures have been taken during the last decade. In the Netherlands nearly all isolated strains are serotyped at the National Salmonella Centre and it is therefore possible to obtain a good idea of the salmonella problem in general. During 1971 the number of strains isolated was higher than ever before and the reasons for the observed increase appear to include the following;

(1) There is a real increase in salmonellosis of man and animals. The large number of isolations during 1971 was favoured by high temperatures during summer and early autumn, including the highest temperature registered this century. This situation promoted the growth and spread of salmonella bacteria which resulted in a record number of cases recorded in man, namely nearly 10,000. It should be emphasized that there are reasons to believe that this number is only between 1 and 5% of the real incidence. Moreover, the influence of imported contaminated foods and feeds and the increasing environmental pollution were of paramount significance.

(2) More interest in the salmonella problem in general.
(3) A special research project in the field of salmonellosis.
(4) Better isolation methods for salmonellae.

Incidence

Man

The incidence in man from 1955–1971 is summarized in Fig. 1, which also shows the number of serotypes isolated from man and the total number of serotypes isolated in the country. Seventy-five per cent of all salmonella cases in man were caused by only 3 serotypes, namely

Salmonella typhimurium	c. 60%
Salmonella panama	c. 10%
Salmonella infantis	c. 6%

It should be mentioned that the percentage of *Salmonella panama* has decreased in recent years. *S. stanley* was third in importance until 1968, but was thereafter replaced by *S. infantis*. The frequency of isolation of the latter serotype is still increasing.

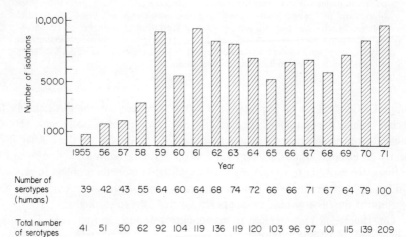

Fig. 1. Salmonella isolations from humans in The Netherlands (1955–1971).

Foods

For man, salmonella contaminated food is the most important source of infection. These foods are mainly products of animal origin such as meat and meat products, poultry and egg products, but also include foods of vegetable origin such as coconut products. The most important sources in the Netherlands are meat and meat products, especially those originating from pigs. To a lesser degree egg products and poultry, where contamination after processing takes place, are important. Contaminated minced meat represents a significant source of infection for man (van Schothorst *et al.*, 1970). To monitor this contamination, minced meat samples from all butcheries in 1 particular town in the Netherlands have been examined once a year since 1959. The results of these investigations are summarized in Table 1.

Table 1

Salmonellae in samples of minced meat collected around July 25 in 1 particular town in the Netherlands

	1966	1967	1968	1969	1970	1971
Number of samples	120	112	101	107	97	95
Number positive	31	46	14	16	21	31
%	25·8	41·1	13·9	15·0	21·6	32·6

Serotypes

S. typhimurium	34·6%	S. livingstone	4·8%
S. panama	15·6%	S. stanley	3·5%
S. infantis	7·8%	Other types (34)	33·7%

One conclusion may be that application of better isolation methods has significantly increased the number of positive results. Each year we find an increasing number of serotypes not previously isolated in the Netherlands. There were for instance in 1971, 50 so-called first isolations registered of which the largest portion, namely 29, originated from reptiles. Nine, however, were isolated from man. In general, the total number of serotypes isolated increases each year.

Slaughter animals

During the last 10 years intensive studies have been carried out of the incidence of salmonellae in normal animals after commercial slaughter. This includes only clinically healthy animals. A summary of the results is given in Tables 2 and 3. It may be concluded from these data that the rate of infection is still increasing.

Animals

Meat from pigs constitutes the most important source of infection for man. Pigs are infected through feeds but also from the environment, especially by insects,

Table 2

Salmonellae in slaughter animals and poultry in The Netherlands

	1960–1962		1969–1971	
	No. of samples	% Positive samples	No. of samples	% Positive samples
Pigs (bacon)	2100	25·3 (faeces 15·2)	700	30·1 (faeces 14·5)
Calves (fattening)	1920	14·5 (faeces 4·9)	1686	(faeces 5·3)
Cattle	1600	0·5	600	0·3
Broilers	1530	17·3 (faeces 4·5	820	18·7 (faeces 6·9)
Hens	770	5·3 (faeces 0·4)	740	7·0 (faeces 1·2)

Table 3

Salmonella serotypes in slaughter animals and poultry in The Netherlands

	1960–1962		1969–1971	
		(%)		(%)
Pigs	S. typhimurium	32·8	S. typhimurium	9·1
	S. bredeney	11·7	S. give	6·1
	S. panama	10·6	S. infantis	22·4
Calves	S. typhimurium	61·4	S. typhimurium	62·9
	S. panama	8·6	S. dublin	35·9
	S. dublin	5·8	S. panama	2·2
Broilers	S. typhimurium	3·8	S. infantis	36·9
	S. stanley	7·6	S. heidelberg	20·1
	S. bareilly	27·4	S. agona	17·3
Hens	S. typhimurium	36·6	S. typhimurium	32·7
	S. bareilly	14·6	S. heidelberg	26·9
	S. oranienburg	12·2	S. infantis	19·2

dust, rodents, birds or man (for instance, through footwear). The animals are practically always clinically healthy. They are carriers which shed salmonellae regularly although in small quantities.

Infection can be prevented or significantly decreased by adequate pelleting of feeds (Edel *et al.*, 1967, 1970) and by thorough measures to prevent contamination from the environment. That such measures have a positive influence is shown by the results of studies in recent years (Tables 4 and 5).

Table 4

Salmonella infection in pigs at farms with pellet feeding and at farms with meal feeding

	Pellet feeding			Meal feeding	
Number of samples	% positive samples	Number of isolated serotypes	Number of samples	% positive samples	Number of isolated serotypes
1356	0·3	3	1764	6·9	15
832	0·1	1	982	1·2	3
1523	1·3	4			
893	0·1	1			
2350	0·2	2			
total 6954	0·4	11 average 2·2	2746	4·9	18 average 9·0

Table 5

Effect of feeding pellets on the prevention and sanitation of salmonella infections in fattening pigs

		Examination after slaughter				
Type of feed	Number of pigs	Probable origin of the infection %				Total salmonella positive pigs (%)
		Juvenile infection	Environ- ment	Feed	Slaughter house	
Meal	60	0·0	45·4	9·1	0·0	54·5
Pellets	60	0·0	34·5	0·0	0·0	34·5
Pellets	120	0·0	77·8	0·0	0·0	77·8
Pellets	348	0·3	74·2	0·0	0·9	75·4
Pellets	279	0·0	0·0	0·0	0·0	0·0

Project Walcheren

In order to study the connection between infections in man, animal and environment, intensive research was carried out on the occurrence of salmonellae in a limited area, namely Walcheren, 1 of the islands in the province of Zeeland in the south-west of the Netherlands. For 1 year (February 1971 to February 1972) all slaughtered pigs were examined for salmonellae. During the period from August–October 1971, faeces from a large number of persons suffering from acute diarrhoea were also examined. In the same period, foods such as minced meat, ice-cream, fish and snacks were investigated. Also examined were flies from fly catchers in piggeries, seagull faeces and chopping-block shavings. From August 1971–February 1972, all effluent waters from sewage purification plants on the island were examined twice a week. The results of these studies are summarized in Table 6.

Table 6

Incidence of salmonellae "Project Walcheren"

Source	Number of samples		Percentage of main serotypes		Phage type of S. typhimurium	
	Examined	Percentage positive			Type	(%)
Pigs	7756	22·3	S. typhimurium	55·8	II 505	59·0
			S. derby V-	9·0	II 501	10·5
			S. infantis	7·1		
Fly catchers	202	1·5	S. typhimurium		II 505	66·7
Faeces of seagulls	60	26·7	S. typhimurium	21·1	II 501	50·0
			S. montevideo	15·7		
			S. infantis	10·2		
Choppingblock shavings	80	6·3	S. typhimurium		II 505	66·7
			S. infantis			
Effluent waters	160	93·8	S. typhimurium	43·0	II 505	61·3
			S. panama	13·0	II 501	1·4
			S. infantis	12·4		
			S. heidelberg	6·4		
Human (patients)	221	20·8	S. typhimurium	56·4	II 505	54·8
			S. panama	10·9	II 501	3·2
			S. infantis	4·3		
			S. heidelberg	4·3		

Causes of Increase in Rate of Infection

Increasing quantities of contaminated foods such as meat from South American countries and feeds of animal and vegetable origin are imported into the Netherlands. Pig feeds especially are responsible for spreading and maintaining infections in pigs. These animals contaminate with their faeces and urine the surroundings of the piggeries, the surface waters and later on the slaughter houses where contamination spreads over carcasses.

Effluent waters of sewage purification plants are nearly always contaminated. Moreover, salmonellae have been isolated regularly in these plants from compost which is used in agriculture, in parks and in private gardens, and which creates a potential danger for man and animals.

Rodents, birds and insects carry over salmonellae and contaminate feeds and foods which again closes the circle. It is clear that this circle is very difficult to break.

Prevention of Infection

The following measures are necessary to prevent the infection by salmonellae of man and animals.

Enforcement of hygienic practices in slaughter houses and meat plants.

Adequate instructions to consumers not to eat raw or insufficiently heated meat products.

Adequate kitchen hygiene.

Prevention of the importation of contaminated foods and feeds.

The most important measure may be to prevent the contamination of meat and meat products which means prevention of infection in animals.

Improvement of environmental sanitation is of the utmost importance. As contaminated surface water plays such an important part in salmonella contamination, every effort should be made to improve the hygienic quality of effluent waters and especially to reduce the number of salmonellae in such waters. An effective contribution by breaking the vicious circle should be improvement of the biological purification in the sewage plants and especially in sewage plants of bio-industries, followed then by the chlorination of effluent waters.

Conclusion

It is an illusion to believe that foods and feeds free from salmonellae can be imported in the near future. Therefore, measures should be taken wherever possible to prevent the growth and spread of salmonella bacteria. For the direct protection of the consumer, improvement of kitchen hygiene is important. Moreover, hygienic standards in the meat industry should be improved. Adequate refrigeration from producer to consumer is important. Post-contamination of heated products should be prevented by good hygienic measures.

Animals, especially pigs, should not receive contaminated feeds, and this can be prevented by adequate pelleting. Spreading of infections in animals on farms should be prevented where possible. It will not be possible in the near future to raise salmonella-free animals on a large scale due especially to the serious contamination of our environment. Therefore, the public should be informed regularly that only refrigeration and heating of foods will be helpful in disease prevention. However, it is possible to raise salmonella-free animals on a small scale. The meat from such animals would be of importance to relatively susceptible groups of people such as those in hospitals, psychiatric institutions, creches and homes for the elderly. In general, measures should be developed to break the vicious circle of salmonella infection.

References

Edel, W., Guinée, P. A. M., Schothorst, M. van. & Kampelmacher, E. H. (1967). Salmonella infections in pigs fattened with pellets and unpelleted meal. *Zentbl. VetMed.* **B14**, 393.

Edel, W., Schothorst, M. van., Guinée, P. A. M. & Kampelmacher, E. H. (1970). Effect of feeding pellets on the prevention and sanitation of salmonella infections in fattening pigs. *Zentbl. VetMet.* **B17**, 730.

Schothorst, van M., Edel, W. & Kampelmacher, E. H. (1970). Voortgezette onderzoekingen over het voorkomen van Salmonella in gehakt, in de maand juli 1965–1969. *Tabd. Diergeneesk* **95**, 279.

Discussion

Christian

Have you an explanation for the very low incidence of salmonellae in slaughtered cattle compared with that in the other animals you have studied?

Kampelmacher

Salmonellosis is in the first case an infection of the young animal (bacon, pig, calf, broiler). Cows eat, at least in the Netherlands, grass. hay and feed-cakes, which have undergone a heat treatment. The intestinal apparatus of the cow (4 stomachs) may be of great influence in reducing salmonellae.

Weitz

Will vaccination of young animals give protection.

Kampelmacher

Not in my opinion since a clinical healthy carrier will not have suffered a septicaemic infection in most cases. I have no data which proves that vaccination has an influence on the carrier state. Have you?

Weitz

No.

Riemann

Your data illustrate the importance of water as a vehicle for the spread of salmonellae in the Netherlands. Are there any practical experiences with respect to the usefulness of chlorination?

Kampelmacher

Yes, this is an important question, since one of the most promising points to break the infection cycle would be the decontamination of effluents. Results with chlorination are excellent so far. But what would chlorination of all effluents mean with regard to the surface water ecology? There are other treatments such as ozone-treatment or gamma irradiation, but to apply them over a large scale would ask for an investment which our generation cannot afford. However, experiments in these directions are currently being undertaken.

Lee

To what extent are animals in Holland exposed to contaminated surface waters? I am thinking particularly of the drinking of such waters by animals.

Kampelmacher

The drinking of water by these animals is not the problem. The problem is the contact from contaminated surface waters via insects, rats, man, etc. to the animal quarters especially piggeries.

Ingram

The percolating filters commonly used for "finishing" effluents normally breed large numbers of specific fly larvae. It would be interesting to identify these and examine them for salmonellae.

Kampelmacher

I did not know of this problem. It would surely be worth having a look into it.

Skovgaard

Your theory that effluents play an important part in spreading salmonellae is certainly important in a country with many rivers, canals, etc. But in Denmark also, the effluent is contaminated but the incidence of salmonellae in the pig population is extremely low. Is this a question of dose level in the effluent or what is your explanation?

Kampelmacher

The dose level may play a role, especially since I believe that the population and animal density is extremely high in the Netherlands. This may explain the rapid build up of infection cycles. What we see today in our country may perhaps be seen in your country if the human and animal population grows in the future.

Weitz

I can see 2 difficulties in salmonella eradication: increasing the size and concentration of farms leads to increased effluent disposal problems and the importation of foods and feeds plays a major role in continuing the cycle of infection especially in relation to the appearance of new serotypes.

The Public Health Significance of Viral Infections of Food Animals

E. P. LARKIN

Virology Branch,
Department of Health, Education and Welfare,
Cincinnati, Ohio,
U.S.A.

Exposure of humans to animals infected by viruses and the possible consumption by humans of virus-containing foodstuffs have received little attention from clinicians and public health officials. Laboratory and field data indicate that a variety of viruses that infect humans may be consistently recovered from animals. Immunological studies demonstrated that the animals were not passive carriers of viruses and were sometimes infected. Restriction of infection to closely related species and the protection endowed by the species barrier are therefore open to question. Systemic infections caused by viruses result in widespread distribution of virus in the animal. The slaughter and sale of carcasses from such animals may subject the food handler and the consumer to extensive viral contact. Limited laboratory data indicate that viruses may persist for several months in meat, milk, cheese, eggs, and shellfish. Viruses of both human and animal origin have been recovered from market foods. The small number of virus particles required to infect a susceptible individual and the extent of contamination reported implicate raw or partially cooked foods as vehicles for virus disease transmission. The significance of such viral contamination of the food supply will only be known as more data become available.

EXPOSURE OF humans to animals infected with viruses and the possible consumption by humans of virus-containing foodstuffs have received little attention from clinicians and public health officials. Large sums of money have been spent on research into viral diseases of animals. These studies emphasize chiefly the control of virus infections from the viewpoint of economics. Diseases responsible for extensive animal losses or which restricted the import or export of animal foods or food animals resulted in intensive research on methods for the control of viral infections. Studies were initiated to protect the farmer from excessive economic losses and to provide the consumer with an adequate supply of animal foods. Little concern has been shown, however, by health officials for possible human infections resulting from contact with, or consumption of, animal products infected by viruses. Available epidemiological data do not indicate that many people are being infected through the consumption of such contaminated foods. However, the probability exists that a variety of virus diseases may be traced to the food supply. The most suspect of all foods consumed are those eaten raw or partially cooked.

258   E. P. LARKIN

Viruses, unlike bacteria, do not grow or increase in numbers in food; but they do propagate in the living animal. The amount of virus produced in the animal varies and is limited by the infectious properties of the virus, the resistance of the host, and the type of infection produced. If viruses are present in or on the food, the type and number present are directly related to the disease potential of the food. The following examples show the degree of contamination that may be encountered in animal tissues: chickens with avian myeloblastosis virus infections have been shown to have titers as high as 1×10^{14} particles/ml of plasma (Beard, 1972); and lymph nodes recovered from cattle with foot-and-mouth disease virus infections have been reported to contain 1×10^5 infectious virus particles/g of tissue (Heidelbaugh & Graves, 1968).

To determine the extent and significance of viral contamination of foods certain basic information is necessary, but this information is difficult to obtain because the relationship between viral disease in animals and man is known in very few instances. This presentation, therefore, is chiefly circumstantial in nature intermixed with known scientific data

Food animals are susceptible to a wide variety of viral infections. The type of infection produced varies from subclinical to various clinical syndromes or to death in the animal. Systemic infections have great public health significance because the virus spreads throughout the various organs and tissues of the animal. Table 1 lists viruses that may produce systemic infections in the bovine. Considerable laboratory and clinical data and the infectious properties to both animals and man are available for some of the viruses listed (Andrewes & Pereira, 1967; Kniazeff, 1969). Nevertheless most viruses recovered from the bovine are of unknown significance from the viewpoints of both animal disease and their possible effects on the human consumer. Similar data are available for the porcine, ovine, avian, and other animals. A number of viruses are infectious for a variety of animals and are not restricted to 1 species. Some viruses infectious for animals are known to be infectious for humans and suspicion is increasing that additional animal viruses may be related to diseases in man.

The animal infectious dose of virus necessary to produce disease is important in discussions on viral diseases of both man and animals. The infectious doses for some viruses are shown in Table 2 (Plotkin & Katz, 1966; Craighead *et al.,* 1960). These data indicate that only a very small number of virus particles may be required to produce infection in the susceptible animal. Probably 1 virus particle is often capable of producing infection. These doses are lower than the infectious doses of bacteria that are reported to be necessary to initiate disease. For example, the minimum infectious dose of *Vibrio parahaemolyticus* for rabbits is $5 \cdot 7 \times 10^4$ bacteria/ml (Twedt, 1972).

Arboviruses have been shown to produce infections resulting in widespread distribution of the virus in the animal (Spradbrow, 1966). Table 3 lists members of the arbovirus group known to infect food animals. It is probable that many of

Table 1

Viruses producing systemic infections in the bovine

Enteroviruses (+?)
Encephalomyocarditis virus +
Foot-and-mouth disease virus +

Arboviruses +

Parainfluenza 3

Reovirus 1, 2 and 3 +

Bovine diarrhoea virus
BVD mucosal disease
Vesicular stomatitis virus +
Rabies virus +

Porcine herpes virus +
Malignant catarrhal fever virus

3 Adenovirus serotypes (?)

Bovine syncytial virus (?)
Bovine leukemia virus (?)

+, Infectious for man; ?, systemic infection probable; (+?), systemic infection and infectious for man probable.

Table 2

Minimal infectious dose of viruses for animals

Virus	Route	Dose*
Yellow fever, monkey	Respiratory	<6 MLD_{50}
Rift valley fever, monkey	Respiratory	<6 MLD_{50}
Rift Valley fever, hamsters	Respiratory	0.5 MLD_{50}
Monkey B, rabbit	Respiratory	<100 TCD_{50}
Monkey B, monkey	Respiratory	<250 TCD_{50}
Newcastle disease, chicken	Respiratory	1 EID_{50}
Infectious bronchitis, chicken	Respiratory	1 EID_{50}
Parainfluenza 3, hamster	Respiratory	1 TCD_{50}
Canine hepatitis, dog	Oral	<6 TCD_{50}
Foot-and-mouth disease, cattle	Tongue	±1 TCD_{50}

* MLD, mouse lethal dose; TCD, tissue culture dose; EID, egg infectious dose.

Table 3

Arbovirus infections of food animals*

Group	Virus	Natural infection (disease)	Natural (serological evidence)	Experimental inoculation (clinical disease)
A	Equine encephalitis			
	Eastern	PH		PH, C
	Western		P	P, C, G
	Venezuela		P	S, G
	Sindbis		C, S, G, B	
	Semliki		C, G, S	
	Middelburg		C, S	S
	Sagiyama		P	
B	West Nile		C, S, G, B	
	Japanese B encephalitis	P	C, S, P, G, B	P, C
	Yellow fever		S, C	
	Louping ill	S, C		S, C, P
	Nairobi	S, G		S, G
	Tick-borne encephalitis in cattle	S, G, P, C		S
	Turkey meningo- encephalitis	T		T
	Wesselsbron	S	C	S, C, P
Ungrouped	Rift Valley fever	S, C		G, S, C
	Bluetongue	S, C, G		S
	Ephemeral fever	C		C
	Near east encephalitis	S, C		S

* P, pig; C, cattle; S, sheep; G, goat; PH, pheasant; T, turkey; B, buffalo.

these viruses will infect an even wider variety of animals but data are not available because of the lack of studies on food animals. Examples of the possible extensive distribution of the virus in organs and tissues of animals are shown in Table 4. Virus was consistently found in most of the samples regardless of the route of inoculation (Eklund & Kennedy, 1962; Gresikova et al., 1962; Pogodina, 1962).

Additional data implicating the food animal in the viral-disease cycle are shown in Table 5. They indicate that animal secretions and discharges are potential sources of infectious virus and may be an important contact route in the spread of disease to other animals or man (Wedum & Kruse, 1969). Some of the diseases listed are usually considered to be restricted to particular sites in the animal and thus isolation from the urine and faeces would not be expected. An increased number of viruses may be found in animal discharges when the aetiology of viral diseases of animals is investigated more extensively.

Table 4

Distribution of virus in organs and tissues of infected animals

Virus*	TBE	TBE	TBE	CTF	LI
Animal	Mouse	Rabbit	Mouse	Mouse	Lamb
Inoculation	IC	IC	Oral	IP	SC
Blood	+	+	+	+	+
Lymph nodes	+	+			+
Spleen	+	+		+	+
Lungs	+	−		+	
Heart	+	−		+	
Liver	+	−	+	+	+
Kidney	+	−		+	
Adrenals	+	−			
Muscle	+	−		+	
Fat	+	−			
Intestine	+	−	+		
Brain	+	−		+	+
Urine			+		
Faeces			+		

* TBE, Tick-borne encephalitis; CTF, Colorado tick fever; LI; louping ill.

The discovery of animal viruses in or on foods is probably considered to be of minor importance to human health because of the protection afforded by the species barrier, but the isolation from food of a virus known to cause disease in the human is of obvious significance. Recently human viruses have been demonstrated in food animals including calves, cows, goats and swine, as well as in other domestic animals such as the horse, dog and cat (Abinanti, 1961; Kalter, 1965; Prier & Riles, 1965; Binn, 1970; Grew *et al.*, 1970; Kiseleva & Berdyliev, 1971). These isolates are listed in Table 6. The same viruses were sometimes isolated from human contacts living in close association with the animals (Grew *et al.*, 1970). Antibody in the blood of some of the animals indicated that infections occurred (Binn, 1970). This extensive reservoir of human viruses in the animal leads one to doubt the effectiveness, or even the existence. of a species barrier and indicates that animal viruses may be agents of subclinical infections or disease in man.

Evidence has accumulated implicating viruses as the probable agents of neoplastic diseases in man and animals. More than 100 different viruses are known to produce cancers of varying types in all the major animal groups. Many of the viruses will replicate and transform human cells in tissue culture (Rauscher, 1971). In Table 7 various neoplasms of food animals and their related viral agents are listed. Viruses of the avian leukosis complex may be present in all sexually mature birds and in most eggs produced by infected hens (Luginbuhl, 1969; Burmester & Purchase, 1970). Most isolations of viruses from fowl indicate that 2 or more strains of the avian leukosis complex are present.

Table 5

Recovery of viruses from urine and faeces of infected food animals

Causative virus or disease	Animal	Urine*	Faeces*
Adenovirus	Cattle	★	+
	Chicken	★	+
	Swine	★	+
African swine fever	Swine	+	+
Avian lymphomatosis	Chicken	★	+
Coxsackie A	Cattle	★	+
Coxsackie B	Cattle	★	+
Eastern equine encephalitis	Chicken	★	+
	Pheasant	★	+
	Rabbit	★	+
Enterovirus			
Avian	Chicken	★	+
Bovine	Cattle	★	+
Swine	Swine	★	+
Foot-and-mouth disease	Cattle	+	+
	Chicken	★	+
	Swine	+	+
Marek's disease	Chicken	★	+
Newcastle virus	Chicken	★	+
	Duck	★	+
	Goose	★	+
	Turkey	★	+
Reovirus	Cattle	★	+
Rinderpest	Cattle	+	+
	Goat	★	+
	Sheep	★	+
Teschen	Swine	★	+
Western equine encephalitis	Chicken	★	+

* +, Virus recovered; ★ no reported data.

Recently a turkey herpesvirus was shown to protect poultry from Marek's disease (Witter, 1971). The method of protection is unknown, but it is probably due to viral interference. Both the Marek and turkey herpesviruses are present in a viable state in the inoculated bird, and a viremia appears to persist for the life of the animal (Witter *et al.*, 1970). A high percentage of chickens is now inoculated with live virus vaccine shortly after hatching. Exposure of the human population to the high concentration of oncogenic avian virus that results from present practices in the poultry industry may be of public health significance.

Viruses of the "C" type have been demonstrated in lymphocytes recovered from the blood and milk of normal and leukemic cattle (Miller *et al.*, 1969). Similar viruses were shown to be present in the blood, organs and mammary glands of leukemic cattle (Nazerian *et al.*, 1968), and virus-like particles were found in high concentrations in milk from these animals (Dutcher *et al.*, 1964).

Table 6

Human viruses isolated from domestic animals *

Virus type		Animal	Infection produced (clinical or serological evidence)
Poliovirus 1		C, D, G, H	C
Coxsackie virus	A5	P	
	A6	D	
	A9	D	
	A10	P	
	A20	D, P	P
	B3	D	D
	B5	D	D
Echo virus 2		D	
6		D, F	D
8		P	
10		C	
19		D, F	F
Reovirus 1, 2, 3		C, D, F	C, D, F
Influenza virus	A2	P, F	P, F
Mumps virus		D	D

* C, Cattle; D, dog; F, feline (cat); G, goat; H, horse; P, pig.

Table 7

Neoplasms in food animals

Food source		Causative agent*
Bovine	Lymphosarcoma	(?)
	Leukemia, leukosis	("C" Type virus)
Avian	Leukosis complex,	"C" Type virus
	Marek's disease	Herpes virus
Pisces	Lymphosarcoma, tumors	(?)
	Carcinoma, etc.	("C" Type virus-pike)
Amphibia	Tumors	Herpes virus
	Liposarcoma	(?)
Mollusca (blue mussel, and pacific oyster)	Leukemia-like disease	(?)
Crustacea (blue crab)	Hemocytic disease	(?)

* (?) Agent unknown; ( ) detected by electron microscopy.

"C" type virus particles have also been demonstrated in leukemic pigs (Frazier *et al.*, 1970).

The high incidence of neoplasms in pike harvested from the North Atlantic (Mulcahy, 1970), and reports of neoplastic disease in freshwater fish in some lakes and streams in the United States (Brown *et al.*, 1972) have caused concern with regard to their public health significance. In addition various tumors and sarcomas have been reported in amphibia (Dawe. 1970) and an extensive outbreak of a leukemia-like disease was reported in the blue mussel and the Pacific oyster off the west coast of the United States (Farley & Sparks, 1970). A haemocytic disease of high incidence was reported in blue crabs harvested from the waters of the Atlantic ocean adjacent to the southern coastal areas of the United States (Newman, 1970).

A recent discovery of enzymes associated with the core of more than 30 viruses known to cause leukemia and sarcomas in chicken, mouse and feline (Todaro, 1971) is of special interest. The polymerase enzymes detected in all leukemia and sarcoma viruses possess similar reactivity and are associated with the synthesis of viral DNA. Similar enzymes have been found in lymphocytes from human leukemia patients, but not in the lymphocytes of normal patients. The ability of some of the leukosis and leukemia viruses to infect cells and animals of various species and the presence of viruses in the tissues, eggs, blood and milk of animals have caused concern about the possibility of adverse effects on humans consuming such food, and the possible role of these particles in human leukemia and cancer.

Human and animal viruses have been isolated from food and food animals (Table 8). The foods include raw and heated milk (Kawakami *et al.*, 1966; Ernek *et al.*, 1968; Gresikova, 1968; Malmquist *et al.*, 1969; Sullivan *et al.*, 1969; Hedger & Dawson, 1970; Gresikova. 1972), butter (Gresikova, 1972; Hedger & Dawson, 1970), beef (Dimopoullos, 1960; Sullivan *et al.*, 1970), oysters (Metcalf & Stiles, 1968; Fugate, 1972; Metcalf, 1972), mussels (Petrilli & Crovari, 1965; Bellelli & Leogrande. 1967; Bendinelli & Ruschi, 1969), and eggs (Delay, 1947; Rubin *et al.*, 1961; Cook, 1971). Only limited viral isolations have been reported, these by only a few laboratories throughout the world. The data presented here clearly demonstrate that viruses of both human and animal origin are present in our foods. The isolation of known human viruses is of obvious public health significance. The finding of parainfluenza 3, the bovine syncytial virus, infectious bronchitis virus, Newcastle disease virus, and the viruses of the avian leukosis complex indicates the great likelihood that other animal viruses are also present.

To determine the extent of the viral contamination of foods and the public health significance of the viruses present in the food supply, additional research is needed in the following areas: (a) the incidence of viral disease in food animals

Table 8

Viral isolates from foods

Encephalitis (tick-borne)	Milk, butter (goat)
Encephalitis (tick-borne)	Milk (cow)
Encephalitis (tick-borne)	Milk (sheep)
Poliovirus 1 and 3	Raw milk (cow)
Parainfluenza 3	Raw milk (cow)
Bovine syncytial virus	Raw milk (cow)
Foot-and-mouth disease virus	Beef, milk, butter (cow)
Poliovirus 1, 2 and 3	Ground beef
Echovirus 6	
Poliovirus 1, Coxsackie virus B2, 3 and 4	Oysters
Echovirus 9	
Echo virus 4 and polio virus 1	Oysters
Poliovirus 1, 2 and 3,	
Coxsackie virus B3, Echovirus 6,	
Reovirus 1, Adenovirus 5,	
Hepatitis candidate virus AR-17	Oysters
Echo virus 5, 6, 8 and 12	
Coxsackie virus A-18	Mussels
Poliovirus 3	Mussels
Echo virus 3, 9 and 13	Mussels
Avian leukosis complex	Eggs
Infectious bronchitis virus	Eggs
Newcastle disease virus	Eggs

and the distribution of virus in the infected animal carcase; (b) the ability of virus to persist in animal tissues and organs after slaughter, storage and distribution; (c) the extent of cross contamination of foods during handling by the food processor or consumer; (d) the infectivity of the virus to the food handler; (e) the ability of the virus to survive the final food processing in the home or food establishment; (f) the infectivity of the virus to the consumer after entry via the oral route; (g) the minimum infectious dose of "animal" viruses for humans; and (h) the development of accurate methods for detecting virus in foods.

These are subjects of increasing importance. Human viruses have been shown to infect domestic animals, but only limited data are available pertaining to the infectivity of animal viruses for humans. There has been a limited number of isolations of viruses from market foods. Methods are now being developed which should improve the rates of isolation of viral contaminants. Epidemiological data indicate that viral diseases of man and animals may be transmitted via the food vehicle. The significance of viruses in foodborne disease will be revealed only when more scientists initiate studies of the extent and type of viral contamination present in the world food supply.

Recommendations

Limited viral isolations have been made from market foods. Improved methods are now being developed which should result in the isolation of additional viral contaminants.

Epidemiological data indicate that viral diseases of man and animals may be transmitted via the food vehicle. How significant a role viruses play in foodborne disease will be revealed only as more interested scientists initiate studies to determine the extent and type of viral contamination present in the world food supply.

References

Abinanti, F. R. (1961). Respiratory and enteric viruses in man and animals. *Pub. Hlth. Rep., Wash.* **76**, 897.

Andrewes, C. & Pereira, H. G. (1967). *Viruses of Vertebrates.* 2nd ed. London: Baillier, Tindall and Cassell.

Beard, J. W. (1972). Personal communication.

Bellelli, M. & Leogrande, G. (1967). Bacteriological and virological research on aquatic animals. *Ann. Sclavo.* **9**, 820.

Bendinelli, M. & Ruschi, A. (1969). Aspects of the pollution of marine waters with isolation of human enterovirus from mussels. *Appl. Microbiol.* **18**, 531.

Brown, E. R., Keith, L. Kwapinski, J. B. G. & Hazdra, J. (1972). Incidence of fish tumors found in a polluted watershed as compared to nonpolluted Canadian waters. *Proc. Am. Ass. Cancer Res.* **13**, 45.

Binn, L. N. (1970). A review of viruses recovered from dogs. *J. Am. vet. med. Ass.* **156**, 1672.

Burmester, B. R. & Purchase, H. G. (1970). Occurrence. transmission and oncogenic spectrum of the avian leukosis viruses. In *Comparative Leukemia Research 1969*, p. 83. (*Bibl. haemat.*, No. 36.) Ed. R. M. Dutcher. Basel/New York: S. Karger.

Craighead, J., Cook, M. & Chanock, R. (1960). Infection of hamsters with Parainfluenza 3 virus. *Proc. Soc. exp. Biol. Med.* **104**, 301.

Cook. J. K. A. (1971). Recovery of infectious bronchitis virus from eggs and chicks produced by experimentally inoculated hens. *J. Comp. Path. Ther.* **81**, 203.

Delay, P. D. (1947). Isolation of avian pneumoencephalitis (Newcastle disease) virus from the yolk sac of 4-day old chicks, embryos, and infertile eggs. *Science N.Y.* **106**, 545.

Dawe, C. J. (1970). Neoplasms of blood cell origin in poikilothermic animals—a status summary. In *Comparative Leukemia Research 1969.* (*Bibl. haemat.*, No. 36.) Ed. R. M. Dutcher. Basel/New York: S. Karger.

Dimopoullos, G. T. (1960). Effects of physical environment on the virus of foot-and-mouth disease. *Ann. N.Y. Acad. Sci.* **83**, 706.

Dutcher, R. M., Larkin, E. P. & Marshak, R. R. (1964). Virus-like particles in cow's milk from a herd with a high incidence of lymphosarcoma. *J. natn. Cancer Inst.* **33**, 1055.

Eklund, C. M. & Kennedy, R. C. (1962). Preliminary studies of pathogenesis of Colorado tick fever virus infection of mice. In *Biology of Viruses of the Tick-borne Encephalitis Complex.* Ed. H. Libíková. New York: Academic Press.

Ernek, E., Kozuch, O. & Nosek, J. (1968). Isolation of tick-borne encephalitis virus from blood and milk of goats grazing in the Tribec Focus Zone. *J. Hyg. Epidem. Microbiol. Immun.* **12**, 32.

Farley, C. A. & Sparks, A. K. (1970). Proliferative diseases of hemocytes, endothelial cells, and corrective tissue cells in mollusks. *Comparative Leukemia Research 1969.* (*Bibl. haemat.* No. 36.) Ed. R. M. Dutcher. Basel/New York: S. Karger.

Frazier, M. E., Ushijima, R. N. & Howard, E. B. (1970). Virus association with ^{90}Sr induced leukemia of miniature swine. In *Comparative Leukemia Research 1969. (Bibl. haemat.* No. 36.) Ed. R. M. Dutcher. Basel/New York: S. Karger.

Fugate, K. (1972). In *International Virology* 2, p. 321. (Proc. 11 int. Congr. Virol., Budapest 1971.) Ed. J. L. Melnick. Basel/New York: S. Karger.

Gresikova, M. (1972). In *International Virology* 2, p. 322. (Proc. 11 int. Congr. Virol., Budapest 1971.) Ed. J. L. Melnick. Basel/New York: S. Karger.

Gresikova, M. (1968). Excretion of tick-borne encephalitis virus in the milk of subcutaneously infected cows. *Acta virol. Prague* 2, 188.

Gresikova, M., Albrecht, P. & Ernek, E. (1962). Study on attenuated virulent louping ill virus. *Biology of Viruses of the Tick-borne Encephalitis Complex,* p. 294. Ed. H. Libíková. New York: Academic Press.

Grew, N., Gohd, R. S., Arguedas, J. & Kato, J. I. (1970). Enteroviruses in rural families and their domestic animals. *Am. J. Epidem.* 91, 518.

Hedger, R. S. & Dawson, P. S. (1970). Foot-and-mouth disease virus in milk: an epidemiological study. *Vet. Rec.* 87, 186.

Heidelbaugh, N. D. & Graves, J. H. (1968). Effects of some techniques applicable in food processing on the infectivity of foot-and-mouth disease virus. *Fd Technol. Champaign.* 22, 120.

Kalter S. S. (1965). Picornaviruses in water. *Transmission of Viruses by the Water Route,* p. 253. Ed. G. Berg. New York: J. Wiley.

Kawakami, Y., Kaji, T., Kume, T., Omuro, M., Hiramune, T., Murase, N. & Matumoto, M. (1966). Infection of cattle with parainfluenza 3 virus with special reference to other infections. I. Virus isolation from milk. *Jap. J. Microbiol.* 10, 159.

Kiseleva, N. V. & Berdyliev, O. M. (1971). Domestic animals (dogs and cats) as carriers of human intestinal viruses. *Zh. Microbiol. Epidem. Immunobiol.* 48, 102.

Kniazeff, A. J. (1969). Viruses infecting cattle and their role as endogenous contaminants of cell cultures. Cell cultures for virus vaccine production. Monograph No. 29, Natl. Cancer Inst., Bethesda, Maryland.

Luginbuhl, R. E. (1969). Viral flora of chick and duck tissue sources. Cell cultures for virus vaccine production. Monograph No. 29, Natl. Cancer Inst., Bethesda, Maryland.

Malmquist, W. A., Van der Maaten, M. J. & Boothe, A. D. (1969). Isolation, immunodiffusion, immunofluorescence, and electron microscopy of a syncytial virus of lymphosarcomatous and apparently normal cattle. *Cancer Res.* 29, 188.

Metcalf, T. G. & Stiles, W. C. (1968). Enteroviruses within an estuarine environment. *Am. J. Epidem.* 88, 379.

Metcalf, T. G. (1972). *International Virology* 2, p. 321. (Proc. 11 int. Congr. Virol., Budapest 1971.) Ed. J. L. Melnick. Basel/New York: S. Karger.

Miller, J. M., Miller, L. D., Olson, C. & Gillette, K. G. (1969). Virus-like particles in phytohemagglutinin-stimulated cultures with reference to bovine lymphosarcoma. *J. natn. Cancer Inst.* 43, 1297.

Mulcahy, M. (1970). Hemic neoplasms in cold blooded animals: lymphosarcoma in the pike *Eso lucius,* p. 644. In *Comparative Leukemia Research 1969. (Bibl. haemat.* No. 36.) Ed. R. M. Dutcher. Basel/New York: S. Karger.

Nazerian, K., Dutcher, R. M., Larkin, E. P., Tumilowicz, J. J. & Eusebio, C. P. (1968). Electron microscopy of virus-like particles found in bovine leukemia. *Am. J. vet. Res.* 29, 387.

Newman. M. W. (1970). A possible neoplastic blood disease of blue crabs. In *Comparative Leukemia Research 1969,* p. 648. (*Bibl. haemat.* No. 36.) Ed. R. M. Dutcher. Basel/New York: S. Karger.

Petrilli, F. & Crovari, P. (1965). Aspects of the pollution of marine waters with particular regard to the situation in Liguria. *C. Ig. Med. Prevent.* 8, 269.

Plotkin, S. A. & Katz, M. (1966). Minimal infective doses of viruses for man by the oral route. In *Transmission of Viruses by the Water Route,* p. 151. Ed. G. Berg. New York: J. Wiley.

Pogodina, V. V. (1962). The course of alimentary infection and development of immunity in tick-borne encephalitis. *Biology of Viruses of the Tick-borne Encephalitis Complex,* p. 275. Ed. H. Libikova. New York: Academic Press.

Prier, J. E. & Riley, R. (1966). Significance of water in natural animal virus transmission. In *Transmission of Viruses by the Water Route,* p. 287. Ed. G. Berg. New York: J. Wiley.

Rauscher, F. J., Jr. (1971). Major opportunities for determination of etiologies and prevention of cancers in man. In *Recent Advances in Human Tumor Virology and Immunology,* p. 3. Ed. W. Nakahara *et al.* Baltimore: Univ. Park Press, orig. publ. by Univ. of Tokyo Press.

Rubin, H., Cornelius, A. & Fanshur, L. (1961). The pattern of congenital transmission of an avian leukosis virus. *Proc. natn. Acad. Sci. USA* **47,** 1058.

Spradbrow, P. (1966). Arbovirus infections of domestic animals. *Vet. Bull., Weybridge* **36,** 55.

Sullivan, R., Fassolitis, A. C. & Read, R. B., Jr. (1969). Isolation of viruses from raw milk, p. 879. Abs. of paper presented at 64th Annual Mtg. of American Dairy Science Association, 22.6.69.

Sullivan, R., Fassolitis, A. C. Read, R. B., Jr. (1970). Method for isolating viruses from ground beef. *J. Fd Sci.* **35,** 624.

Todaro, G. J. (1971). RNA dependent DNA polymerases in RNA viruses, tumor cells and normal cells. *Proc. Fifth Internat. Sym. on Comparative Leukemia Res.* (In press.) Basel/New York: S. Karger.

Twedt, R. M. (1972). Personal communication.

Wedum, A. G. & Kruse, R. H. (1969). *Assessment of risk of human infection in the microbiological laboratory* (2nd ed.). Misc. Pub. 30, Dept. of the Army, Ft. Detrick, Md.

Witter, R. L. (1971). Marek's disease research—history and perspectives. *Poult. Sci.* **50,** 333.

Witter, R. L., Solomon, J. J., Champion, L. R. & Nazerian, K. (1970). Long-term studies of Marek's disease infection in individual chickens. *Avian Dis.* **14,** 346.

Discussion

Harrigan

How useful is epidemiological evidence of a circumstantial nature in incriminating foods as a vector in transmission of virus diseases? For example, what is the significance of the report of a few years ago that the distribution in the U.S.A. of areas of high incidence of human leukaemia could be correlated quite well with the marketing areas for bovine milk?

Larkin

Epidemiological data are definitely very useful in incriminating viruses as causative agents in food-borne disease. A good example is in the case of infectious hepatitis where foods have been incriminated from the epidemiological data alone. In regards to correlation between consumption of bovine milk and human leukaemia the epidemiological data is insufficient to statistically show that such a relationship actually exists. However, several European workers believe that there may be a correlation and eradication programmes in several countries of herds with bovine leukosis may produce an answer within the next several years. However, at the present time no information is available to demonstrate that human leukaemia and consumption of leukotic animals or their products are in any way interrelated.

Brachman

Epidemiological investigations, properly conceived and conducted, should be able to stand by themselves. The epidemiologist should be able to define the epidemic, develop the clinical spectrum of disease, identify the mode of transmission, the vehicle of infection, calculate the attack rate, predict the future course of the disease and outline control measures. The agent may be known or unknown before the laboratory confirms the epidemiologists findings. The laboratory performs valuable assistance to the epidemiologist. John Snow has demonstrated the extent to which the epidemiologist can practice his profession without laboratory identification of the aetiologic agent.

Riemann

Are all the viruses which have been isolated so far from foods destroyed by heating?

Larkin

Probably the viruses found in foods will be inactivated by pasteurization times and temperatures. There are reports in the literature that tick-borne encephalitis is slightly more resistant and may persist at these temperatures and times. There are also reports that infectious hepatitis will withstand boiling water but I have doubts as to whether this is true.

Rehm

What is the real danger for man when cattle are infected with rabies virus?

Larkin

The danger is quite evident if the individual has an open cut on his hand or arm and attempts to relieve the bovine which has a constriction of the throat. There has been an outbreak in cattle in Kentucky, U.S.A. recently, involving c. 25–30 animals. There have been no recorded cases of human rabies contracted from the cattle.

Reinius

The food virology programme of W.H.O. involves collection of data on detection methods, identification of viruses, persistence of viruses in foods and of the epidemiology of food-borne diseases due to viruses.

Elliott

Do you know of any work that would indicate the times and temperatures necessary to destroy foot and mouth disease virus in meats, both cured and uncured?

Larkin

Data is available in the literature on this subject. I don't remember the actual times and temperatures required but I believe that Hedger & Dawson (*Vet. Rec.* 83, 186) found that the HTST pasteurization parameters were sufficient to inactivate the virus in milk.

Wood

Unpasteurized milk is an established viral vector and several common viruses have been associated with aseptic meningitis in children. Do you think that it is possible that Prof. Sinell's observations concerning *Str. agalactiae* in unpasteurized milk and infant meningitis could be explained by a viral agent?

Larkin

A number of viruses discussed have been shown to cause aseptic meningitis. If the patient observed had aseptic meningitis it could have been produced as you suggest.

Simmons

Firstly I don't believe that meningitis caused by *Str. agalactiae* would be confused with a viral meningitis. The reaction in the cerebrospinal fluid would be quite different in the 2 conditions. In a bacterial infection the CSF would contain polymorphs and the glucose would be reduced; in a virus infection the

CSF would contain lymphocytes and the glucose concentration would remain within normal limits. One example of a virus transmitted from food animals to man is the Newcastle disease virus which I understand is quite a problem in broiler workers.

Ingram

If the viruses concerned are common in animals and their products it would seem useful to make epidemiological investigations among humans connected with animals, e.g. in slaughterhouses (cf. *Brucella*), or in foot-and-mouth disease situations. Is it the case that there are unusual viral problems among such people? Until this is established it seems wiser not to put undue effort into other aspects of the question.

Larkin

The subject of my paper was restricted to the significance of viral infections of food animals. If you consider the virus coming from secondary sources such as: contaminated water, insect and rodent faeces and urine, the human food handler, etc., you enlarge the sources of contamination that must be considered. Epidemiological studies would be quite useful especially serological ones, that would determine whether antibody was present in the human to a number of known animal viruses. It is essential that laboratories develop the competence to recover viruses from foods otherwise we will be restricted to circumstantial evidence which may or may not give us the proof we need. If the virus is isolated from the food the question is solved.

Riemann

I agree with Prof. Ingram and I suggest the group of food-borne diseases with unknown causes as a natural area for investigation of the possible involvement of viruses.

Larkin

We have plans to attempt to isolate viruses from foods involved in food-borne disease outbreaks. However, at present we are developing and comparing different methods for efficiency and simplicity. We have used these methods to show that viruses are present in our foods. What we need is more laboratories working in this field to provide more data as to the extent and type of viral contamination present in the food supply.

Telford

In the field of Public Health considerable emphasis is placed on keeping food premises free from rats, mice and cockroaches. No speaker this afternoon has dealt with the part these play in the conveyance of disease. Is it of bacterial or viral origin or both?

Session 3

Special Laboratory Techniques

Chairman: Dr Betty C. Hobbs

Staphylococcal Food Poisoning with Special Reference to the Detection of Enterotoxin in Food

R. J. GILBERT AND ANTONETTE A. WIENEKE

Food Hygiene Laboratory,
Central Public Health Laboratory,
Colindale Avenue,
London NW9 5HT, England

Food poisoning caused by enterotoxigenic staphylococci continues to be a problem in this and many other countries. Cultures of *Staphylococcus aureus* from 94 food poisoning incidents in the United Kingdom have been examined for their ability to produce enterotoxins A, B, C, D and E. Enterotoxin was produced by means of a sac culture technique and detected serologically by the slide gel double-diffusion method. Representative cultures from 89 of the 94 incidents (95%) produced enterotoxins A, B, C, D or E: 76 (81%) of the cultures produced enterotoxins A, D or both A and D.

Two methods which have been used in this laboratory for the extraction of enterotoxin from foods implicated in staphylococcal food poisoning are described. Enterotoxin has been detected serologically in 16 of 29 food samples from 25 separate incidents: 11 samples contained enterotoxin A, 3 contained D, 1 both A and B, and 1 both A and D. The method in current use for examining food can be divided into 4 steps: (i) separation of the enterotoxin from insoluble food constituents; (ii) separation from soluble constituents using a carboxy-methylcellulose column; (iii) concentration of the eluted extract to 0·2 ml; and (iv) examination of this extract by means of gel-diffusion.

Introduction

STAPHYLOCOCCAL FOOD poisoning is an acute gastro-enteritis resulting from the ingestion of food containing the enterotoxin formed by certain strains of *Staphylococcus aureus*. At least 5 serologically distinct enterotoxins, A, B, C, D and E, are now recognized and all have been implicated in food poisoning outbreaks.

Since 1949 an attempt has been made by the Public Health Laboratory Service to obtain as much information as possible about all outbreaks and sporadic cases of food poisoning in England and Wales. In the period 1949–1971 over 2000 incidents of staphylococcal food poisoning were reported. Although the figures for the total number of cases are incomplete they probably total 15–18,000. Table 1 gives the number of reported incidents and cases for the period 1962–1971. Staphylococcal food poisoning accounts for about 2% of all reported food poisoning incidents and about 5% of the reported cases in England and Wales. In other countries, however, staphylococcal food poisoning is more

Table 1

Staphylococcal food poisoning in England and Wales (1962–1971)

Year	Number of reported incidents*	Number of reported cases	References
1962	143	629	Report (1963)
1963	74	214	Report (1964)
1964	107	426	Vernon (1965)
1965	74	899	Vernon (1966)
1966	54	262	Vernon (1967)
1967	29	461	Vernon (1969)
1968	63	309	Vernon (1970)
1969	c. 46	c. 500	+
1970	c. 55	c. 560	+
1971	c. 16	c. 330	+

* General outbreaks, family outbreaks and sporadic cases; + Cumulative totals from published data in the Annual Report of the Chief Medical Officer of Health and unpublished data from the Epidemiological Research and Food Hygiene Laboratories.

common and in some, such as the U.S.A. (World Health Organization, 1969; Brachman *et al.*, 1973), New Zealand (McDougall, 1970) and Hungary (Ormay, 1971), it is responsible for more outbreaks than either *Salmonella* spp. or *Clostridium welchii*.

Table 2 shows the foods responsible for 175 outbreaks of staphylococcal food poisoning in England and Wales. Cold meats, including cured products such as ham, tongue and corned beef, and cooked poultry were the foods most frequently implicated but some outbreaks were associated with dairy products, fish, cooked seafoods and canned vegetables.

The main mode of infection of food by staphylococci is from the hands of the food handler or from his nose or septic lesions via his hands to the food, followed by storage of the food at temperatures which encourage multiplication of the organisms and production of enterotoxin. The cow is another source of enterotoxin-producing staphylococci and raw milk and raw milk products such as cream and cheese have given rise to outbreaks in many countries.

Bacteriological evidence is essential for confirmation of outbreaks of staphylococcal food poisoning. In most outbreaks, large numbers of *Staph. aureus* can be isolated from suspected foods and from faecal specimens; counts of *Staph. aureus* in foods from 39 incidents varied from $7\cdot5 \times 10^5$ to 9×10^9/g (Gilbert *et al.*, 1972). Phage-typing enables strains isolated from victims, suspected foods and suspected food handlers to be correlated. Most staphylococci implicated in food poisoning are lysed by phages of group III or groups I and III and most are resistant to penicillin. Enterotoxin-producing coagulase-negative staphylococci have been implicated in 2 outbreaks (Omori & Kato, 1959; Breckinridge & Bergdoll, 1971).

Table 2

Foods responsible for 175 outbreaks of staphylococcal food poisoning (1962-1971)

Food responsible	Number of outbreaks	Total	%
Meat and poultry			
Cold meat—beef, lamb, pork	6		
Cold poultry—chicken, turkey	21		
Reheated meat	4		
Reheated poultry	3		
Ham and boiled bacon	33		
Canned pork products	6		
Corned beef	13	128	73
Pressed beef	1		
Tongue	15		
Meat pies—various	15		
Sausages and sausage rolls	5		
Potted meat, paté, meat paste, brawn and rissoles	6		
Dairy products			
Trifle, cream cakes and custard	15		
Cheese	2		
Milk—raw	1	20	
Milk jelly	1		
Butter sauce	1		
Fish and seafoods			
Canned fish	6		
Processed and made-up fish	6	17	27
Frozen prawns	4		
Canned prawns	1		
Vegetables			
Canned vegetables, e.g. peas	7	8	
Potato	1		
Other foods			
Egg rolls	1	2	
Hollandaise sauce	1		

The staphylococcal enterotoxins have been purified by Professor M. S. Bergdoll and colleagues at the Food Research Institute, University of Wisconsin and the late Dr E. P. Casman and colleagues at the Food and Drug Administration, Washington, D.C. Demonstration of their antigenicity has made it possible to use antigen-antibody tests for the serological detection of enterotoxin in culture supernatants. Serological methods are more sensitive and less expensive than methods involving the feeding of monkeys or intraperitoneal injection into kittens. Several methods for detecting enterotoxin have been developed, including immunofluorescence, haemagglutination inhibition and reversed passive haemagglutination, but the most widely used techniques are those employing gel-diffusion. Efforts to detect enterotoxin in foods have been concentrated primarily on procedures for extracting and concentrating the enterotoxin from foods. Techniques found useful include acid precipitation and chloroform extraction, specifically adapted for milk and cheese (Read, Bradshaw, Pritchard and Black, 1965), gel filtration through Sephadex G-100 (Casman & Bennett, 1965), chromatography with Amberlite CG 50 ion exchange resin (Hall *et al.*, 1965) and chromatography with carboxymethyl-cellulose (Casman & Bennett, 1965; Zehren & Zehren, 1968).

Comprehensive reviews on the enterotoxins and on staphylococcal food poisoning have been prepared by Bergdoll (1967, 1970), Casman (1967), Angelotti (1969) and Minor & Marth (1971, 1972).

The experimental data outlined below is a combination of published (Simkovicova & Gilbert, 1971; Gilbert *et al.*, 1972) and unpublished work from this laboratory.

Materials and Methods

Cultures

Three hundred and twenty one cultures of *Staph. aureus* were received from public health and hospital laboratories. They were obtained from 94 food poisoning incidents during the years 1953–1972: for each incident there was good clinical and bacteriological evidence that staphylococcal food poisoning had occurred. *Staph. aureus* had been isolated, usually in large numbers, from 1 or more foods and from faecal specimens. In many cases isolations had also been made from specimens of vomit, from swabs of food equipment and from the hands and nose of food handlers. All the strains examined were coagulase-positive.

Foods

Twenty-nine foods including samples of cooked meat and poultry, canned and frozen cooked prawns, cream cakes, trifle and cheese from 25 of the food poisoning incidents were submitted for enterotoxin tests.

Phage-typing

Cultures were phage-typed in the Cross Infection Reference Laboratory at Colindale using the international set of 22 typing phages (Report, 1967).

Biochemical and antibiotic sensitivity tests

A representative strain from 92 of the incidents was tested for deoxyribo-nuclease activity on DNase agar (Oxoid), phosphatase activity on phenolphthalein phosphate agar and colonial pigmentation at $37°$ on cream agar (O'Connor *et al.*, 1966). Sensitivities to antibiotics were determined by a disk-diffusion method with Multodisks (Oxoid, code No. 11-14C).

Control enterotoxins and antisera

Enterotoxins A, B, C, D and E, all partially purified, and their specific antisera were kindly supplied by Professor M. S. Bergdoll (A, B, C and E) and Dr R. W. Bennett (D).

Production of enterotoxin by cultures of Staph. aureus

Cultures were tested for their ability to produce enterotoxins A, B, C, D and E in double-strength brain heart infusion (Difco) by means of a sac-culture technique (Donnelly *et al.*, 1967).

Extraction of enterotoxin from foods

Two methods were used for the extraction of enterotoxin, the first for 17 foods submitted during 1969 and 1970 and the second for 12 foods submitted from 1971–July 1972. Both methods have been described in detail (Gilbert *et al.*, 1972) but a schematic presentation is given in Fig. 1. Method 1 is a simple extraction and concentration procedure. Method 2 is a slight modification of the method of Casman (1967) and includes a step for the separation of enterotoxin from other soluble constituents of the food by means of a carboxymethyl-cellulose column.

Experiments have been carried out on the recovery of enterotoxins from various foods to which measured amounts of toxin had been added. The estimated recovery of enterotoxins A, B or C, using method 1 for the extraction, varied between 20 and 55% of the quantity added with a mean value of 33%. The estimated recovery of enterotoxin A from foods using method 2 for the extraction was 20%.

Gel immunodiffusion test for enterotoxin

The slide gel double-diffusion method of Wadsworth (1957) and Crowle (1958) with minor modifications was used for the detection of enterotoxin in culture filtrates and in foods.

Results

Examination of cultures from food poisoning incidents

The supernatant fluids from sac cultures of 321 strains of *Staph. aureus* isolated from 94 food poisoning incidents were tested for enterotoxins A, B and C, and 1

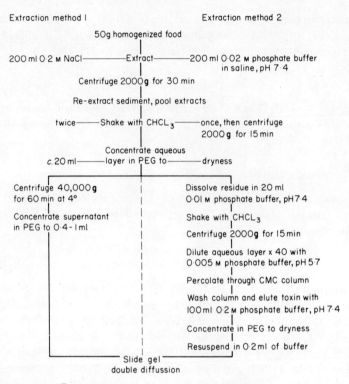

Fig. 1. Two methods used for the extraction of enterotoxin from foods implicated in staphylococcal food poisoning.

strain from each incident was tested for enterotoxins D and E. On completion of the phage-typing and enterotoxin tests an analysis of the results indicated the strain most probably responsible for each incident. A single strain usually from the implicated food was chosen to represent each of the 94 incidents. The relationship between enterotoxin production and phage group of these strains is shown in Table 3. Enterotoxin A either alone or with enterotoxins B, C or D was produced by 73 strains (78%). Enterotoxin C was produced by 1 strain and enterotoxin D by 7 strains: 6 strains produced both enterotoxins C and D. One strain produced enterotoxin E and 1 strain both enterotoxins C and E. Five strains did not produce enterotoxins A, B, C, D or E but culture filtrates of each produced an emetic response in monkeys typical of staphylococcal food poisoning (Professor M. S. Bergdoll, personal communication). Eighty-eight (94%)

Table 3

Relation between phage group and enterotoxin production of strains of
Staph. aureus *from 94 food poisoning incidents*

Phage group	Number of strains	Number producing enterotoxins A–E	Number of strains producing enterotoxin*									
			A	B	C	D	E	A&B	A&C	A&D	C&D	C&E
I	3	3	2		1							
III	60	57	34			4	1	1		15	2	
IV	1	1				1						
I/III	28	27	9			2		1	2	9	4	
Misc.	1	1										1
Non-typable	1	0										
Totals	94	89	45	0	1	7	1	2	2	24	6	1

* Three strains were not tested for enterotoxin E production.

strains were lysed by phages of group III and 28 of these were also lysed by one or more phages of group I.

Several of the biochemical and antibiotic sensitivity tests were carried out in 2 laboratories (R.J.G. at Colindale and Dr Susan E. J. Young at the Public Health Laboratory, Luton) and almost identical results were obtained. All 92 strains tested were DNase-positive and phosphatase-positive. Eighty-five of the strains produced orange or buff colonies on cream agar, 4 produced yellow colonies and 3 white colonies. The 4 strains forming yellow colonies produced enterotoxins A or A and D whereas the 3 strains forming white colonies produced enterotoxins C, C and D or C and E. In the antibiotic sensitivity tests 76 of the 92 strains gave results characteristic of penicillinase production. Two of the strains also showed resistance to streptomycin, tetracycline and erythromycin; both strains were lysed by phage 85 only and both produced yellow colonies on cream agar.

Detection of enterotoxin in foods implicated in food poisoning incidents

Twenty-nine foods implicated in 25 food poisoning incidents were examined for the presence of enterotoxins A, B and C and 14 of these for enterotoxin D. Enterotoxin was detected in 9 of 17 foods using method 1 for the extraction of toxin and in 7 of 12 foods from the more recent outbreaks using method 2. Eleven foods contained enterotoxin A, 3 contained D, 1 contained both A and B and 1 both A and D (Table 4). In each instance *Staph. aureus* was present in large numbers from 5×10^6–2×10^9/g of food and the enterotoxin detected was the same as that produced by growth of the organisms in sac-cultures. In 1 outbreak, however, only enterotoxin D was detected in the food, whereas strains isolated from the food and from patients produced enterotoxins C and D. In 3 outbreaks the strains produced enterotoxins A and D; the corresponding foods were positive for enterotoxin A, but were not tested for enterotoxin D.

Attempts were made to determine the approximate amount of enterotoxin in some of the foods. The amount of enterotoxin A or A and B in 12 foods from separate incidents varied from <0·01–0·025 µg/g (Table 4).

Enterotoxin was not detected in 13 foods. Cultures of *Staph. aureus* from 9 of these foods and from specimens of patients produced enterotoxins A, A and D, or C. The number of staphylococci in 5 of these 9 foods was less than 1×10^6/g; the 4 foods with more than 1×10^6 staphylococci/g were all examined for enterotoxin by the first extraction method. The strains isolated from 4 of the negative foods did not produce enterotoxins A, B, C or D. Subsequent tests have shown that 3 of the strains isolated from foods implicated in 1 outbreak produced enterotoxin E; unfortunately no food extracts were left for further investigation.

Discussion

All the food poisoning strains examined were obtained from incidents that had been carefully investigated. Enterotoxins A, B, C, D or E were produced by strains from 89 of 94 incidents (95%); enterotoxins A, B, C and D were produced by strains from 77 of 80 incidents (96%) in the U.S.A. (Casman *et al.*, 1967). Enterotoxins A and A and D together are the predominant types causing food poisoning in both countries and also in Canada (Toshach & Thorsteinson, 1972). Cultures producing enterotoxin C or both C and D have also been implicated in several outbreaks in England, the U.S.A. and Canada. Cultures producing enterotoxin B either alone or with A, C or D are implicated less frequently.

The results in Table 3 confirm the findings of numerous workers that a staphylococcus incriminated in food poisoning is most likely to be lysed by phages of group III or phages of both groups I and III. However, it is impossible to predict whether a particular strain belonging to phage group III or groups I and III is capable of producing enterotoxin. Numerous cultures in our collection, that have been isolated from a wide variety of routine food samples, nasal swabs and wound infections, are lysed by phages of group III but do not produce enterotoxins A, B or C. Conversely we have some strains that produce enterotoxin which are lysed by phages of groups I, II, IV or the miscellaneous group and many which are untypable strains.

Two methods were used for the extraction of enterotoxin from foods incriminated in food poisoning. Method 2 was more satisfactory than method 1 because the final extract contained less of the soluble constituents of the food

Table 4

Detection of enterotoxin in foods implicated in 16 incidents of staphylococcal food poisoning (1969–1972)

Extraction method used (see text)	Incident	Food examined	Amount examined (g)	Count of Staph. aureus (g)	Enterotoxin produced by cultures	Enterotoxin detected in food Type	Amount (µg/g)
1	1	Prawns (canned)	25	40×10^6	A and B	A and B	A – 0·03 B – 0·03
	2	Cold chicken	100	55×10^6	A and D	A (D not tested)	0·02
	3	Cold chicken	54	150×10^6	A	A	0·04
	4	Trifle	52	150×10^6	A and D	A	0·05
	5	Tongue and beef	65	200×10^6	A and D	A (D not tested)	0·04
	6	Ham	50	250×10^6	A	A	0·05
	7	Ham	50	450×10^6	A	A	0·08
	8	Cold chicken	51	200×10^6	D	D	*
	9	Ham	27	1000×10^6	C and D	D	*
2	10	Prawns (frozen)	71	6×10^6	A	A	0·02
	11	Prawns (canned)	17	5×10^6	A	A	0·09
	12	Vanilla cake	36	150×10^6	A	A	<0·01
	13	Torta cream cake	31	1000×10^6	A	A	0·02
	14	Ham and potato	4	4000×10^6	A	A	0·25
	15	Brawn	50	600×10^6	A and D	A and D	Not determined
	16	Veal, ham and egg pie	23	2000×10^6	D	D	*

* Not calculated because concentration of reference enterotoxin unknown.

which interfered with the gel diffusion test and it could be concentrated to a smaller volume because it was less viscid.

Most of the food samples submitted for enterotoxin tests weighed between 30–100 g. In 1 incident, however, enterotoxin A was detected in a sample of ham and potato weighing only 4 g (Table 4). Although this result was particularly encouraging there is still a need for a more simple and shorter technique for detecting minute amounts of enterotoxin in food. In view of the current interest in the staphylococcal enterotoxins and the fact that entero-toxins and antisera are now commercially available it would seem likely that much progress will be made during the next decade. For example, further development of rapid and sensitive techniques such as immunofluorescence (Genigeorgis & Sadler, 1966) and reversed passive haemagglutination (Silverman *et al.*, 1968) are likely to replace gel diffusion procedures for the detection of enterotoxin in food if adequate refinements can be made in the initial extraction procedures.

The work described in this paper would have been impossible without the cultures of *Staph. aureus* and samples of food that were received from over 50 public health and hospital laboratories in the United Kingdom. Since 1969 enterotoxin tests have been carried out on cultures from about 80% of the reported general and family outbreaks of staphylococcal food poisoning in this country.

Acknowledgements

We are grateful to Professor M. S. Bergdoll, Food Research Institute, University of Wisconsin for providing enterotoxins A, B, C and E and their antisera and to Dr R. W. Bennett, Food and Drug Administration, Washington, D.C., for providing enterotoxin D and its antiserum. We would also like to thank the staff of Cross Infection Reference Laboratory for phage-typing the cultures and Dr Betty C. Hobbs for her constant encouragement.

Part of this work was made possible by a generous grant from the Wellcome Trust.

References

Angelotti, R. (1969). Staphylococcal intoxications. In *Food-borne Infections and Intoxica-tions*, p. 359. Ed. H. Riemann. New York: Academic Press.

Bergdoll, M. S. (1967). The staphylococcal enterotoxins. In *Biochemistry of some Food-borne Microbial Toxins*. Eds R. I. Mateles & G. N. Wogan, p. 1. The Massachusetts Institute of Technology: The M.I.T. Press.

Bergdoll, M. S. (1970). Enterotoxins. In *Microbial Toxins,* Vol. 3, p. 265. Eds T. C. Montie, S. Kadis & S. J. Ajl. New York: Academic Press.

Brachman, P. S., Taylor, A., Gangarosa, E. J., Merson, M. H. & Barker, W. H. (1973). Food poisoning in the U.S.A. In *Proceedings of the 8th International Symposium. The Microbiological Safety of Food.* p. 143. London: Academic Press.

Breckinridge, J. C. & Bergdoll, M. S. (1971). Outbreak of food-borne gastro-enteritis due to a coagulase-negative enterotoxin producing staphylococcus. *New Engl. J. Med.* **284**, 541.

Casman, E. P. (1967). Staphylococcal food poisoning. *Hlth Lab. Sci.* **4**, 199.

Casman, E. P. & Bennett, R. W. (1965). Detection of staphylococcal enterotoxin in food. *Appl. Microbiol.* **13**, 181.

Casman, E. P., Bennett, R. W., Dorsey, A. E. & Issa, J. A. (1967). Identification of a fourth staphylococcal enterotoxin, enterotoxin D. *J. Bact.* **94**, 1875.

Crowle, A. J. (1958). A simplified micro double-diffusion agar precipitin technique. *J. Lab. clin. Med.* **52**, 784.

Donnelly, C. B., Leslie, J. E., Black, L. A. & Lewis, K. H. (1967). Serological identification of enterotoxigenic staphylococci from cheese. *Appl. Microbiol.* **15**, 1382.

Genigeorgis, C. & Sadler, W. W. (1966). Immunofluorescent detection of staphylococcal enterotoxin B. II. Detection in foods. *J. Fd Sci.* **31**, 605.

Gilbert, R. J., Wieneke, A. A., Lanser, J. & Simkovicova, M. (1972). Serological detection of enterotoxin in foods implicated in staphylococcal food poisoning. *J. Hyg., Camb.* **70**, 755.

Hall, H. E., Angelotti, R. & Lewis, K. H. (1965). Detection of the staphylococcal enterotoxins in food. *Hlth Lab. Sci.* **2**, 179.

McDougall, J. (1970). Symp. Food Toxin. Massey University, N.Z. Cited from *Br. Fd J.* (1971) **73**, 70.

Minor, T. E. & Marth, E. H. (1971). *Staphylococcus aureus* and staphylococcal food intoxications. A review. Part I. *J. Milk Fd Technol.* **34**, 557.

Minor, T. E. & Marth, E. H. (1972). *Staphylococcus aureus* and staphylococcal food intoxications. A review. Part II. *J. Milk Fd Technol.* **35**, 21; Part III, **35**, 77; Part IV, **35**, 228.

O'Connor, J. J., Willis, A. T. & Smith, J. A. (1966). Pigmentation of *Staphylococcus aureus. J. Path. Bact.* **92**, 585.

Omori, G. & Kato, Y. (1959). A staphylococcal food-poisoning caused by a coagulase negative strain. *Biken J.* **2**, 92.

Ormay, L. (1971). Some problems concerning bacteriological standards for food. *Elelm. Ipar* **25**, 74.

Read, R. B., Pritchard, W. L., Bradshaw, J. & Black, L. A. (1965). *In vitro* assay of enterotoxins A and B from milk. *J. Dairy Sci.* **48**, 411.

Read, R. B., Bradshaw, J., Pritchard, W. L. & Black, L. A. (1965). Assay of staphylococcal enterotoxin from cheese. *J. Dairy Sci.* **48**, 420.

Report (1963). Food poisoning in England and Wales, 1962. A report of the Public Health Laboratory Service. *Mon. Bull. Minist. Hlth* **22**, 200.

Report (1964). Food poisoning in England and Wales 1963. A report of the Public Health Laboratory Service. *Mon. Bull. Minist. Hlth* **23**, 189.

Report (1967). Report of the Subcommittee on phage-typing of staphylococci of the International Committee on nomenclature of bacteria, Moscow, July 1966. *Int. J. Syst. Bact.* **17**, 113.

Silverman, S. J., Knott, A. R. & Howard, M. (1968). Rapid, sensitive assay for staphylococcal enterotoxin and a comparison of serological methods. *Appl. Microbiol.* **16**, 1019.

Simkovicova, M. & Gilbert, R. J. (1971). Serological detection of enterotoxin from food-poisoning strains of *Staphylococcus aureus. J. Med. Microbiol.* **4**, 19.

Toshach, S. & Thorsteinson, S. (1972). Detection of staphylococcal enterotoxin by the gel diffusion test. *Can. J. publ. Hlth* **63**, 58.

Vernon, E. (1965). Food poisoning in England and Wales, 1964. *Mon. Bull. Minist. Hlth* **24**, 321.

Vernon, E. (1966). Food poisoning in England and Wales, 1965. *Mon. Bull. Minist. Hlth* **25**, 194.

Vernon, E. (1967). Food poisoning in England and Wales, 1966. *Mon. Bull. Minist. Hlth* **26**, 235.

284   R. J. GILBERT AND ANTONETTE A. WIENEKE

Vernon, E. (1969). Food poisoning and *Salmonella* infections in England and Wales, 1967. *Publ. Hlth, Lond.* **83**, 205.
Vernon, E. (1970). Food poisoning and *Salmonella* infections in England and Wales, 1968. *Publ. Hlth, Lond.* **84**, 239.
Wadsworth, C. (1957). A slide microtechnique for the analysis of immune precipitates in gel. *Int. Archs Allergy appl. Immunol.* **10**, 355.
World Health Organization (1969). Surveillance summary of foodborne disease outbreaks: United States of America, 1968. *Wkly Epidem. Rec.* **44**, 349.
Zehren, V. L. & Zehren, V. F. (1968). Examination of large quantities of cheese for staphylococcal enterotoxin A. *J. Dairy Sci.* **51**, 635.

Discussion

Hobbs

Are there any instances of foods in which enterotoxin has been demonstrated in the absence of isolations of *Staph. aureus*?

Gilbert

There have been no instances in our laboratory. Enterotoxin has been detected only in foods from outbreaks where the count of *Staph. aureus* was greater than 6 million/g of food.

Bergdoll

One instance of food poisoning with dried milk and 1 with malted milk powder in which no staphylococci were found have been brought to our attention. On examination of these products enterotoxin D was found in each case. Dr Gilbert reported finding no food poisoning outbreaks from enterotoxin B. Until this summer we also had not recorded any, but we have had 3 outbreaks in Milwaukee (Wisconsin) due to enterotoxin B. This was the only enterotoxin produced by the organisms isolated from the implicated food and the only toxin in the foods. One outbreak was from devilled egg which contained 1 µg enterotoxin B/g of egg. This is a very high level.

Olson

With respect to the discussion of enterotoxin levels at which food might be condemned, it is suggested that in view of the magnitude of numbers of enterotoxigenic staphylococci necessary to produce detectable toxin (by the methods in use at this time), a food may be considered unacceptable long before enterotoxin might be demonstrated. Good manufacturing practices and proper food handling would not permit staphylococcal populations to reach anywhere near the level necessary to form detectable levels of enterotoxin.

Sutton

Is there any correlation between the level of enterotoxin in the food and the morbidity rate and severity of symptoms? Also does the nature of the enterotoxin have any effect on symptoms?

Gilbert

Yes, there is some correlation between the level of enterotoxin in the food and the morbidity and severity of symptoms. However, one must remember that some people may be much more sensitive to the enterotoxins than others and there may be a considerable variation in the amount of enterotoxin in various parts of a contaminated food, i.e. the toxin may not be evenly distributed. The symptoms of staphylococcal food poisoning and their severity are probably the same for enterotoxins A, B, C, D and E. This is supported by tests in monkeys and in human volunteers in the U.S.A.

Elliott

Dr Gilbert's reference to enterotoxin detection in a food with 750,000 *Staph. aureus*/g is of some interest. Can we have further information; for example, might some cells have died before analysis?

Gilbert

There is a slight misconception here. Enterotoxin A has been detected in samples of canned prawns and frozen prawns from separate outbreaks where the counts of *Staph. aureus* were 5×10^6 and 6×10^6/g. The frozen prawns were thawed and eaten several hours later in the form of a prawn cocktail and it would seem likely that the staphylococci increased in numbers during this time interval (relative to the count on a sample of the original frozen prawns).

Barnes

Was any direct microscopic count made on the low staphylococcal count prawns which caused food poisoning? If the staphylococci had multiplied prior to freezing some would have died on freezing and the numbers dying would probably depend upon the temperature of the frozen storage and the length of time stored.

Gilbert

No direct microscopic count was made on the frozen prawns.

Moss

There are 5 cultures remaining in your collection of *Staph. aureus* implicated in outbreaks of food poisoning which you have shown not to produce enterotoxins A–E. Does not the presence of these 5 isolates imply that you may need to look further for agents other than the known enterotoxins in foods?

Gilbert

Cultures from 89 of the 94 outbreaks produced enterotoxins A, B, C, D or E. Other enterotoxins will no doubt be isolated in the future. Cultures from the 5 outbreaks that did not produce known enterotoxins were sent to Professor Bergdoll for further tests. Culture filtrates produced a typical emetic response in monkeys indicating that enterotoxin was present.

Enterotoxin Detection*

M. S. BERGDOLL

*University of Wisconsin,
Madison,
Wisconsin,
U.S.A.*

The quantity of staphylococcal enterotoxin required to make a sensitive person ill is believed to be less than 1 μg and therefore any method that is recommended for the detection of this toxin must be sufficiently sensitive to detect such small amounts. The method of Casman and Bennett or a modification thereof is at present the only reliable procedure available for demonstrating enterotoxin in food. However, the technique is time consuming and is not sensitive enough to find small amounts of enterotoxin. What is required is a rapid and sensitive detection procedure which can be used with unconcentrated food extracts. Two such methods, reversed passive haemagglutination and radio immunoassay, are under investigation and this paper outlines some of the advantages and disadvantages of these techniques.

THOSE WHO ARE interested in testing foods for staphylococcal enterotoxin can be divided into 3 groups, namely: public health agencies, control agencies and food processors. All of these groups are interested in the safety of foods, but each has a somewhat different concern. The public health agencies are primarily concerned with investigating food poisoning outbreaks—after the fact concern, so to speak. The control agencies are also interested in investigating food poisoning outbreaks, but, in addition, they are concerned with protecting the consumer against potentially hazardous foods. The food processor is primarily concerned with preventing hazardous foods from reaching the consumer. The task of public health laboratories is somewhat easier than the others because chances that food from outbreaks contains enterotoxin are good. If staphylococci are isolated, as they were in all the instances reported by Gilbert & Wieneke (1973)—although this is not always so—it is likely that they are enterotoxigenic and with our help they were able to demonstrate that fact. It is necessary to examine an occasional strain by some biological test, such as the monkey feeding test, to demonstrate enterotoxigenicity. Gilbert and Wieneke were not able to detect enterotoxin in all of the foods examined, but this was primarily because the samples were too small. However, if the organisms are enterotoxigenic, it maybe concluded that enterotoxin was present in the food.

* Contribution from the College of Agricultural and Life Sciences, University of Wisconsin, Madison, Wisconsin, U.S.A.

The position with the regulating agencies is the same as that for public health laboratories for food poisoning outbreaks, but it is different when suspect foods are examined. In this situation, even if the staphylococci isolated produce enterotoxin, it cannot be assumed that the food contains enterotoxin or at least in sufficient quantities to be of concern. Thus it is necessary to use methods which are sufficiently sensitive to detect very small quantities of toxin. This is also true for the food processor.

At the present time, it is my judgment that the Casman & Bennett (1965) method or some modification thereof such as that described by Gilbert & Wieneke (1973) (modified by us also) is the only reliable method for detecting enterotoxin in foods. Currently it is considered to be the reference method.

Even so, it is far from ideal because it is time-consuming and cumbersome and many people have difficulty achieving the desired sensitivity. What sensitivity is required? At least, we should be able to detect less enterotoxin than will make a sensitive person ill. We do now know how much this is, but I believe it to be less than 1 μg. In any case the most that is allowable is less than the least the regulatory agencies can detect in their laboratories. In the United States Food and Drug Administration laboratories, this amount would be less than 0·2 μg enterotoxin/100 g of food. I do not believe that the physical limits of the method will allow for detecting much less than this amount and the quantity may vary somewhat depending on the type of food examined. The methods are not quantitative and it is pointless to try to make them so, other than to obtain a rough estimate of the amount of toxin present. As a guide it should be known how little enterotoxin can be added to 100 g of the food in question to detect it by the Casman and Bennett method or some modification thereof.

The success of any method depends to a great extent on the technique used to detect the toxin in the extract. I would suggest that the minimal practical limit of the microslide technique is $c.$ 0·1 μg/ml with the knowledge that many people have difficulty in achieving this level of sensitivity. The length of time for the development of precipitation lines is usually 1–5 days with 1 day minimum at 37°. The Casman and Bennett method could be shortened by using a faster method of detection such as immunoelectrophoresis, proposed by Chugg (1972). This can be accomplished in 1 h with a sensitivity equivalent to that of the microslide technique (Gasper, 1972). Since the major drawback to the Casman and Bennett method is the concentration procedure—concentration of the extract from 100 g of food to 0·2 ml—there is need to find a sensitive detection procedure which can be used with unconcentrated food extracts. Two such methods are currently under investigation, namely, the reversed passive haemagglutination and the radioimmunoassay techniques (Johnson *et al.,* 1971; Dickie, 1970).

Silverman *et al.* (1968) proposed the reversed passive haemagglutination method for detection of enterotoxin in foods. This method is based on the absorption of enterotoxin antibodies to tannic acid-treated sheep red blood cells followed by agglutination of the sensitized cells with enterotoxin. The sensitivity of the method is 0·0015 μg enterotoxin/ml; thus it should be possible to detect about 0·1 μg enterotoxin/100 g of food in 1 volume/weight extract of the food (e.g. 10 ml/10 g) with concentration. The time required to obtain results is 1 or 2 days. Several laboratories are using this technique at the present time as a screening method. We have been working with it and a similar procedure, that is, coupling the antibodies to the red blood cells with bis-diazotized benzidine and we have encountered some difficulties (Reiser & Conaway, unpublished data, 1972). Firstly, the antibodies in some antibody preparations, particularly those from enterotoxin A, do not adsorb or couple to the red blood cells. The method is doomed if this baffling problem cannot be solved. Secondly, sensitized cells may be agglutinated by enterotoxin-negative food extracts, particularly meat extracts, thus giving false positive reactions. Thirdly, sometimes the buffer system causes agglutination of the control cells, thus making it impossible to carry out the reaction. The second problem has been solved for the most part by trypsin treatment of the food extract, to digest interfering proteins. The third problem may have been solved by making the phosphate saline buffer freshly each day with deionized distilled water. Certain procedures must be followed in preparing the food extract to obtain good results with different types of foods.

Much time has been spent on this method and perhaps it should be abandoned, but we realize that it might be the only sensitive method available to many laboratories if the difficulties can be solved. We will not recommend it, however, until we are satisfied that the results cannot be misinterpreted.

The radioimmunassay technique is carried out as follows: (a) antibodies are adsorbed to the inside surface of plastic tubes; (b) a mixture of radio-iodinated enterotoxin is mixed with the unknown and placed in the tube; (c) after a suitable reaction time, usually overnight, the mixture is removed and the radio-activity remaining in the tubes is determined; and (d) the amount of toxin present in the unknown is calculated from the amount of radioactive toxin taken up. The sensitivity of the method is similar to that of the reversed passive haemagglutination method (0·0015 μg/ml), but work carried out so far indicates that analysis can be done on unconcentrated and untreated food extracts (Johnson et al., 1972). We have not yet worked with this method, but predict that it will become the method of choice for those laboratories that can use it. The disadvantages are: (a) facilities for handling radioactive materials must be available, including counting equipment; (b) each laboratory must prepare its own radioactive toxin; and (c) fairly highly purified toxin must be available, probably at least 90% pure. At the present time, (c) is a drawback because the purified enterotoxins are not readily available and D has not yet been purified.

The last 2 methods are not generally available for use but, it is hoped that they will be available within a reasonable time. They would greatly simplify the detection of enterotoxin in foods even if they could be used only as screening methods for food processors.

It must be emphasized that, at the present, we cannot guarantee that any food is free of enterotoxin, because not all of the enterotoxins have been identified. There is some consolation in the knowledge that most food poisoning outbreaks investigated have been caused by identified enterotoxins, but not all, as Gilbert & Wieneke (1973) showed. We need 1 test that is specific for all the enterotoxins, but the possibilities of finding such a method are difficult to predict.

Our line of approach to this problem is to utilize some common structure of the enterotoxins for development of a specific reagent. To do this we need a detailed knowledge of the structure of the enterotoxins. For a beginning, we have determined the amino acid sequence of enterotoxin B (Huang & Bergdoll, 1970) and we are working on A. Our best hope centers around the cystine loop since each enterotoxin apparently contains only 1 cystine molecule. It is our hope that it will be the same for all enterotoxins and that it can be used to develop a specific reagent.

References

Casman, E. P. & Bennett, R. W. (1965). Detection of staphylococcal enterotoxin in food. *Appl. Microbiol.* **13**, 181.

Chugg, L. (1972). Rapid, sensitive methods for enumeration and enterotoxin assay for *Staphylococcus aureus.* Ph.D. Thesis, Oregon State University.

Dickie, N. (1970). Detection of staphylococcal enterotoxin. *Can. Inst. Fd Technol. J.* **3**, 143.

Gasper, E. (1972). Personal communication.

Gilbert, R. J. & Wieneke, A. A. (1973). Staphylococcal food poisoning with special reference to the detection of enterotoxin in food. In *Proceedings of the 8th International Symposium. The Microbiological Safety of Food.* p. 273. London: Academic Press.

Huang, I-Y. & Bergdoll, M. S. (1970). The primary structure of staphylococcal enterotoxin B. III. The cyanogen bromide peptides of reduced and amino-ethylated enterotoxin B, and the complete amino acid sequence. *J. Biol. Chem.* **245**, 3518.

Johnson, H. M., Bukovic, J. A., Kauffman, P. E. & Peeler, J. T. (1971). Staphylococcal enterotoxin B: Solid-phase Radioimmunoassay. *Appl. Microbiol.* **22**, 837.

Johnson, H. M., Bukovic, J. A. & Kauffman, P. E. (1972). Staphylococcal enterotoxins A and B: Solid-phase radioimmunoassay in food. Abstracts of the Annual Meeting of the Amer. Soc. for Microbiol. p. 23.

Silverman, S. J., Knott, A. R. & Howard, M. (1968). Rapid, sensitive assay for staphylococcal enterotoxin and a comparison of serological methods. *Appl. Microbiol.* **16**, 1019.

Discussion

Riemann

Is there any new information about the pharmacological effect of enterotoxins?

Bergdoll

There is no recent information. The studies of Sugiyama indicate that the effect of enterotoxin for ordinary food poisoning is a direct one on the intestinal tract. Other effects can be observed when large doses are injected such as high temperature and low blood pressure. The low blood pressure is probably due to peripheral pooling of the blood. An occasional death from staphylococcal food poisoning is probably due to these secondary reactions. The kidneys are the major organ for removal of the toxin from blood and an accident in our laboratory resulted in an effect of the toxin in the individual's kidneys.

Ingram

If the quantity of toxin which at present can be detected chemically substantially exceeds that needed to cause food poisoning it is not surprising that Dr Gilbert should occasionally fail to demonstrate enterotoxin in cases where poisoning occurs. If on the other hand, chemical detection were possible at levels much below those producing poisoning, we should need to reconsider the common phraseology "Toxin should be absent . . ." Could Dr Bergdoll comment on minimum detectable quantities?

Bergdoll

The human dose is probably less than 1μg. The minimum that can be detected in foods by the Casman and Bennett method is 0·2 μg enterotoxin/100 g of food. At the present time if the operator can detect this amount in his control system but cannot in the sample, then the sample could be considered safe for consumption. Smaller amounts can be detected by more sensitive methods and it is hoped that eventually a minimal level for food rejection will be set.

Monty

Do the different types of enterotoxin differ in their resistance to heat?

Bergdoll

The purified enterotoxins are more vulnerable to heat than is toxin in foods. Mr Denny of The National Canners Association, U.S.A., has shown that good canning procedures will destroy enterotoxin. Enterotoxin B appears to be more heat resistant than A with C intermediate.

Barrow

A fatal case of food poisoning following consumption of cockles occurred last year. Large numbers of enterotoxin A—producing staphylococci and marine bacteria were isolated from the gut and from samples of the cockles. Similar staphylococci were isolated from food handlers, refrigerators, etc. and although this occurred in the holiday season and large amounts of these cockles were sold no other case came to light. The representative strains of marine bacteria were kept and when tested recently it was found that the only Kanagawa positive strain of *V. parahaemolyticus* in our collection was isolated from this fatal case. Possibly there was a synergistic effect from the 2 pathogens?

Olson

Regarding the question on enterotoxin detection in foods in which no viable staphylococci could be demonstrated, in butter samples involved in an extensive outbreak in the U.S. about a year ago, no viable staphylococci could be detected yet enterotoxin A was found.

Sinell

In addition to the methods summarized by Professor Bergdoll, our experiments using the reversed haemagglutination technique were carried out over a long period without success. The direct haemagglutination method described by Johnson (1967) gave more reliable results. This method, however, needs also a highly purified enterotoxin in relatively large quantities, although the microtitre technique was applied. Most reproducible results could be obtained when the antigen was fixed upon the surface of the erythrocytes using glutaraldehyde as described by Avrameas.

Bioassay Methods for Mycotoxins

B. JARVIS AND M. O. MOSS

British Food Manufacturing Industries Research Association,
Randalls Road,
Leatherhead,
Surrey, England

Department of Biological Sciences,
University of Surrey,
Guildford,
Surrey, England

During the past few years, many laboratory bioassay methods for mycotoxins have been developed including the use of vertebrates and invertebrates, fertile eggs, bacteria, and cells grown in tissue culture. The advantages and disadvantages of the more generally applicable laboratory assay methods are discussed, in relation to the purpose for which the method is required, i.e., screening cultures for toxigenesis, confirmation of physico-chemical analyses, etc. Results are presented on comparative studies using oral, subcutaneous, intraperitoneal, intradermal and dermal routes of administration of mycotoxins to laboratory animals and on the fertile egg test and antibacterial tests. The general applicability of these methods as routine test procedures for detection of mycotoxins in feed, food and culture medium systems are discussed.

SINCE THE discovery of the aflatoxins in the early 1960s world-wide interest has been focused on the toxic metabolites of microfungi. The occurrence of mycotoxins as contaminants of foods and feedingstuffs may render the food unsafe, and in the U.S.A. the Food and Drug Administration considers these compounds to be of major significance as contaminants (Fischbach & Rodricks, 1972). Mycotoxins show a wide range of acute and chronic toxic manifestations including carcinogenic, oestrogenic and teratogenic effects and there is a considerable variation in response to individual toxins by different species of animal.

For control purposes both physico-chemical and biological test methods have been developed for mycotoxins. Bioassay methods are used for 3 purposes.

(1) To screen fungal cultures for their ability to produce mycotoxins and to examine foods and feedingstuffs for the presence of potentially toxic materials. In this context it is interesting to note that many isolated organisms were shown to be capable of producing mycotoxins long before the organisms were implicated as possible causal agents of diseases of unknown aetiology.

(2) To follow the bioproduction and isolation of new mycotoxins until

physico-chemical assay methods have been developed for the specific detection and quantification of the particular toxin. It is obviously essential to have a method whereby the toxicity of fungal extracts may be assessed and laboratory bioassay methods fulfil an important function in this area.

(3) To confirm results obtained in physico-chemical analyses of specific toxins. It is of interest that in the U.S.A. legal proceedings in respect of aflatoxin contamination of food materials require biological confirmation of the chemical analyses (Fischbach & Rodricks, 1972). The purpose of the present paper is to review some of the bioassay methods used for mycotoxin detection and to present comparative data for the applicability of some of these methods in screening organisms for toxigenicity. We do not propose to discuss assay methods used in studies of mycotoxin carcinogenicity, teratogenicity, oestrogenicity or for any purpose other than acute toxicity.

Whole Animal Bioassay Systems

Bioassay systems may be divided conveniently into 2 distinct categories, namely whole animal and laboratory bioassay systems. As in any toxicological work, various modes of administration of mycotoxins may be used. Each method has certain advantages and disadvantages according to the purpose for which it is used. The most obvious system is the use of feeding trials in selected species of animal using either naturally or artificially contaminated foodstuffs. Artificial contamination of the animals' normal diet with culture filtrates, culture extracts or solutions of purified metabolites or by admixture of fungal mycelium or laboratory infected natural substrates with the diet has been used by many workers (Tsunoda, 1970; Forgacs & Carll, 1962). This approach was successful in confirming of disease syndromes observed in the field where mould contaminated feed was implicated as a possible causal agent (Krogh, 1972). It has been used also in the toxicological screening of microfungi cultivated as possible sources of biomass (Daunter, 1972) and for the screening of fungi isolated from contaminated foods and feeds (Tsunoda, 1970).

Oral intubation of concentrated culture filtrates or of aqueous suspensions of dried solvent extracts of mycelium or culture filtrates is sometimes used (Ueno, 1971; van Warmelo et al., 1971; Wilson, 1971). This method suffers from the disadvantage of requiring a high degree of experimental expertise for administration of test materials but has an advantage over feeding trials in that the amount of material taken by an animal can be controlled. Less time is generally required for preparative work than in the production of special diets. Intraperitoneal (IP) and subcutaneous (SC) injections are frequently used in toxicological work (Ichinoe et al., 1970; Ueno, 1970). Provided that satisfactory injection regimes are used these procedures are suitable for the study of inherently toxic compounds, i.e. those which are not rendered more toxic by passage through the digestive tract. Most known mycotoxins probably fall into this category but it must be emphasized that our understanding of the toxicity of a group of compounds will be influenced by the methodology used to study it.

Many workers have demonstrated dermal toxicity when certain mycotoxins are applied on or in the depilated skin of sensitive laboratory animals (Joffe, 1960; Forgacs & Carll, 1962; Ueno, 1971). Animal species used in dermal toxicity studies include mice, rats, guinea pigs and rabbits and dermal effects have been reported for mycotoxins isolated from many species of *Aspergillus, Cladosporium, Fusarium, Mucor, Penicillium* and *Stachybotrys.* The simplicity of the dermal patch test obviously makes it attractive as a preliminary screening test but caution is always necessary in interpreting results since not all mycotoxins produce dermal reactions and the level of sensitivity of dermal tests is usually fairly low (Gedek, 1972). The intradermal test requires a high degree of expertise and interpretation of the results is often difficult. Many otherwise atoxic compounds cause dermal ulceration when injected into the skin and the carrier solvent frequently has a pronounced effect on the result obtained (Grasso, P. pers. comm.).

In addition to the general advantages and disadvantages of specific methods it must be borne in mind that the species of animal selected for a screening programme will inevitably be a compromise. No single animal will be sensitive to the wide range of mycotoxins which might be of interest as contaminants of food and feedingstuffs. Economic factors are likely to reduce the range of species available to the majority of laboratories. The choice of solvent used as a vehicle for administration of crude extracts or purified toxins may also affect the apparent toxicity and these effects may themselves depend upon the species of animal. Thus rubratoxin appears to be 10 times more toxic to mice when given by the IP route in dimethylsulphoxide than it does when propylene glycol is used as the solvent. No such solvent effect can be demonstrated in the rat for this particular toxin (Wogan *et al.,* 1971).

When using animal tests on crude extracts of moulded material, fungal mycelium or culture medium it is essential to remove all traces of residual organic solvents which were used for the extraction. Such solvents may range from simple alcohols such as methanol to halogenated hydrocarbons such as chloroform, and their presence in even trace amounts may interfere with the interpretation of results. These general criteria apply equally to the other bioassay systems to be described.

Alternative Bioassay Systems

A disadvantage to many laboratories working on mycotoxins is the requirement for experimental animal facilities. For this reason and also to increase the throughput of samples and reduce the quantity of material required for the test other bioassay systems have been developed. Many different test systems have been studied for both screening and confirmatory purposes and these include the use of avian egg embryos (Verret *et al.*, 1964; Gedek, 1972), brine shrimps (Brown *et al.*, 1968), zebra fish (Abedi & McKinley, 1968), protozoa (Hayes, W., pers. comm.; Wyatt & Townsend, 1973), micro-organisms (Lillehoj *et al.*, 1966; Clements, 1968) and cell lines grown in tissue culture, e.g. (Gabliks *et al.*, 1965; Engelbrecht, 1971; Gedek, 1972). The advantages of these various methods are summarized in Table 1.

Table 1

Advantages and disadvantages of laboratory bioassay methods for mycotoxins

	Test system				
	Egg embryo	Brine shrimp	Micro-organism	Tissue culture	Protozoa
Level of sensitivity	μg	ng	μg	ng	μg
Specificity	Low	Low	Var.	Var.	High
Expertise	Mod.	Low	Mod.	High	Mod./high
"Time"	High	Low	Low	Low	Low
Operator fatigue	Low	V. high	Mod.	High	Low
Solvent effects	Var.	High	High	High	?

In the egg embryo test both the age of the embryo and the route of administration affect the sensitivity. In the experience of both ourselves and other workers the air cell route of administration provides a more sensitive response than does yolk sac injection. The egg embryo test has low specificity since many chemicals, including organic solvents, produce lethal effects. A detailed protocol for use of the egg test as confirmation of chemical analyses of aflatoxin has been published (Brown, 1970). Whilst only a moderate degree of technical expertise is required and operator fatigue is comparatively low, a considerable number of eggs are required for each confirmatory bioassay undertaken.

The brine shrimp *(Artemia salina)* assay, developed for aflatoxin testing, is more sensitive to the effects of aflatoxins than to other mycotoxins such as ochratoxin (Brown, 1970). However, it requires little technical expertise and provides results within a few hours but operator fatigue is high and the method is extremely sensitive to extraneous solvent.

The use of micro-organisms as test organisms suffers from the major disadvantage of very low specificity. These procedures can be used with confidence when working with pure mycotoxins but are completely unsuitable for preliminary studies on mycotoxins. The degree of expertise required is the same as that for microbiological assays of antibiotics but even so, in purely chemical laboratories problems with the establishment of such methods may arise. In our own experience many mycotoxins have no antimicrobial effect and the selection of test organisms is dependent upon the nature of the mycotoxin. Non-lethal cytotoxic effects can also be seen in many microorganisms exposed to sublethal doses of aflatoxins and certain other mycotoxins (Lillehoj *et al.*, 1966; Brown, 1970).

Of the other bioassay procedures probably the most sensitive is the use of human or animal cell lines grown in tissue culture. The procedure is more sensitive than other bioassay methods (and chemical methods) (Gedek, 1972) but suffers from the potential disadvantage that food components may adversely affect the cells. The expertise and facilities required for maintenance of cells in tissue culture and the degree of operator fatigue in assessing the response of treated cells are disadvantages for the general introduction of this procedure as a routine method. Nevertheless, this is potentially a most valuable tool and further study of its applications would appear to be desirable.

Many other bioassay methods have been developed which have not been discussed. In our opinion they are methods which are not suitable for general use because of the requirements for specialized equipment or techniques. In omitting them we have no intention of disparaging their application in the specialized areas for which they were developed.

Results of Comparative Screening Tests

Of the various methods outlined above, several have been used at different times by ourselves as screening procedures for cultures which have been implicated as the possible causative organisms in mycotoxicosis outbreaks or for organisms isolated from spoiled foods. These procedures have been used also for the examination of certain fungal metabolites.

Table 2 summarizes results obtained some years ago at The Tropical Products Institute when a study was made of microfungi isolated from mould contaminated feeds and feed constituents implicated in field outbreaks of mycotoxicoses. The egg embryo test was used as a primary screen for cultures grown on different media and under various cultural conditions and for solvent extracts from such cultures. The extracts were tested also by the dermal patch and intradermal methods. The toxicity of the culture filtrates to the egg embryos generally correlated with the subsequent isolation from the cultures of specific mycotoxins. In many other instances no evidence for toxicity was observed using either the egg embryo or the dermal toxicity tests, and no toxic metabolites were isolated from the cultures. One aspect of the early work on the fertile egg test was the assessment of the optimum age of embryo for use as a test system. Whilst non-incubated eggs were the most sensitive the problems

Table 2

Use of the egg embryo test for screening purposes (Jarvis & Townsend, unpublished data, 1963)

Organism	Strain	Medium	% Mortality after 6 days		Mycotoxin subsequently isolated
			Yolk sac	Air sac	
A. flavus	T.12	CZ	100	100	Aflatoxin B$_1$
A. clavatus	MR072	CZ/RT	0	100	Patulin
A. chevalieri	MR058	CZ	0	0	Xanthocillin X*
		RT	100	100	
P. rubrum	MR043	CZ/RT	100	100	Rubratoxin B†
P. martensii	MR068	CZ	40	100	Penicillic acid‡
		RT	20	80	
P. islandicum	MR066	CZ/RT	0	100	?
Gliocladium sp.	MR074	CZ	0	100	?
		RT	0	20	
Inoculated		CZ	10	0	
controls		RT	10	10	
Un-inoculated controls	–	–	0	0	

Media: CZ – Czapek; RT – Raulin Thom.

* Coveney, Peck & Townsend (1966); † Townsend, Moss & Peck (1966); ‡ Moss, M. O. Unpublished data.

associated with non-viable eggs resulted in high mortality levels in the control groups. With 4 day embryos the viability is obvious and the embryos are more sensitive to mycotoxins than are older (e.g. 10 day) embryos. As mentioned previously, the air cell route of administration was found to be more sensitive than injection into the yolk sac and posed fewer technical problems, for example the need to work under strictly aseptic conditions.

Many workers have produced data for the LD$_{50}$ values of different mycotoxins in egg embryos using both the air cell and the yolk sac routes of administration and our results (Table 3) are generally in agreement with those reported by Brown (1970) and Gedek (1972). We differ, however, from Gedek in our assessment of the lethality of rubratoxin B to the egg embryo. When rubratoxin B is administered in propylene glycol a significant decrease in the LD$_{50}$ value is observed compared with that of culture filtrates containing known quantities of rubratoxin (Table 3). Dose response curves for aflatoxins B$_1$ and G$_1$ and for rubratoxin B are presented in Fig. 1.

The potentiating effect of solvents on the toxicity of rubratoxin B has been reported previously (Wogan *et al.*, 1971). When administered by the IP route the LD$_{50}$ of rubratoxin in mice is reduced from 2·6 mg/kg in propylene glycol to 0·27 mg/kg in dimethylsulphoxide. Wilson & Wilson (1962) observed a dermal effect from crude culture filtrates but failed to demonstrate any effect with

Table 3

Acute toxicities of mycotoxins in the egg embryo test

Toxin	Single dose LD$_{50}$ *	
	Air cell µg/egg	Yolk sac µg/egg
Aflatoxin B$_1$	0·03	0·05
G$_1$	1·05	NT
Ochratoxin A	9·80	NT
Sterigmatocystin	1·10	NT
Rubratoxin B (in PG)	0·25	55·00
Rubratoxin B (in water)	48·00	>100

* 4 day embryo; dose 0·04 ml; NT—Not tested.

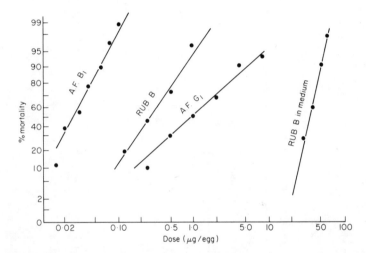

Fig. 1. Dose response curves at 21 days in the egg embryo test for mycotoxins administered via the air cell route.

purified rubratoxin B. In our studies we have obtained similar results using aqueous solutions of crystalline rubratoxin and crude culture filtrates known to contain rubratoxin; but when crystalline rubratoxin was dissolved in propylene glycol or dimethylsulphoxide a pronounced dermal effect was observed which was considerably in excess of the effect caused by the solvent alone. Only when a potentiator is used to carry the rubratoxin into the skin of the test animal is a dermal reaction observed. Similar effects might conceivably occur with other dermally atoxic mycotoxins.

Comparative results of acute and dermal toxicity for a number of strains of microfungi isolated from various sources are presented in Table 4. As expected

Table 4

Comparative results using different bioassay methods for screening fungi

| Organism | Strain | Medium | Mouse | | | | Guinea pig | | Egg embryo |
			Dermal patch	Oral	IP	SC	Dermal patch	IP	
A. flavus	T.12	YES	–	–	–	–	++	+++	+++
A. flavus	NCFT 64	YES	–	–	+	+	–	++	++
A. oryzae var. effusus	CMI18142	YES	–	+	+	+	+	+	+++
P. rubrum	MR043	RTM	++	+++	+++	++	+	++	+++
Monascus ruber	BFR027	RT	–	–	–	–	NT	NT	+++
M. acetoabutans	CMI159918	RT	–	++	–	–	NT	NT	+
Cl. cladosporioides	AD$_1$	CAROB	–	–	–	–	–	–	++
A. amstelodami	BFR	YES	–	–	–	–	–	–	–

Toxicity: +++ Highly toxic; ++ Moderately toxic; + Mildly toxic; – Atoxic.
Media: YES – Yeast extract sucrose; RT – Raulin Thom; RTM – Raulin Thom + malt extract; CAROB – Carob extract.

an aflatoxin-producing strain of *Aspergillus flavus* had no toxic effect on mice but acute toxicity was observed in both the guinea pig and the egg embryo. A tremorgenic toxin-producing strain of *A. flavus* caused convulsions in mice and was lethal to guinea pigs. Egg embryos were very sensitive to the tremorgen. Similar but less pronounced results were observed for a kojic acid-producing strain of *A. oryzae* var *effusus* (cf. Wilson, 1971). In the case of a rubratoxin-producing strain of *Penicillium rubrum* mild dermal reactions were obtained in both mice and guinea pigs and pronounced toxicity was observed in the other test systems. A strain of *Monascus ruber* isolated from a commercial food product caused a high mortality in egg embryos but was not toxic to mice. Unfortunately results are not yet available in other test animals. Similar, but less striking, results were observed with a strain of *Moniliella acetoabutans* an organism known to produce spoilage of pickled foods (Dakin & Stolk, 1968).

Recent work by Daunter (1972) demonstrated the toxicity to mice of mycelium from *Cladosporium cladosporioides,* an organism selected for biomass production on a carob-extract medium on the basis of its growth characteristics and its reported atoxicity (Ichinoe *et al.,* 1970). Cultivation of this organism on ordinary laboratory media failed to reveal any toxigenicity but when cultivated on the carob medium used by Daunter a toxic effect was observed with culture filtrates in both egg embryos and mice after initiation of autolysis of the mycelium. No evidence for toxicity in culture filtrates was found with young cultures. Solvent extracts of the mycelium of this organism and feeding experiments are being made. Results are not yet available. Martin *et al.* (1971) have reported a markedly toxic reaction from 3 out of 6 strains of *Clad. cladosporioides* when tested by the "standard duckling test" which may be a more sensitive test species.

Other strains of fungi which we have isolated from mould-spoiled foods have failed to produce a toxic response in any of the systems used. Amongst the organisms tested have been strains of *Aspergillus amstelodami, A. ochraceus, A. terreus, A. niger* and *Sporendonema sebi.*

Conclusions

From the data presented we suggest that the most generally useful screening technique for mycotoxin production by organisms isolated from foods is the egg embryo test used in conjunction with IP administration of filtrates or extracts to mice and to a second test species such as the guinea pig. The use of dermal patch testing, whilst providing rapid results in some instances, may lead to many potentially toxic organisms being discarded as atoxic. The use of intradermal tests is also considered unsatisfactory for routine procedures.

An idealized protocol for mycotoxin screening is presented in Fig. 2. For any sample of food or feed implicated as a possible cause of disease then the *a priori*

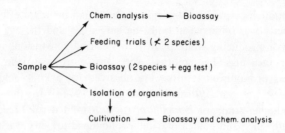

Fig. 2. Idealized protocol for mycotoxin screening.

test should be chemical analysis for known mycotoxins such as aflatoxin, ochratoxin, patulin, etc., followed by confirmation of presumptive positive results using a suitable bioassay procedure, e.g. egg embryos or tissue culture. When sufficient material exists it is obviously essential to undertake feeding trials in at least 2 animal species, although where a veterinary mycotoxicosis has occurred the choice of test animal will obviously be governed by the species implicated in the field outbreak.

If insufficient feed is available for feeding trials and if the presence of specific toxins is not confirmed in a known toxic food it is necessary to isolate the fungi from the food material and to cultivate these on suitable substrates. Such cultures may then be screened using at least 2 different animal species and extracts may be tested by some suitable micromethod, e.g. the egg embryo test. In undertaking such screenings it is desirable to cultivate the organism under different nutritional and physical conditions since the production of metabolites by fungi is known to depend upon a wide range of parameters including the composition of the medium, the degree of aeration and the temperature (Jarvis, 1971; Hayes, 1972).

Acknowledgements

We are indebted to Dr P. C. Spensley, Director of The Tropical Products Institute, for persmission to publish the data in Table 2. The invaluable assistance of Miss A. C. Rhodes and Miss A. Peggs in the current research programme is acknowledged. A more detailed account of this work will be published elsewhere.

References

Abedi, Z. H. & McKinley, W. P. (1968). Zebra fish eggs and larvae as aflatoxin bioassay test organisms. *J. Ass. off. anal. Chem.* **51**, 902.

Brown, R. F., Wildman, J. D. & Eppley, R. M. (1968). Temperature-dose relationships with aflatoxin on the brine shrimp. *J. Ass. off. anal. Chem.* **51**, 905.

Brown, R. F. (1970). Some bioassay methods for mycotoxins. In *Proc. 1st U.S.–Japan Conference on Toxic Microorganisms,* p. 12. Ed. M. Herzberg. Washington: U.S. Department of the Interior.

Clements, N. L. (1968). Rapid confirmatory test for aflatoxin B$_1$ using *Bacillus megaterium. J. Ass. off. anal. Chem.* **51** 1192.

Coveney, R. D., Peck, H. M. & Townsend, R. J. (1966). Recent advances in mycotoxicoses. In *Microbiological Deterioration in the Tropics.* S.C.I. Monograph No. 23. London: Society of Chemical Industry.

Dakin, J. C. & Stolk, A. C. (1968). *Moniliella acetoabutans:* some further characteristics and industrial significance. *J. Fd Technol.* **3,** 49.

Daunter, B. (1972). The evaluation of fungi using tower fermenters. M.Sc. Thesis. University of Aston-in-Birmingham.

Engelbrecht, J. C. (1971). The effects of aflatoxin B$_1$ and sterigmato–cystin on two different types of cell culture. In *Mycotoxins in Human Health,* p. 215. Ed. I. F. H. Purchase. London: Macmillan.

Fischbach, H. & Rodricks, J. V. (1972). Current efforts of the U.S. Food and Drug Administration to control mycotoxins in foods. Abstracts I.U.P.A.C. Symposium Control of Mycotoxins, p. 31.

Forgacs, J. & Carll, W. T. (1962). Mycotoxicoses. *Adv. vet. Sci.* **7,** 273.

Gabliks, J., Schaeffer, W., Friedman, L. & Wogan, G. (1965). Effect of aflatoxin B$_1$ on cell cultures. *J. Bact.* **90,** 720.

Gedek, B. (1972). Biologischer Nachweis von Mykotoxinen. *Zentbl. VetMed. B.* **19,** 15.

Hayes, A. W. (1972). Environmental and nutritional factors affecting production of rubratoxin by *Penicillium rubrum* STOLL. Abstracts, I.U.P.A.C. Symposium Control of Mycotoxins, p. 9.

Ichinoe, M., Udagawa, S-I., Tazawa, M. & Kurata, H. (1970). Some considerations on a biological method for the detection of mycotoxins in Japanese foods. In *Proc. 1st U.S.–Japan Conference on Toxic Micro-organisms,* p. 191. Ed. M. Herzberg. Washington: U.S. Department of the Interior.

Jarvis, B. (1971). Factors affecting the production of mycotoxins. *J. appl. Bact.* **34,** 199.

Joffe, A. Z. (1965). Toxin production by cereal fungi causing alimentary toxic aleukia in man. In *Mycotoxins in Foodstuffs,* p. 77. Ed. G. N. Wogan. Cambridge, Mass.: M.I.T. Press.

Krogh, P. (1972). Natural occurrence of ochratoxin A and citrinin in cereals associated with field outbreaks of swine nephropathy. Abstracts. *I.U.P.A.C. Symposium Control of Mycotoxins,* p. 19.

Littehoj, E. B., Ciegler, A. & Hall, H. H. (1967). Aflatoxin B$_1$ uptake by *Flavobacterium auranticum* and resulting toxic effects. *J. Bact.* **93,** 464.

Martin, P. H. D., Gilman, G. A. & Keen, P. (1971). The incidence of fungi in foodstuffs and their significance based on a survey in the Eastern Transvaal and Swaziland. In *Mycotoxins in Human Health,* p. 281. Ed. I. F. H. Purchase. London: Macmillans.

Townsend, R. J., Moss, M. O. & Peck, H. M. (1966). Isolation and characterisation of hepatotoxins from *Penicillium rubrum. J. Pharm. Pharmac.* **18,** 471.

Tsunoda, H. (1970). Micro-organisms which deteriorate the stored cereals and grains. *Proc. 1st U.S.–Japan Conference on Toxic Microorganisms,* p. 143. Ed. M. Herzberg. Washington: U.S. Department of the Interior.

Ueno, Y. (1971). Toxicological and biological properties of Fusarenon–X, a cytotoxic mycotoxin of *Fusarium nivale* Fn–2B. In *Mycotoxins in Human Health,* p. 163. Ed. I. F. H. Purchase. London: Macmillans.

Verret, M. J., Marliac, J. P. & McLaughlin, J. (1964). Use of the chick embryo in the assay of aflatoxin toxicity. *J. Ass. off. anal. Chem.* **47,** 1003.

van Warmelo, K. T., Marasas, W. F. O., Adelaar, T. F., Kellerman, T. S., van Rensburg, I. B. J. & Minne, J. A. (1971). Experimental evidence that lupinosis of sheep is a mycotoxicosis caused by the fungus *Phomopsis leptostromiformis* (Kühn) Bubák. In *Mycotoxins in Human Health,* p. 185. Ed. I. F. H. Purchase. London: Macmillans.

Wilson, B. J. (1971). Recently discovered metabolites with unusual toxic manifestations. In *Mycotoxins in Human Health*, p. 223. Ed. I. F. H. Purchase. London: Macmillans.

Wilson, B. J. & Wilson, C. H. (1962). Extraction and preliminary characterization of a hepatotoxic substance from cultures of *Penicillium rubrum. J. Bact.* **84,** 283.

Wogan, G. N., Edwards, G. S. & Newberne, P. M. (1971). Acute and chronic toxicity of Rubratoxin B. *Toxic. appl. Pharmac.* **19,** 712.

Wyatt, T. D. & Townsend, R. J. (1973). The bioassay of Rubratoxins A & B using *Tetrahymena pyriformis. J. gen. Microbiol.* (In press.)

Discussion

Kampelmacher

Do you have an explanation for the striking differences in yolk sac and air sac injection in your fertile eggs? And what about the evaluation if you find 0% lethality with the yolk sac route and 100% lethality by the air sac route?

Jarvis

The embryonic blood circulation system is directly in contact with the allantoic membrane of the air sac and I think it is probable that the toxin is more readily distributed to the embryo from the air sac than occurs when it is injected into the yolk sac. In the latter case one probably gets a localized build up of toxin which may not be absorbed for some considerable time. At this stage the embryo may be much less sensitive to the effects of the toxin. Regarding the second question, I would personally take the results of the air sac route in preference to the yolk sac data as indicative of the possible toxic effects of the material administered. In practice we gave up using the yolk sac route some time ago.

Rehm

What reactions did you look for in the oral test on mice? Did you examine for infiltration of liver or fatty liver or other changes?

Jarvis

The initial test was to determine the post administration reaction of the animals to the test material. Post mortem examinations were carried out on all the test animals whether they died or were destroyed. Gross pathological changes were recorded but histopathology on all the tissues has not been undertaken to date.

Rehm

Why did you not use tests with protozoa such as *Tetrahymena*?

Jarvis

There is always a limit to the number of tests which one can undertake. At other times we have also used antimicrobial assays, zebra fish and brine shrimp assays. In general we have not found these to be reliable, reproducible methods for mycotoxins. I am aware of the *Tetrahymena* system used by Townsend but we have not tried it for other toxins.

Rehm

We have found it to be most useful for assay of byssochlamic acid.

Woodbine

Townsend *et al.* have developed a very sensitive method for rubratoxin assay using *Tetrahymena pyriformis*. It was reliable and very quick but needs considerable experience to operate (for example, care in the medium used).

DISCUSSION 305

Jarvis

Wally Hayes in Alabama has also worked on this system for rubratoxin. If one is doing a large number of assays on one toxin it is undoubtedly useful but it still suffers the drawback (in toxicological terms) of using a test system which is far removed from the avian or mammal which might be at risk from mycotoxins. In the toxicological testing of food additives it is accepted that one must use a number of species of mammals. It seems likely to me that more attention will not be directed to the mycotoxin problem until toxic effects are demonstrated in animals with mycotoxins other than aflatoxin isolated from foods.

Barrow

One of the penalties of being a Public Health Laboratory Director is that we not infrequently receive samples such as mouldy pies, some of which are said to have been associated with illness. More often than not the actual reason for submission of these samples by the Public Health authorities is to decide whether or not to take legal action against the retailer or producer. Should we pay more attention to these samples? Should we culture and identify the moulds concerned? Could such foods indeed cause illness?

Jarvis

It is difficult at present to say to what extent such spoiled foods might be the causes of diseases of unknown aetiology. There is evidence from other countries (largely circumstantial) that consumption of mouldy foods has produced illness (see Purchase, 1971). Scott in Canada has isolated mycotoxin-producing moulds from mouldy products implicated in outbreaks of food poisoning but we and many other workers have isolated such organisms from a wide range of mouldy foods not implicated in such outbreaks. I personally consider that considerably more attention should be focused on the possible mycotoxin link in incidents of food poisoning where mouldy foods are reputed to have been involved. If you receive such products I should be very happy to collaborate in examining them for the presence of possible mycotoxins and to identify the moulds present.

Hobbs

Is there any evidence that the fungi used industrially for cheese produces toxins either in pure culture or in cheese?

Jarvis

It is well documented that such commercial strains do not produce aflatoxins but I do not think they have been examined for production of other mycotoxins. One aspect about which I am concerned is the use of moulds for biomass production or for the production of food additives. In the former case I know of instances where a relatively minor change in the cultural conditions resulted in the production of mycotoxins which were not produced under the normal cultural parameters. Even if such biomass products were used for animal feed, problems with residues in the carcass could arise. I think we need to keep a careful watch on these aspects.

Nelson

Have any authenticated cases of human mycotoxicosis been reported?

Jarvis

There was a recent incident of acute aflatoxicosis in Germany resulting from consumption of peanuts and I believe that alcohol may have been a contributory factor. Outbreaks of mycotoxicoses have occurred in Japan and elsewhere. Dr Rehm or someone else from Germany may have more information.

Dénombrement des *Clostridium* à Partir de Produits Alimentaires Pre-reduits—Caracterisation de *Clostridium perfringens*

H. BEERENS,* L. FIEVEZ ET CH. ROMOND

C.E.R.T.I.A.—Villeneuve d'Ascq (France),
Faculté de Médecine Vétérinaire, Université de Liège-Bruxelles (Belgium)
Institut Pasteur, Lille, France

Samples of dry sausages were homogenized aerobically or in an atmosphere of 99% nitrogen and 1% hydrogen and the influence of these 2 procedures on the isolation of *Clostridium* sp. and of *Cl. perfringens* was studied. The efficiency of 3 media—a sulphite agar (V.F) and 2 liquid media (Rosenow and R.C.M. Oxoid)—were compared for the enumeration of the clostridia. The broth cultures and all black colonies on V.F. medium were streaked on to blood agar for identification of the *Clostridium* sp. and *Cl. perfringens*. Four μg/ml of lysozyme was added to all media to suppress the growth of organisms of the *Bacillus* group, which otherwise obscured the anaerobes. Experiments with 19 20 g samples of sausage showed that homogenizing under anaerobic conditions almost halved the colony count of *Clostridium* sp. when on V.F. medium, giving 59 colonies from anaerobic and 105 colonies from aerobic homogenates. When the liquid media were used, counts from anaerobic and aerobic homogenates were similar, being 217 and 201 with Rosenow broth and 96 and 100 with R.C.M. broth. Homogenization under anaerobic conditions reduced the colony count of *Cl. perfringens* isolated from 12 20 g samples of sausage from 59 to 30 when V.F. medium was used for the enumeration, from 7 to 5 with Rosenow broth and from 7 to 4 with R.C.M. broth. The results indicated that it was disadvantageous to homogenize samples with the exclusion of oxygen. Rosenow broth was the most efficient medium for the enumeration of *Clostridium* sp. and V.F. medium for the enumeration of *Cl. perfringens*. The addition of lysozyme to the media had no effect on the count of *Clostridium*. Rosenow broth was slightly more efficient than R.C.M. broth for the enumeration of *Cl. perfringens* on chicken skin.

LES BACTERIES anaérobies ne supportant pas l'oxygène de l'air, nous recherchons dans ce travail s'il n'est pas préférable d'homogénéiser les prises d'essai de produits en atmosphère réductrice, lors de la préparation des suspensions mères en vue de l'analyse bactériologique. La germination des spores de germes anaérobies devrait en être facilitée. Nous comparons ensuite la valeur des milieux solides à celle des milieux liquides pour le dénombrement des *Clostridium* en général et de *Clostridium perfringens* en particulier.

---

* C.E.R.T.I.A. Centre d'Etudes et de Recherches Technologiques des Industries Alimentaires, 369, rue Jules Guesde, 59-Villeneuve d'Ascq (Flers-Bourg), France.

Materiel et Methodes

Matériel

L'intérêt de la préparation des suspensions mères en atmosphère réductrice est recherché en utilisant des saucissons secs comme matériel expérimental.

L'étude comparative de la valeur des milieux solides et des milieux liquides, pour le dénombrement des spores de *Clostridium* et la recherche de *Clostridium perfringens* porte sur deux matières premières différentes: la peau des carcasses de poulet d'une part, et des saucissons secs d'autre part. Toutefois, pour la peau de poulet, l'étude ne concerne que les milieux liquides et *Clostridium perfringens* exclusivement.

Méthodes

Exemple des peux de poulet

On prélève de chaque côté du bréchet, 50 cm^2 de peau et on les introduit dans un flacon stérile préalablement taré. On ajoute à ces 100 cm^2 de peau une quantité de tryptone-sel telle qu'on amène le poids à 100 g. On broie à l'Ultra-Turrax pendant 20 secondes. Dans ces conditions, 1 ml de cette suspension correspond à 1 cm^2 de peau. On pasteurise 10 min à 80°, suivant la méthode habituelle pour ne conserver que les spores bactériennes. On dilue enfin cette suspension mère au 1/10 et au 1/100.

Nous comparons, pour le dénombrement des spores de *Clostridium perfringens,* deux milieux liquides: milieu de Rosenow et milieu R.C.M. Oxoïd.

Les milieux liquides régénérés sont ensemencés de la façon suivante: 5 tubes reçoivent 1 ml de suspension mère; 5 autres 1 ml de la dilution au 1/10 et enfin les 5 derniers, 1 ml de la dilution au 1/100 (technique du nombre le plus probable).

Tous les tubes sont incubés à 37° pendant 3 jours. On recherche alors la présence de *Clostridium perfringens.* Dans ce but la culture de chaque tube de milieu de Rosenow ou de R.C.M. susceptible de contenir des *Clostridium* est isolée sur gélose V.L. sang coulée en boîtes de Pétri. L'incubation se poursuit pendant 72 h à 37°. Les différents types de colonies sont repiqués dans des milieux de Rosenow.

On note le caractère du milieu de Rosenow après 3 jours d'incubation à 37°.

L'identification de *Clostridium perfringens* se réalise de façon classique en recherchant la production d'indole et de lécithinase, l'activité du germe sur la gélatine, le lait cystéiné, différents sucres. Les résultats de cette galerie sont vérifiés par la méthode de Willis, neutralization spécifique de la lécithinase par l'immunsérum anti-perfringens.

Exemple des saucissons secs

Avec les saucissons secs, deux types de suspensions-mères sont préparés: l'un en atmosphère normale, l'autre en atmosphère réductrice.

Suivant la méthode courante: Prélèvement de 20 g de l'échantillon, addition de 180 ml de tryptone-sel désaérée, mixage au Turmix pendant 1 min, puis transvasement aseptique de la suspension dans un tube de 20 pour pasteurization 10 min à 80°.

Suivant la seconde technique, réalisée en atmosphère réductrice (azote 99%, hydrogène 1%), le prélèvement, la dilution, le mixage s'opèrent comme précédement, mais on a soin avant l'opération, de faire arriver dans le bocal du mixer un courant de mélange réducteur stérile, afin d'éliminer entièrement l'air. On entretient le courant gazeux pendant tout le broyage. Les manipulations sont poursuivies comme précédemment en évitant au maximum le contact de la suspension et de l'air.

On compare dans cette partie la valeur du V.F. sulfite à celle de 2 milieux liquides, celui de Rosenow et le R.C.M. ensemencés pour une série avec la suspension normale et pour une autre avec la suspension réduite.

Pour chaque type de suspension, nous ensemençons suivant la technique de Mac Grady les quantités suivantes: 6 tubes reçoivent chacun 5 ml de suspension soit 3 fois 1 g de matière première analysée. 1 ml introduit dans 3 tubes correspond à 3 fois 1/10 de g et 0,1 ml de la même suspension dans 3 tubes de milieu permet la numération des spores dans 3 fois 1/100 de g; les milieux liquides ains ensemencés sont incubés à 37° pendant 3 jours. Le V.L. sulfite est surveillé soigneusement dès l'apparition de colonies noires, car très rapidement le tube noircit totalement, rendant toute lecture impossible. Très souvent, pour cette raison, on ne peut laisser l'incubation se poursuivre 3 jours.

Il est nécessaire de rechercher dans chaque tube de milieu liquide, la présence de *Clostridium* et éventuellement de *Clostridium perfringens* et de vérifier que les colonies noires en V.F. sulfite sont bien dues à des *Clostridium* et combien d'entre elles sont des *Clostridium perfringens*.

Une première série d'essais nous montre l'importance de la contamination par des *Bacillus* et nous sommes alors immédiatement confrontés à cette difficulté: Isoler des *Clostridium* en présence de nombreux *Bacillus*. A partir de colonies noires en V.F. sulfite, nous essayons la purification des *Clostridium* par épuisement dans des tubes milieu profond S.S.P.S. (soytone, sulfite, polymyxine, sulfadiazine). Le solde de cette série d'essais est un échec. Nous pensons, pour permettre la culture des spores de *Clostridium perfringens* devenues lysozyme dépendantes, à ajouter 4µg de lysozyme par ml de milieu selon Sébald (Cassier & Sebald, 1969). Nous sommes alors rapidement surpris de trouver en V.F. sulfite des colonies noires beaucoup moins contaminées par des *Bacillus*. Nous étendons alors l'addition de lysozyme 4 µg/ml à tous les milieux liquides ou solides et même à ceux retenus pour l'isolement.

Nous pratiquons donc maintenant comme suit: les colonies noires en V.F. sulfite lysozyme sont passées en V.L. profond lysozyme. En présence de germes anaérobies, on isole sur V.L. sang lysozyme coulé en boîte de Petri. On vérifie le type respiratoire de chaque type de colonies et en même temps, on les repique sur milieu de Willis + lysozyme, afin de rechercher *Clostridium perfringens*. Les milieux liquides (Rosenow et R.C.M.) additionnés de lysozyme sont isolés sur V.L. sang lysozyme. Chaque type de colonies est analysé comme précédemment.

Résultats

L'utilisation de ces méthodes nous amène aux résultats suivants: Nous les donnons dans 2 tableaux; l'un concernant le travail sur les poulets (1) l'autre se rapportant aux saucissons secs (2).

Tableau 1

Nombre de spores Cl. perfringens *par* cm^2 *de peau prélèvée sur les carcasses des 35 poulets contaminés*

Nombre le plus probable avec les milieux:	Nombre de poulets	Nombre de spores/cm^2
	11	0
	14	0,20
Rosenow	3	0,45
	6	0,78
	1	1,30
	Total: 35	Total: 10,13
	11	0
	17	0,20
	3	0,45
R.C.M.	1	0,78
	1	1,30
	1	2,30
	1	7,90
	Total: 35	Total: 17,03

Présence de spores de *Cl. perfringens* à la surface de 100 carcasses de poulets.

(1) présence en Rosenow et R.C.M. simultanément: 13 poulets
(2) présence en Rosenow et absence en R.C.M.:   11 poulets
(3) présence en R.C.M. et absence en Rosenow: 11 poulets
(4) absence en Rosenow et R.C.M.:   65 poulets

soit 35 poulets contaminés sur 100.

Tableau 2

No.	V.F. Reduit Clostridium	V.F. Reduit Cl. perfringens	V.F. Non reduit Clostridium	V.F. Non reduit Cl. perfringens	Rosenow Reduit Clostridium	Rosenow Reduit Cl. perfringens	Rosenow Non reduit Clostridium	Rosenow Non reduit Cl. perfringens	R.C.M. Reduit Clostridium	R.C.M. Reduit Cl. perfringens	R.C.M. Non reduit Clostridium	R.C.M. Non reduit Cl. perfringens
1	1	Avant utilisation du Lysozyme	2		0,4		1,1		1,5		2,3	
2	2		10		9,3		0		4,3		4,3	
3	0		0		0		0,4	0,4	0		0	
4	1		2		9,3		2,3		0,4		0,4	
5	0		0		0		0		0		0	
6	0		1		0,7		1,6		0		0	
7	0		0		0,4		1,1		0,4		0,4	
8	5,3	4,3	2	1,3	3,6	0,7	12	0	15	0,4	12	0
9	0,9	0,3	0,9	0	1,5	0,4	2,1	0	1,1	0,4	0,7	0,4
10	3,6	0,3	3,2	0,3	2,3	0,4	9,3	0,4	12	0	29	0,4
11	5,1	2,1	3,3	1,2	7,5	0,4	12	0,9	15	0	4,3	0,9
12	3,6	0,9	3,3	1,2	0,9	0,4	9,3	0,4	15	0,9	15	1,1
13	0,3	0	0,6	0	1,5	0	6,4	0	1,1	0	0,9	0
14	1,9	0	1,2	0,3	21	0	21	0	9,3		2,9	
15	5,1	0	7,6	0	24	0	9,3	0	9,3	0	9,3	0
16	3,3	0	1,2	0	4,3	0	0,7	0	0,7	0	0,9	0
17	3,3	0	3	0	110	0	110	0,4	5,3	0	0,4	0
18	20	20	50	50	4,3	2,4	9,3	4,3	9,3	2,1	9,3	4,3
19	3	2	14	5	0,9	0	9,3	0,4	0,9	0	4,3	0
20			2	0			0,9	0			0,9	0
21			0	0			0,4	0			0,4	0
22			3	0			2,3	0,4			0,9	0
23			0	0			0,4	0			0,4	0
24			0	0			0,7	0			0,4	0
25			4				4,3				9,3	
26			0	0			0	0			0,3	0
27			4	0			4,3	0			4,3	0
28			0,5	0			0,7	0			0,4	0
29			1				0,7	0			0,4	0
30			0	0			0	0			0,4	0
31			0,9	0			0,4	0			0,4	0
32			0,3	0,3			2,3	0,4			1,5	0,4

Discussion

L'intérêt de l'homogénéisation en atmosphère réductrice fait l'objet de la première partie de ce travail.

Il faut considérer l'effet de la réduction sur le dénombrement des *Clostridium* totaux d'abord, et sur celui du *Clostridium perfringens* ensuite.

L influence du mixage en atmosphère d'azote et d'hydrogène sur la numération de l'ensemble des *Clostridium* apparaît dans le tableau n° 3. Dans le cas du V.F. sulfite + lysozyme, nous trouvons au total 105,3 spores, après mixage en atmosphère normale contre seulement 59,4 après réduction. Cette manipulation a, de toute évidence, un effet néfaste. Par contre, en milieu de Rosenow ou R.C.M., ses conséquences sont moins apparentes, les chiffres trouvés étant peu différents mais malgré tout en faveur de l'homogénéisation en atmosphère normale.

Tableau 3

Milieux	Nombre total de spores de *Clostridium* après mixage en atmosphère	
	Réductrice	Normale
V.F. sulfite	59,4	105,3
Rosenow	201,9	217,2
R.C.M.	!00,6	96,4

Tableau 4

Milieux	Nombre total de spores de *Clostridium perfringens* après mixage en atmosphère	
	Réductrice	Normale
V.F. sulfite	29,9	59,3
Rosenow	4,7	6,8
R.C.M.	3,8	7,1

Sur le dénombrement des spores de *Clostridium perfringens* (tableau n° 4), le rôle de la réduction est absolument désastreux puisqu'il provoque une perie de près de 50% des spores—29,9 contre 59,3 en V.F. sulfite + lysozyme 3,8 contre 7,1 en R.C.M. En milieu de Rosenow, les différences sont un peu moins sensibles mais la variation est en faveur du mixage, en absence d'agent réducteur. Parmi les *Clostridium* dénombrés, *Clostridium perfringens* apparît comme le plus sensible à cette modification de technique. C'est en V.F. sulfite qu'on trouve la plus forte quantité de *Clostridium perfringens,* mais c'est aussi dans ce milieu que se fait sentir davantage l'influence de la réduction (voir tableau n° 3).

Contrairement à notre hypothèse de départ, l'homogénéisation des échantillons de produits alimentaires, dans une atmosphère d'azote et d'hydrogène n'est pas bénéfique au dénombrement des spores de bactéries anaérobies mais bien plus, elle est néfaste à la numération de celles de *Clostridium perfringens.*

L'étude comparative de la valeur des milieux liquides et des milieux solides dans le dénombrement des *Clostridium* totaux et de *Clostridium perfringens* en particulier fait l'objet de la deuxième partie de ce travail.

Elle est envisagée en utilisant comme matériel expérimental, la peau des carcasses de poulet et les saucissons secs.

Sur 100 carcasses examinées, seules 35 possèdent à leur surface des spores de *Cl. perfringens*. Le tableau n° 1 rapporte le nombre de spores trouvées, par cm^2 de surface. Les deux milieux retenus, Rosenow et R.C.M. sont à peu près équivalents avec un léger avantage, plus apparent que réel, au R.C.M.

Lorsqu'on utilise le saucisson sec comme matériel expérimental, on bute immédiatement sur la difficulté bien connue que nous signalions précédemment = isoler des *Clostridium* d'un milieu contenant de nombreux *Bacillus*.

Les recherches basées sur l'emploi de milieux classiques sans lysozyme ne nous ont apporté que des déboires. Les 20 analyses tentées se répartissent en 14 échecs et 6 résultats interprétables dont 4 seulement ont permis l'isolement de *Clostridium,* les 2 autres étant négatives. Dans les mêmes conditions mais en présence de lysozyme, les résultats de 18 échantillons sur 20 sont interprétables. Lorsqu'on tente un isolement en V.F. sulfite, les colonies noires montrent presque toujours une association de *Clostridium* et de *Bacillus*. Fort heureusement, l'utilisation du lysozyme dans tous les milieux, à raison de 4 μg/ml, change complèment les données du problème. Les *Bacillus* sont considérablement inhibés dans leur développement. Généralement, les colonies noires en V.F. suflite + lysozyme ne sont plus des associations *Clostridium-Bacillus*. En présence de lysozyme, sur V.L. sang, les *Bacillus* ne donnent plus, en atmosphère anaérobie, de colonies envahissantes; bien plus, elles sont souvent petites. Aussi, nous nous sommes posé la question de savoir si le lysozyme, bien que favorable à la germination des spores de *Clostridium perfringens* n'allait pas modifier considérablement les résultats des numérations. Le tableau 5 exprime la réponse à cette question. Ce sont les résultats de 7 dénombrements effectués en V.F.

Tableau 5

Echantillons	Nombre de spores de *Clostridium*	
	Avec lysozyme	Sans lysozyme
1	6	11
2	3	2
3	13	17
4	11	11
5	5	2
6	25	11
7	4	4
	Total 67	Total 58

sulfite, avec et sans lysozyme. Mis à part deux exemples, le chiffre trouvé avec lysozyme est égal ou supérieur à celui noté dans les conditions ordinaires. Le nombre des comparaisons est peut-être faible, mais il est limité par la difficulté d'isolement des *Clostridium* en absence de lysozyme. Puisqu'il ne semble pas diminuer la valeur des résultats enregistré et étant donné les grandes facilités qu'il apporte nous ne pouvons que conseiller son emploi dans le dénombrement des *Clostridium* dans les saucissons secs, et sans doute à partir d'un milieu fortement contaminé par des *Bacillus*.

Cette difficulté surmontée, nous pouvons alors reprendre notre comparaison de la valeur de V.F. sulfite + lysozyme et des milieux liquides + lysozyme (Rosenow et R.C.M.). Le tableau 6 résume les résultats de ces investigations. Dans la numération des *Clostridium* totaux, les chiffres obtenus qui sont de toute évidence les plus élevés en milieu de Rosenow, sont plus faibles en R.C.M. ou V.F. sulfite, mais peu différents l'un de l'autre. Il faut remarquer que les colonies noires en V.F. sulfite ne sont constituées que des germes sulfito-réducteurs alors qu'on isole des milieux liquides, tous les *Clostridium* réducteurs ou non.

Tableau 6

Nombre d'échant. de saucissons secs	Nombre total de spores					
	V.F. sulfite		Rosenow		R.C.M.	
	Cl. totaux	*Cl. perfringens*	*Cl.* totaux	*Cl. perfringens*	*Cl.* totaux	*Cl. perfringens*
32	121		234,6		116,4	
25		59,6		7,6		7,5

Enfin, lorsque parmi les *Clostridium*, on recherche particulièrement *Clostridium perfringens*, d'un intérêt certain en bactériologie alimentaire (Sebald & Cassier, 1970), le milieu de choix est de toute évidence le V.F. sulfite + lysozyme. En effet, dans 25 échantillons de saucisson sec, on dénombre 59,6 spores de *Clostridium perfringens* avec ce milieu sulfité, alors que les autres milieux n'en révèlent que respectivement 7,6 et 7,5.

Conclusions

De ces différentes recherches, on peut tirer quatre conclusions principales:

(1) L'homogénéisation en atmosphère réductrice azote hydrogène des échantillons de produits alimentaires en vue du dénombrement des spores de

bactéries anaérobies, est à déconseiller et même à proscrire si l'on s'intéresse principalement à *Clostridium perfringens.*

(2) Pour les numérations de *Clostridium* et surtout en milieux fortement contaminés par des *Bacillus,* il est recommandable d'ajouter du lysozyme à raison de 4 µg/ml, ce qui facilite la germination des spores de *Clostridium perfringens* et en général l'isolement des anaérobies sporulés.

(3) L'exemple des saucissons secs nous permet d'affirmer que pour la recherche et le dénombrement des spores de *Clostridium,* il faut choisir les milieux liquides. Par contre, pour effectuer la même opération avec *Clostridium perfringens,* les milieux au sulfite semblent nettement préférables. Ce fut le cas du V.F. sulfite pour les saucissons secs.

(4) Les résultats de l'étude comparée des milieux liquides pour la recherche des spores de *Clostridium perfringens* ne nous autorisent pas à accorder la préférence, d'une manière systématique, au milieu de Rosenow ou au milieu R.C.M.

Bibliographie

Cassier, M. et Sebald, M. (1969). Germination lysozyme-dépendante des spores de *Clostridium perfringens*–ATCC 3624 après traitement thermique. *Annls Inst. Pasteur, Paris* **117,** 312.

Sebald, M. et Cassier, M. (1970). Toxi-infections alimentaires dues à *Clostridium perfringens. Bull. Inst. Pasteur* **68,** 7.

Discussion

Ingram

Am I right in understanding that the numbers shown in your slides represent totals of MPN counts over all the samples (30 or so) represented? Though the differences appear substantial, some of them would scarcely be significant if the figures represented individual MPN counts.

Fievez

The numbers shown in my slides represent totals of MPN counts over all the samples.

Skovgaard

You asked for an explanation of the lysozyme effect towards *Bacillus* spp. I would offer the explanation that it interferes with oxygen uptake. Lysozyme inhibits Gram negative but not Gram positive rods. It will not inhibit the microaerophilic lactobacilli or streptococci. So it probably interferes with oxygen metabolism of the obligately aerobic *Bacillus* but the anaerobic and microaerophilic organisms are lysozyme resistant.

Ashton

In addition to inhibiting *Bacillus* spp., lysozyme may permit germination of heat-damaged spores by replacing an inactivated lytic enzyme system. Such heat-damaged spores would not be able to germinate in the absence of lysozyme and hence would not be detected. The phenomenon of lysozyme-dependent

outgrowth of heat damaged *Cl. perfringens* spores has been well described in the
literature.

Rehm

I do not agree with Dr Skovgaard, because the action of lysozyme is well
known. It splits the 1,4 glucoside linkages between *N*-acetyl-glucosamine and
muramic acid, especially in Gram-positive bacteria.

Roberts

The effect of lysozyme on the $\beta 1 : 4$ link of murein should affect *Bacillus*
as well as *Clostridium* spp. The effect of lysozyme on *Cl. perfringens* is not a
general phenomenon—it is effective only on heat sensitive spores and has no
effect on heat-resistant spores.

I should like to ask Dr Fievez whether he has determined that all spores in a
pure culture of *Cl. perfringens* form black colonies in VF—sulphite agar, i.e. that
non-black colonies are absent?

Fievez

No.

The Use of Fluorescent Antibodies and Other Techniques for the Detection of *Clostridium botulinum*

ANN BAILLIE, J. S. CROWTHER AND A. C. BAIRD-PARKER

Unilever Research Laboratory, Colworth/Welwyn,
Colworth House,
Sharnbrook,
Bedford, England

The fluorescent antibody technique was examined as a method for detecting whole cells of *Clostridium botulinum*. Fluorescein labelled antibodies prepared in rabbits against whole cells of *Cl. botulinum* types A, B, C, E and F, were used to stain overnight cultures of 158 strains of *Cl. botulinum* and 54 strains of other clostridia. Proteolytic types A, B and F (but not non-proteolytic types B and F) stained with sera prepared against type A, whilst non-proteolytic types B, E and F stained with sera prepared against non-proteolytic types B, E and F. Types C and D stained with sera prepared against C and E. Few cross-reactions were found between *Cl. botulinum* and other species of clostridia, but some of the non-toxic organisms resembling *Cl. botulinum* gave positive results. Electroimmunodiffusion was examined as a method for detecting *Clostridium botulinum* toxin. When culture supernatants were electro-phoresed in agarose gels containing botulinum anti-toxin, precipitation lines were produced. These preliminary experiments will be described and the future of the method for the rapid detection of botulinum toxin will be discussed.

Introduction

THERE ARE 2 distinct approaches to the laboratory detection of *Clostridium botulinum*, namely to look for intact cells or spores using cultural or microscopic techniques, or to look for soluble toxins using *in vitro* or *in vivo* tests. Several studies have been designed to evaluate the fluorescent antibody technique as a method for detecting intact cells or spores of *Cl. botulinum* (Kalitana, 1960; Bulatova & Kabanova, 1960; Georgala & Boothroyd, 1966). The earlier reports were sometimes conflicting, but these were rationalized by Walker & Batty (1964) and Lynt *et al.* (1971) who showed that *Cl. botulinum* could be divided into 3 groups on the basis of their fluorescent antibody cross-reactions: type A and proteolytic types B and F; types C and D; type E and non-proteolytic types B and F.

The fluorescent technique is of little value, however, for examining foods in which the bacterial cells have disintegrated through lysis or mechanical damage. Such situations demand a fast, reliable test to detect soluble botulinum toxin. The accepted method is the mouse toxicity test, which, although simple, has disadvantages. It is slow, taking up to 3 days to confirm the results, and non-specific deaths may occur due to protein shock and other causes.

There is, therefore, a need for a more rapid, specific test to replace mouse toxicity. Several methods have been considered, including haemagglutination, bentonite flocculation (Johnson *et al.*, 1966) and immunodiffusion (Sugiyama *et al.*, 1967; Vermilyea *et al.*, 1968). More recently, Miller & Anderson (1971) have described an electro-immunodiffusion technique to assay purified *Cl. botulinum* type A toxin.

In this paper we describe the use of the fluorescent antibody technique to survey a collection of some 200 strains of *Cl. botulinum* and other clostridia, and also an example of the use of fluorescent antibodies to enumerate *Cl. botulinum* in a sample of food. We outline preliminary studies on the application of the electro-immunodiffusion test to detect *Cl. botulinum* toxin.

Materials and Methods

Some 212 strains of clostridia were obtained from various sources. They comprised: *Clostridium botulinum* type A (50 strains), type B (18 strains), type C (15), type D (1), type E (67), and type F (7); *Clostridium sporogenes* (14), *Cl. tetani* (7), *Cl. perfringens* (2), *Cl. septicum* (3) and *Cl. oedematiens* (2). Twenty-four non-toxic cultures resembling *Cl. botulinum* and 2 unknown cultures were also studied.

Detection of Cl. botulinum *cells by fluorescent antibody*

Fluorescent sera

Six different fluorescein labelled antisera were tested. Antisera to *Cl. botulinum* types A, C and E were obtained commercially (Wellcome Reagents Limited, Beckenham, Kent). Antisera to 2 strains of non-proteolytic type B and 1 strain of non-proteolytic type F *Cl. botulinum* were kindly prepared by Dr P. D. Walker at the Wellcome Research Laboratories, Beckenham.

Staining method

Cells from 16 h cultures in Reinforced Clostridial Medium (RCM) were harvested by centrifugation and resuspended in sterile saline. Smears were made on clean glass slides (0·8–1 mm thick), air-dried and fixed in acetone for 10 min. One drop of the appropriate antiserum was spread evenly over the smear. Smears were placed in large Petri dishes containing moistened filter paper and allowed to stain for 30 min. Excess antiserum was removed by washing with several changes of phosphate buffered saline, pH 7·1. Smears were blotted gently and mounted in buffered glycerol pH 9·0.

Fluorescent microscopy

Slides were examined using a Reichert Neopan microscope with an HBO 200 mercury vapour lamp and a ×40 or a ×100 glycerin immersion objective. The filters for normal working were a BG 12 (3 or 4·5 mm thick) primary, with an OG 1 (1·5 mm) as secondary. When critical UV illumination was required, the primary filter was a combined UG 1 (2·5 mm) + BG 12 (2·5 or 3 mm) in conjunction with a GG 9 (1 mm) secondary filter. Both dark ground and bright field condensers gave satisfactory results.

Detection of Cl. botulinum *toxin*

Mouse toxicity test

0·3 ml of each supernatant was mixed with 0·3 ml 2% (w/v) streptomycin sulphate (Glaxo Laboratories Limited, Greenford), and 0·5 ml of the resulting mixtures were injected intraperitoneally into 20 g mice. Any toxic reactions were checked using type specific *Cl. botulinum* antitoxins.

Electro-immunodiffusion test

Cl. botulinum *antitoxins*. Eleven commercially available *Cl. botulinum* anti-toxins were obtained: monovalent antisera types A–E (Institut Pasteur, Paris); polyvalent antiserum types A–F (State Serum Institute, Copenhagen); mono-valent antisera types A and B (Wellcome Reagents Limited, Beckenham, Kent); monovalent type E and trivalent types A, B and E (Connaught Medical Research Laboratories, Willowdale, Ontario); and monovalent British Standard Antitoxin types A–F (Medical Research Council, Hampstead Laboratories, Holly Hill, London).

In addition *Cl. botulinum* type A antitoxin (hereafter referred to as Colworth antitoxin) was prepared in the laboratory by immunizing rabbits with formalized semipurified *Cl. botulinum* type A toxin (kindly supplied by Dr P. D. Walker).

Electro-immunodiffusion. Culture supernatants of selected strains of *Cl. botulinum* were electrophoresed in agarose gels following the method of Miller & Anderson (1971). Clean lantern slide cover glasses (8·1 × 8·1 cm) were coated with 2% Ionagar (Oxoid) and dried. The slides were overlayered with a mixture of 0·5 ml of the appropriate antiserum and 12·5 ml 1% agarose. The latter was prepared in barbitone acetate buffer (Oxoid, 0·05 μ, pH 8.6) containing 0·37 g calcium lactate/l. When the gels had set, wells were cut and filled with *c*. 5 μl culture supernatant. The slides were placed horizontally in a water-cooled electrophoresis tank (Shandon) and electrical connections between the slides and the electrophoresis buffer made with absorbent lint. A potential of 150 V was applied and the current adjusted to *c.* 5 A/cm. After 30 min, slides were removed from the electrophoresis tank, washed overnight in normal saline, rinsed in several changes of distilled water and dried in a warm air oven.

The slides were immersed for 20 min in stain containing Amido Black (Bayer) 0·25 g, mercuric chloride 2·5 g, thiazine red 0·05 g, glacial acetic acid 3·0 ml and distilled water 97 ml (A. W. Anderson, pers. comm.). Excess stain was rinsed off with tap water and the slides were decolourized in 2% acetic acid until the background was clear (5–10 min).

Detection of Cl. botulinum *in food*

A sample of fresh meat inoculated with spores and vegetative cells of *Cl. botulinum* type A and stored at room temperature was macerated in gelatin phosphate buffer (pH 7·0) and the numbers of *Cl. botulinum* estimated in pasteurized (60° for 30 min) and unpasteurized samples by 5 tube MPN counts in Differential Reinforced Clostridial Medium (Freame & Fitzpatrick, 1971). Tubes were incubated at 30° and positive (blackened) tubes were sampled the day after blackening appeared. Presence of *Cl. botulinum* in positive tubes was determined by the fluorescent antibody technique for intact cells and by the detection of toxin in cell-free supernatants by the mouse test.

Results

Staining reactions of cultures of Cl. botulinum

The staining reactions of the cultures are shown in Table 1. The strains of *Cl. botulinum* could be divided into 3 groups on the basis of the fluorescent antibody reaction.

Group 1. Type A and proteolytic types B and F. These organisms stained with type A antiserum but not usually with antisera against type E or non-proteolytic types B or F.

Group 2. Types C and D. These stained with antisera against types C and E but not antiserum against type A.

Group 3. Type E and non-proteolytic types B and F. These stained with type E antiserum and non-proteolytic type B and F antisera, but not with type A or C antisera.

Five non toxic strains resembling *Cl. botulinum* stained with antiserum to *Cl. botulinum* type A, and 3 strains with type C antiserum. Three stains stained with non-proteolytic type B antiserum and 2 with antiserum to non-proteolytic type F. One strain of *Cl. septicum* stained with antisera to non-proteolytic types B and F.

Table 1

Fluorescent antibody reactions of 212 strains of clostridia tested with antisera to
Cl. botulinum

Organism	No. of strains tested	No. giving positive reaction with antiserum type				
		A	B(NP)*	C	E	F(NP)*
Cl. botulinum type A	50	50	3	5	0	4
Cl. botulinum type B	11	11	0	2	1	0
Cl. botulinum type B (NP)	7	1	7	0	6	7
Cl. botulinum type C	15	0	1	15	15	4
Cl. botulinum type D	1	0	1	1	1	1
Cl. botulinum type E	67	0	60	2	67	65
Cl. botulinum type F	4	4	0	0	0	0
Cl. botulinum type F (NP)	3	0	3	0	3	3
Cl. sporogenes	14	1	0	0	0	0
Cl. tetani	7	0	0	0	0	0
Cl. septicum	3	0	1	0	0	1
Cl. oedematiens	2	0	0	0	0	0
Cl. perfringens	2	0	0	0	0	0
Non-toxic	19	3	2	0	0	1
Non-toxic (NP)	5	2	1	3	1	1
Unknown	2	0	–	0	0	–

* NP, non-proteolytic strains.

Detection of Cl. botulinum *in fresh meat*

Figure 1 illustrates the use of the fluorescent antibody technique to detect *Cl. botulinum* type A in MPN counts of a macerate of fresh meat inoculated with the organism. For the unpasteurized sample, the number of toxic tubes rose slowly and by day 12 of this incubation gave a count of only $7{\cdot}0 \times 10^6$ cells/g meat. In contrast, the fluorescent count rose rapidly and by day 5 had reached a nearly maximum count of $1{\cdot}2 \times 10^8$ cells/g meat. When the sample was pasteurized the toxic and fluorescent counts rose at the same rate and reached the same maximum value of $7{\cdot}9 \times 10^5$ cells/g.

Electro-immunodiffusion

Table 2 summarizes the results of electrophoresing culture supernatants of each of the toxigenic types of *Cl. botulinum* in agarose gels containing each of the 11 antitoxins. A typical electroimmunodiffusion gel is shown in Fig. 2.

Type A antitoxins

Wellcome A antitoxin gave precipitin lines with types A and B cultures and with proteolytic type F. The British Standard type A and Colworth antitoxins

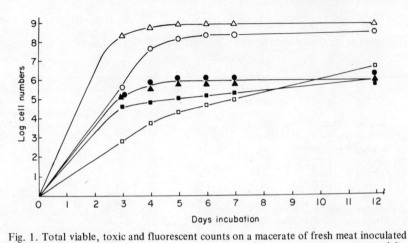

Fig. 1. Total viable, toxic and fluorescent counts on a macerate of fresh meat inoculated with *Cl. botulinum* type A. △——△, Total count; ○——○, fluorescent count; □——□, toxicity count; ▲——▲, ●——●, ■——■, pasteurised.

Table 2

Electroimmunodiffusion reactions of supernatants of cultures of Cl. botulinum *and* Cl. sporogenes *tested against* Cl. botulinum *antitoxins*

Antitoxin		Types of culture tested								*Cl.*
Type	Source	A	B	B(NP)*	C	D	E	F	F(NP)	*sporogenes*
A	Wellcome	+†	+	+	−‡	−	−	+	−	−
	Pasteur	−	−	NT§	−	−	−	−		−
	British standard	+	+	NT	−	−	−	+	−	+
	Colworth	+	+	NT	−	−	−	+		+
	Wellcome	+	+	NT	−	−	−	−		−
B	{ Pasteur	+	+	NT	−	−	−	+		+
	{ British Standard	+	+	NT	−	−	−	+		+
C	{ Pasteur	−	−		−	−	−	−		−
	{ British Standard	−	−		+	+	−	−		+
D	{ Pasteur	−	−		+	+	−	−		−
	{ British Standard	−	−		+	+	−	−		−
	{ Pasteur	−	−		−	−	+	−		−
E	{ British Standard	−	−	+	−	−	+	−	+	−
	{ Connaught	+	+		+	+	+	+	+	
Poly valent A–F	State Serum Institute	+	+	+	+	+	+	+	+	+
ABE	Connaught	+	+	+	+	+	+	−	+	

* NP, non-proteolytic; † +, precipitin lines detected; ‡ −, no precipitin lines detected; § NT, tested.

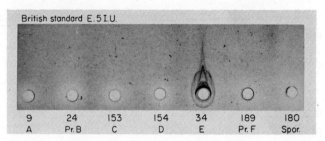

Fig. 2. Electroimmunodiffusion of culture supernatants of various strains of *Cl. botulinum* in an agarose gel containing British Standard *Cl. botulinum* type E antitoxin at a concentration of 5 i.u/ml. Strains: 9, type A; 24, proteolytic type B; 153, type C; 154, type D; 34, type E; 189, proteolytic type F; 180, *Cl. sporogenes*.

behaved similarly but also gave lines with *Cl. sporogenes*. Antiserum from the Pasteur Institute failed to show precipitin lines with any of the cultures tested.

Type B antitoxins

Wellcome, Pasteur and British Standard type B antitoxins gave precipitin lines with culture supernatants of type A and proteolytic type B cultures. Pasteur and British Standard sera also gave lines with proteolytic type F and *Cl. sporogenes*.

Type C and D antitoxins

British Standard types C and D and Pasteur type D antitoxins gave lines with types C and D only.

Type E antitoxins

British Standard and Pasteur antitoxins gave lines with type E strains. The former also gave lines with strains of non-proteolytic types B and F.

Polyvalent antitoxins

Connaught trivalent A, B and E antitoxin gave lines with strains of types C, D and F as well as with A, B and E. Polyvalent A–F antitoxin gave lines with all toxigenic types of *Cl. botulinum* and with *Cl. sporogenes*.

Discussion

The results of the fluorescent antibody study described here agree well with previous studies by Walker & Batty (1964) and Lynt *et al.* (1971) and show that strains of *Cl. botulinum* fall into 3 groups with respect to their fluorescent staining. It is noteworthy that all 158 strains of *Cl. botulinum* could be detected in cultures by the use of at least 1 of the commercially available fluorescent antisera.

The fluorescent antibody technique was of especial value in the examination of enrichment cultures of inoculated foods, since intact cells of *Cl. botulinum* could be detected before the cultures became toxic for mice.

An ideal test for *Clostridium botulinum* toxin must have several necessary qualities. It must be simple, fast, specific, sensitive, quantitative and inexpensive. The electro-immunodiffusion technique of Miller & Anderson (1971) seems to possess all these qualities. The reliability of the method depends entirely on having specific, precipitating antitoxins which do not cross-precipitate with other toxigenic types of *Cl. botulinum* or with any other organism. None of the commercially available sera matched these requirements but the British Standard type E antitoxin seemed to be the most useful. It is not known which of the precipitin lines, if any, correspond to the toxin. Some of the lines may represent non-toxic antigens such as toxin precursors, cell or phage proteins or enzymes involved in cell, toxin or phage synthesis. These may be present in cultures in higher concentrations than the toxin or may be better precipitating antigens than the toxin, and thus could prove to be better parameters of the growth of *Cl. botulinum*.

In conclusion, we feel that there is good indication that the electro-immunodiffusion technique described here could be developed to provide a simple, fast, quantitative, specific method for detecting *Cl. botulinum*.

Acknowledgements

We are greatly indebted to Dr P. D. Walker, Wellcome Research Laboratories, Beckenham, for supplying sera and toxins. We acknowledge Professor A. W. Anderson for his kind help and advice. We thank all persons who have given cultures from time to time.

References

Bulatova, T. I. & Kabanova, Y. A. (1960). Identification of the botulism pathogen with luminescent sera. *J. Microbiol. Epidem. Immunobiol.* **31**, 18.

Freame, Barbara & Fitzpatrick, B. W. F. (1971). The use of Differential Clostridial Medium for the isolation and enumeration of clostridia from food. In *Isolation of Anaerobes.* Eds D. A. Shapton & R. G. Board. London: Academic Press.

Georgala, D. L. & Boothroyd, Margery (1966). Fluorescent staining techniques applied to *Cl. botulinum* types A, B and E. In *Botulism 1966.* Eds M. Ingram & T. A. Roberts. London: Churchill.

Johnson, H. M., Brenner, R., Angelotti, R. & Hall, H. E. (1966). Serological studies of types A, B and E botulinal toxins by passive haemagglutination and bentonite flocculation. *J. Bact.* **91**, 967.

Kalitana, T. A. (1960). The detection of *Clostridium botulinum* by means of luminescent antibodies. *Bull. Biol. Med. exp. U.R.S.S.* **49**, 81.

Miller, Carol & Anderson, A. W. (1971). Rapid detection and quantitative estimation of type A botulinum toxin by electroimmunodiffusion. *Infect. Immunol.* **4**, 126.

Lynt, R. K., Solomon, H. M. & Kautter, D. A. (1971). Immunofluorescence among strains of *Clostridium botulinum* and other clostridia by direct and indirect methods. *J. Fd Sci.* **36**, 594.

Sugiyama, H., von Mayerauser, B., Gogat, G. & Heimsch, R. C. (1967). Immunological reactivity of trypsinised *Clostridium botulinum* type E toxin. *Proc. Soc. exp. Biol. Med.* **126**, 690.

Vermilyea, B. L., Walker, H. W. & Ayres, J. C. (1968). Detection of botulinumtoxins by immunodiffusion. *Appl. Microbiol.* **16**, 21.

Walker, P. D. & Batty, Irene (1964). Fluorescent studies in the genus *Clostridium* II. A rapid method for differentiating *Clostridium botulinum* types A, B and F, types C and D and type E. *J. appl. Bact.* **27**, 140.

Discussion

Sutton

Have you attempted to use this method to detect botulinum toxin in foods or in patients' serum?

Baillie

We have not yet tried this, but we hope so to do when we have increased the sensitivity of the method.

Riemann

We have also experienced—working with fluorescent antibody microscopy—the necessity of using antiserum prepared against highly purified toxins. The most sensitive *in vitro* test for botulinum toxins is probably the reverse passive haemagglutination test described by Sakaguchi; this test depends also upon the availability of monospecific antisera.

Roberts

Have you observed any difference in "pattern" in the precipitin lines produced by *Cl. sporogenes* and *Cl. botulinum*? Regarding the sensitivity, Prof. A. W. Anderson at Oregon State University still has a student working with purified toxin and hyperimmune rabbit serum, but the method is slightly less sensitive than is the mouse test.

Baillie

We have not yet examined sufficient strains of *Cl. sporogenes* to determine if there is a consistent difference in pattern obtained with this organism and the proteolytic strains of *Cl. botulinum*.

Insalata

Our work has shown that commercially available mono-specific *Cl. botulinum* antisera consistently show cross-reaction and cross-protection. Is this a possible explanation for the "false positive" reactions on fluorescent antibody method for detection of *Cl. botulinum*?

Baillie

It is likely that the different toxigenic types of *Cl. botulinum* have some shared surface antigens which would result in cross-reactions in fluorescent antibody testing.

Significance and Detection of Histamine in Food

C. IENISTEA

Institute of Hygiene and Public Health,
Bucharest,
Romania

The author has searched the literature and compiled data on the significance and detection of histamine in food. Histamine in food may be of physiological origin or it may result from bacterial decarboxylation, and, when it occurs in larger amounts, it may be harmful to humans. When foodstuffs containing histidine are infected by large numbers of bacteria forming histidine decarboxylase and stored under circumstances in which the latter are able to multiply, favourable conditions are created for enzymic action. Among the foods that may contain histamine in large amounts are some fish and fish products and some kinds of cheese. It appears likely that histaminolytic bacteria play an important part in food containing large amounts of histamine. Several instances of histamine poisoning are described in the literature but there is little knowledge of fish poisoning or of the possible role that spoilage organisms play in producing illness. In the author's opinion, histamine poisoning of alimentary origin must be included in the microbial food poisoning chapter, and not with the chemical outbreaks, as it is caused by the presence in histidine-containing foods of large numbers of bacteria forming histidine-decarboxylase. Methods of detection of histamine in foods are also mentioned.

Introduction and Historical Survey

FOOD POISONING due to histamine is infrequently mentioned in official statistics of food poisoning, and papers concerning this subject have, as a rule, been published in periodicals related to various specialities. In some countries, for example in France (Pointeau-Pouliquen, 1958), it is included with food poisoning. In the U.S.A. and some other countries, it is recorded with food-borne illness such as scombroid fish poisoning and is classified as poisoning by chemicals (Center for Disease Control, 1968, 1969, 1970). Since almost all cases of this type occur after the ingestion of fish, they are sometimes described under the heading of fish poisoning of uncertain bacteriological origin (Bryan, 1969).

The present paper is a compilation of data concerning the significance of the presence of histamine in food as related to the composition and bacterial contamination of the food.

A few years after Windau & Vogt (Knut, 1950) had synthesized histamine, Pauly (1904) described a method for the quantitative determination of imidazole compounds, and it was subsequently shown that the bacteria causing

327

putrefaction are liable to decarboxylate histidine with conversion to histamine. Histamine was isolated from soy beans undergoing putrefaction and from a drink called *Tamari shoyu* (Yoshimura, 1910).

Chemical investigations revealed the presence of histidine and histamine in tunny fish muscle (Suzuki & Yoshimura, 1909).

Chemical methods (Ackermann, 1910; Bertrand & Berthelot, 1913; Mellanby & Twort, 1912/1913), as well as biological methods based on the contraction of the isolated guinea-pig uterus (Dale & Laidlaw, 1910/1911) or ileum (Guggenheim & Loeffler, 1916), have been used for the detection of histamine.

Particular attention has been focused on histamine-forming bacteria; they were found to include several strains of *Escherichia coli* (Mellanby & Twort. 1912/1913) as well as a variety of *Aerobacter aerogenes* (Bertrand & Berthelot, 1913) termed *Bacillus aminophilus intestinalis*. This organism was detected in the intestine of patients suffering from *dementia praecox* (Bayard, 1916), and a comparative study (Jones, 1918) was made of the number of histidine decarboxylating bacteria in the faeces of such patients and of normal subjects.

The faeces of healthy human subjects were found to contain from 0–5 μg histamine/ml (Myrhman & Tomenius, 1939). In patients suffering from asthma the amount is increased, probably because of quantitative or qualitative changes in the intestinal bacteria flora.

Subsequent investigations were concerned with aerobic (Koessler & Hanke, 1919) or anaerobic (Zunz, 1919) histidine–decarboxylating and histamine-forming bacteria, and a substance causing contraction of the isolated guinea-pig intestine was demonstrated in cultures of *Clostridium perfringens* (Kendall & Schmidt, 1926).

Significance of Histamine in Food

(a) Histidine and histamine content of food

Histamine is normally present in various tissues and viscera (Table 1), and no potential hazards are related to this fact (Boyer *et al.,* 1956; Hillig, 1956; Plagnol & Aldrin, 1963).

Various foods, such as cabbages and cucumbers that have undergone lactic fermentation, canned tomatoes (Diemair 1941; Gleichmann, 1934; Keil & Kritter, 1935), several kinds of cheese (Dolezalek, 1956; Feldberg & Schilf, 1930; Geiger, 1955), some fish such as herring, sardine, tunny, raw salmon (Geiger, 1955), as well as shark muscle tissue (Oishi, 1953), also contain histamine in amounts that are harmless to man.

The histaminogenic potential of food is related to its histidine content, to the presence of bacterial histidine decarboxylase and to the environmental conditions.

Table 1

Histamine contents of some tissues and viscera *

Skin (fowl, fish)	1000–140,000 μg/100 g
Tunny fish skin	2000 μg/100 g
Muscle tissue	200–600 μg/100 g
Tunny fish muscle tissue	1000–4000 μg/100 g
Intestine; intestine preparations	10,000 μg/100 g
Liver (fowl)	100–3000 μg/100 g
Lung	1500–5000 μg/100 g
Blood (swine)	0·9–1·5 μg/100 g
Blood (rabbit)	10 μg/100 g
Blood (fowl)	50 μg/100 g

* From Dumont (1959); Henry (1960) and Peeters (1963).

Proteins contain about 2–3% histidine. The histidine content of blood is 6%, whereas meat, milk, eggs and cereals contain only small amounts of this amino-acid (Belitz & Schormüller, 1965). Fish, especially tunny, herring, mackerel and sardines, have high histidine contents. Tunny fish contains 1·5–2·5% histidine (Parrot & Nicot, 1965; Peeters, 1963; Verge & Boyer, 1958) and is therefore considered potentially histaminogenic. Histidine may be converted by decarboxylation to histamine before any sign of protein breakdown becomes apparent and this fact is relevant to the causation of disease.

Histamine formation in the *Scombridae* is not associated with apparent premonitory changes (Savi, 1957), and tunny fish becomes unsuitable for consumption before bacterial proliferation has reached the stage at which it will cause spoilage detectable by organoleptic examination (Koessler & Hanke, 1919). In stale fish histamine may be formed in the presence of bacteria such as *Achromobacter histaminum.* During storage the amount of histamine increases rapidly as confirmed by experimental test (Geiger, 1955).

When fish has acquired a faint disagreeable smell the histamine content may be as high as 3000 μg/100 g (Sager & Horwitz, 1957), and when the smell becomes strongly offensive histamine may be present in an amount of 25,000 μg/100 g.

Great differences in histamine content have been recorded among tunny fish from the same batch, the concentrations ranging from 2800–400,000 μg/100 g in some investigations (Boyer *et al.,* 1956) and from 12,000–464,000 μg/100 g in others (Ferencik *et al.,* 1961). Similar findings were made in fresh and canned sardines (Boyer & Jacob, 1961; Vidal & Mouret, 1957) and are presumed to be related to the degree of contamination of the specimens (Ramel *et al.,* 1965). There is also some evidence that the histamine content may vary in different parts of 1 fish (Boyer *et al.,* 1956; Hillig, 1956; Plagnol & Aldrin, 1963).

Not all of the histidine is converted to histamine (Hughes, 1959). This is partly due to the fact that spoiled food may contain various histamine producing bacteria (Katae & Kawaguchi, 1957, 1959), some of which, such as *Ach. histaminum* and *Proteus mirabilis*, will form larger amounts of histamine than will others.

Pseudomonas fluorescens has been shown to break down histidine (Meyer, 1968) and strains of *Betabacterium buchneri* have been described that form histamine in fish marinade (Meyer, 1962, 1968; Winkle, cited by Meyer, 1968). Histamine becomes demonstrable in fish pickle when the viable count of bacteria exceeds 10^7/ml (Meyer, 1968).

The possible presence of histamine in sterilized cans is due to its heat resistance when the temperature at which sterilization is carried out is too low to destroy the amine (Gounelle, 1960; Gounelle & Pointeau-Pouliquen, 1961). Histamine is partially destroyed in 3 h at $102°$, and 90 min at $116°$ in 250 g cans.

Apart from fish, few foodstuffs have been examined for their histamine content. Cheese may contain large amounts of amines produced by bacteria acting on the amino acids of casein (Asatoor *et al.*, 1963). Histamine may be present and has been demonstrated in several varieties of cheese, such as Romadour, Olomouo (Olmützer), Gruyère and Brynza (Dolezalek, 1956) as well as Trappist, Tilsit and Roquefort (Swiatek & Kisza, 1959; Uuspaa & Torsti, 1951). In meat products it is either absent (Cantoni *et al.*, 1970) or present in comparatively small amounts, i.e. 130–500 μg/100 g in some sausages and 400–1550 μg/100 g in crude dried sausage. If preparation is faulty the histamine content of such products may increase to 10,000 μg/100 g (Henry,1960).

One particular instance that has been mentioned (Henry, 1960) concerns therapeutic liver preparations to be taken by mouth, for which the 1949 Codex specified a maximum admissible histamine content of 300 μg/g dry substance.

Blackwell *et al.* (1969) investigated yeast preparations and recorded histamine contents up to 2800 μg/g. The same workers demonstrated that histamine-free preparations may be obtained by modifying the conditions of preparation.

The presence of histidine decarboxylases and histamine formation were recorded also in higher plants (Werle & Raub, 1948). Wine may contain histamine in amounts up to 1000 μg/100 ml (Ough, 1971). Histamine may likewise be present in must containing large amounts of nitrogen compounds.

(b) The role of histidine decarboxylating bacteria

A great variety of bacteria play a decisive part in the production of histamine in food. Feldberg & Schilf (1930) considered the occurrence of histamine in nature to be due largely to the existence of these bacteria.

Histidine decarboxylating bacteria have been isolated from the intestine of healthy human subjects, of patients and of animals, as well as from food of normal appearance or displaying signs of spoilage. Histidine decarboxylating activity varies according to the bacterial species and strain, and depends on environmental conditions (Ienistea, 1971). In addition, some bacteria display a certain degree of specificity as regards the amino acids present in the substrate. Strains forming large amounts of histamine and strains producing insignificant amounts of this substance have been described (Kimata & Tanaka, 1954a, b).

The number of histamine-forming bacteria may influence to a certain extent the amount of histamine occurring in bacterial suspensions or food. Experimental investigations in tunny fish (Shifrine $et\ al.$, 1959) showed that alteration commenced after 3–4 days of storage at room temperature or at 30°. At that time the muscle tissue contained $4\cdot6 \times 10^8$–$6\cdot3 \times 10^9$ bacteria/g and the histamine content in various samples ranged from 0–2·0%.

The present author (Ienistea $et\ al.$, 1960) did not find a direct relationship between bacterial numbers and the histamine content in all instances and considered that the age of the cells may influence their enzyme activity. The number of bacteria in the suspensions investigated was about 1×10^9/ml which is close to that found in altered food. Under these circumstances the amount of histamine produced by various enteropathogenic and nonpathogenic strains of $E.\ coli$ ranged from 0·5–28·5 mg/100 ml with 1 strain, $E.\ coli$ 0126 B16, producing 83·5 mg histamine /100 ml.

It has been shown that in some Gram-positive bacteria grown on artificial media histidine decarboxylase rapidly disappears, whereas in others it may remain active for periods of up to 10 years (Koessler $et\ al.$, 1928). In appropriate culture media histamine production starts after 14–16 h of incubation, persists for about 4–5 days, and subsequently decreases (Gale, 1940). Young cells display comparatively low levels of activity (Gale, 1940).

For histamine to be produced by bacteria, the presence of amino acids, such as glycine, leucine, arginine, alanine (Feldberg & Schilf, 1930), asparagine and cystine (Eggerth, 1939), as well as sugars is required. Histidine decarboxylation occurs in an acid medium, within a pH range from 5·0–8·0 with an optimum at pH 5·0–5·5 (Eggerth, 1939). The appearance of histamine in the culture medium is associated with a decrease in pH (Hanke & Koessler, 1922; Koessler $et\ al.$, 1928).

Temperature influences enzyme activity, i.e. bacterial suspensions are more active at 27° than at 37° (Eggerth, 1939, 1940). In Japanese fish (Kimata & Kawai, 1953) histamine formation was found to occur at 20°. In tunny fish held at 30° for 2 days, 20% of the histidine was converted to histamine (Van Veen & Latuasan, 1950).

Several groups of bacteria are known to produce histamine and some authors

(Koessler, Hanke & Sheppard, 1928; Scheibner, 1968; Baumgarten, 1970) have pointed out the higher the incidence rate of such micro-organisms in the genera *Escherichia* and *Salmonella*.

Within the family *Enterobacteriaceae*, histamine-forming bacteria have been found in the genera *Escherichia, Aerobacter, Salmonella, Shigella* and *Proteus*. Other species possessing this ability belong to the genera *Achromobacter, Pseudomonas, Streptococcus, Lactobacillus, Betabacterium* and *Clostridium*.

Strains of *Cl. perfringens* have been investigated in this respect (Kendall & Gebauer, 1930, 1931; Kendall & Schmitt, 1926; Knut, 1950), and it has been shown that histamine is commonly present in cultures of this micro-organism and that B and C toxins cause the isolated guinea-pig intestine to contract (Prévot *et al.*, 1967). Histamine was also detected in glucose-VF broth cultures of type C beta *Cl. botulinum* and of *Cl. sordellii* (Prévot & Thouvenot, 1962). Type F *Cl. perfringens*, which causes necrotic enteritis in man, forms large amounts of histamine (Koslowski *et al.*, 1951), while strains of *Cl. sporogenes* produce small amounts (Billy, 1962). Among the bacteria known to be commonly involved in fermentations, the lactobacilli were stated by some authors to exert a strong decarboxylating action on histidine (Guirard & Snell, 1954; Rodwell, 1953). However, this was not confirmed in cheese undergoing lactic fermentation during ripening (Schormüler & Hutch, 1957).

(c) Histamine in foodstuffs involved in food poisoning

The histamine content of fish has been related by some authors (Boyer *et al.*, 1956; Ramel *et al.*, 1965) to the effects produced, as follows:

normal	less than 5000 μg/100 g
increased; may cause mild disorders in susceptible subjects	5000–10,000 μg/100 g
toxic	10,000–100,000 μg/100 g
highly toxic	over 100,000 μg/100 g

These figures are supported by findings made in cases of food poisoning in which the amounts found in tunny fish were 180,000–400,000 μg/100 g (Boyer *et al.*, 1956), up to 320,000 μg/100 g (Kriska *et al.*, 1960), over 200,000 μg/100 g (Ferencik *et al.*, 1961), 100,000–500,000 μg/100 g (Legroux *et al.*, 1946*a, b*; Legroux, Levaditi & Second, 1946; Legroux, Levaditi & Bovet, 1946), and 500,000 μg/100 g (Dumont, 1959).

Incidents have occasionally followed the ingestion of spiny lobster, shrimps or shellfish (Verge & Boyer, 1958). Large amounts of histamine have been found in fillets of anchovies or herrings (37,000–93,750 μg/100 g) (Peeters, 1963), in dried fish (400,000 μg/100 g) (Van Veen & Latuasan, 1950) and in *Nuoc-Mam*

(700 μg/ml) Cousin & Noyer, 1944). In 1 instance of food poisoning (Douglas *et al.*, 1967), 85,000 μg/100 g were found in Gouda cheese.

(d) Destruction of histamine by bacteria

Bacterial histamine must be included among the enzymic factors which influence the amount of histamine present in food at any time. Histamine is destroyed by the histaminase present in various tissues, especially in the kidney and small intestine.

Compared with the data on histamine producers, less information is available concerning histaminolytic bacteria. Such enzyme activity has been described in several Gram negative species, in particular in *Ps. aeruginosa* (Werle, 1940; Ahlmark, 1944; Knut, 1950), *E. coli* and *Cl. fesseri* (Ahlmark, 1944). It has been pointed out that various bacteria, such as *Serratia marcescens, Pr. vulgaris* and *Sarcina flava*, are able to destroy the histamine formed by *Ach. histaminum* or *Pr. morganii* (Katae & Kawaguchi, 1957). However, histaminase activity was not demonstrable in a number of *Salmonella* strains (Knut, 1950).

Histaminolytic bacteria are likely to play an important part in food containing large amounts of histamine, resulting in the establishment of an equilibrium between histamine production and destruction (Ienistea, 1971).

3. Detection of Histamine in Food

Biological and chemical methods are available for qualitative and quantitative determinations.

(a) Biological methods

These methods are based on the *in vitro* contraction of the guinea-pig uterus or intestine in the presence of histamine. They are frequently used for the assay of small amounts up to 0·1 μg. Valuable information may be obtained in this manner, but the procedure is too fastidious to yield reproducible results (Parrot & Rense, 1954; Tabor, 1955; Henry, 1960).

The classical biological method using the isolated terminal ileum of the guinea-pig has been referred to by many investigators (Barsoum & Gaddum 1935; Code & McIntire, 1956; Huidobro, 1956; Ienistea *et al.*, 1960; Kendall & Schmidt, 1926; Quevauviller, 1959; Suchet, 1959).

Intramuscular injection in the guinea-pig of an extract of material presumed to have caused food poisoning has been used (Legroux, Levaditi & Second, 1946). Some authors (Ferencik *et al.*, 1961) injected aqueous extracts of fish into the peritoneal cavity of guinea-pigs and under these circumstances death occurred at a dose level of 4·0–4·5 mg histamine/kg body weight.

Extracts may be introduced into the stomach of guinea-pigs by means of a catheter. The lethal dose is 150–200 mg histamine/kg body weight.

Biological methods have also been used for the assay of histamine in bacterial cultures (Mellanby & Twort, 1912/1913; Hanke & Koessler, 1924; Kendall & Schmitt, 1926; Koessler et al., 1928; Kendall & Gebauer, 1930; Hirai, 1933; Edlbacker et al., 1937; Eggerth, 1939; Werle, 1941; Rocha & Silva, 1944; Halpern & Walthert, 1945; Ehrismann & Werle, 1947; Werle & Raub, 1948; Ienistea et al., 1960).

Scratch tests using the suspected material have been carried out by (Douglas et al., 1967). If histamine is present a strong weal-and-flare reaction is induced which parallels the reaction to 0·1% histamine hydrochloride.

(b) Chemical methods

As histamine is highly resistant to acid conditions, a method has been devised involving hot acid hydrolysis after treatment of the material with trichloroacetic acid (Barsoum & Gaddum, 1935; Code, 1937; Peeters, 1963; Ramel et al., 1965). Another method of acid extraction is described by Douglas et al., (1951). According to Ferencik et al. (1961), the best method for extraction from fish homogenates is by means of n-amyl alcohol in a normal hydrochloric acid solution. The Warburg apparatus was used by Gale (1940, 1941a, b) to investigate histidine decarboxylation in several strains of E. coli.

Colorimetric tests are based on Pauly's reaction. Such tests have been described and used by several authors (Koessler & Hanke, 1919; Koessler & Hanke, 1920; Eggerth et al., 1939). They are appropriate for use with samples containing over 40 µg histamine/g (Ferencik et al. 1961).

Paper chromatography using ninhydrin is a more sensitive method yielding superior results. It has been used for histamine assays in cheese (Dolezálek, 1965), fish (Ferencik et al., 1961) and fish pickles (Meyer, 1961).

Another method of histamine assay in canned tunny fish involves extraction, subsequent purification by absorption on an ion exchange column and elution, coupling of the histamine to a diazo compound, purification of the coloured solution and photometric reading (Sager & Horwitz, 1957).

Electrophoresis (Ferencik et al., 1961) and high-voltage electrophoresis in a thin layer of silica gel (de Koning, 1969) have been used for histamine assays in fish and in cheese. In cultures of Cl. perfringens histamine has been assayed by ion electrophoresis using acetic ninhydrin as the developing solution (Blas & Sarraf, 1960); up to 8 mg histamine/g of dry substance were detected in this manner.

A sensitive and specific method using radioactive isotopes has been developed for the assay of histidine decarboxylase activity (Levine & Watts, 1966).

Recently, assays by fluorometric methods were carried out in various foodstuffs (de Koning, 1969; Ough, 1971).

4. Histamine Toxicity

The upper limit of histamine tolerance in a human subject of 70 kg body weight is about 5–6 mg. Deleterious effects in relation to the amount of histamine ingested at one meal are as follows (Henry, 1960; Peeters, 1963).

Mild poisoning	8–40 mg histamine
Disorders of moderate intensity	70–1000 mg histamine
Severe incidents	1500–4000 mg histamine

It was found by Anrep *et al.* (1944) that an increased dietary supply of histamine resulted in increased excretion of the substance. Protecting mechanisms in the body include processes of acetylation and histaminopexis. The proteins of the blood serum are considered to exert a histaminolytic action (Henry, 1960) and the adrenals also play an important part.

For obvious symptoms of poisoning to appear, comparatively large amounts of histamine must overcome the intestinal barrier and gain access to the blood. This may occur after the ingestion of a very large amount of histamine.

Experiments in guinea-pigs have shown that a latent period of about 10–15 min precedes the onset of disorders. During an initial phase histamine is maintained in the intestine. Excess histamine subsequently passes the intestinal barrier and invades the body. Mucin has been shown to exert a histaminopexic action (Parrot & Nicot, 1965), which, however, cannot prevent the passage of excessive amounts of histamine through the intestinal barrier.

Experiments in guinea-pigs (Koessler & Hanke, 1919) have shown that when 100 mg histamine dihydrochloride are administered orally, 61·8% of the substance is no longer demonstrable in the intestine after 2 h.

The existence of an histamine barrier at the level of the digestive tract has been demonstrated *in situ* (Kendall & Varney, 1927; Feigen & Campbell, 1946) in stomach and intestine preparations. Under certain circumstances diamines may enhance the toxic action of histamine by facilitating the passage of the latter through the intestinal barrier (Parrot *et al.*, 1947; Boyer & Jacob, 1961). When putrescine was administered before histamine the toxicity of the latter was enhanced, and the animals were killed by amounts considerably below the normal lethal dose (Parrot & Nocot, 1965). Apparently the association with histamine of some diamines, such as putrescine, cadaverine and spermine, exerts a deleterious effect on the animal body.

5. Disorders due to the Presence of Histamine in Foods

(a) Food poisoning due to histamine in man

Disorders of this type have been described in both adults and children. Their incidence rate cannot be precisely stated since many cases escape diagnosis because methods of histamine assay are not available in every laboratory.

Statistics concerning histamine poisoning have been published for France (Pointeau-Pouliquen, 1958), Japan (Aiiso, 1958), and the U.S.A. (Center for Disease Control, 1968, 1969, 1970).

Several outbreaks of histamine poisoning in families and institutions are described in the literature, occurring especially after the ingestion of tunny fish (Legroux et al., 1946a, b; Legroux, Levaditi & Bovet, 1946; Legroux, Levaditi & Second, 1946; Legroux et al., 1947; Boyer et al., 1956; Jovanović, 1957; Sapin Jaloustre & Sapin Jaloustre, 1957; Spuzić et al., 1957; Verge & Boyer, 1958; Dumont, 1959; Ferencik et al., 1961; Gounelle & Pointeau-Pouliquen, 1961; Meyer, 1968), herring (Bisbini et al., 1960; Hayashi, 1955), and canned sardines (Vidal & Mouret, 1957; Popović et al., 1960) as well as of some Pelamidae caught along the coasts of the Black Sea (Meyer, 1968).

In Indochina the preparation called Nuoc-Mam, which is obtained by the slow hydrolysis of various fish species, contains large amounts of histamine and may cause poisoning (Legroux, Levaditi & Bovet, 1946).

A few years ago Douglas et al. (1967) reported an instance of histamine poisoning that was caused by the ingestion of 100 g of Gouda cheese. The cheese had been stored for 2 years and contained 85 mg histamine/100 g of dry material.

(b) Other disorders caused in humans by histamine of dietary origin

Histamine was also incriminated as a causal agent in summer diarrhoea of children (Mellanby, 1915/1916; Neumann, 1949). In such conditions Cl. perfringens was found in large numbers in the intestine of the patients (Kendall, 1926; Kendall & Schmitt, 1926). This micro-organism produces histamine, and the occurrence of decreased blood pressure among the symptoms recorded was considered to be related to this fact. The effect of specific therapy in such cases might be explained by the excretion or neutralization of histamine (Neumann, 1949).

Histamine-forming strains of E. coli may likewise have a causal role in children's enteritis (Malachowska & Misiorek, 1957). The isolation of such strains from several patients revealed a correlation between the clinical symptoms and the amount of histamine produced. The strains isolated from severe cases produced 100 μg of histamine/ml of bacterial suspension.

(c) Disorders caused by related substances of dietary origin

Apart from histamine, some histamine-like substances are presumed to be possible causes of disease. Thus, saurine has been mentioned in Japan (Kawabata *et al.*, 1955; Meyer, 1968; Bryan, 1969). It is produced in fish by some *Proteus* strains.

Other instances of scombroid fish poisoning have been reported (Cooper *et al.*, 1964; Listick & Condit, 1964; Center for Disease Control, 1968, 1969, 1970, 1971). They occurred after the ingestion of tunny fish, mackerel, bonito fish or a Taiwanese preparation called *Mahi Mahi*. In one case, Listick & Condit (1964) isolated strains of *Pr. morganii* and *Pr. vulgaris* from a few frozen mackerel. According to some authors (Bryan, 1969) information concerning fish poisoning and the part played by spoilage organisms is still inadequate.

6. General Remarks Concerning the Significance of Histamine, as Present in Food

In the opinion of the present author the investigations carried out so far emphasize the following points.

(i) The presence of large amounts of histamine in food results from bacterial enzyme activity and demonstrates that extensive bacterial growth has occurred recently or at some time in the past.

(ii) The number of histamine-forming bacteria present, either of a single species or in synergistic associations, may to a certain extent influence the amount of histamine formed in food.

(iii) Some of the histamine-producing bacteria are known to be pathogens, whereas the pathogenic action of others is suspected but has not yet been conclusively proved. The author considers that bacteria belonging to the latter group may be included among the pathogens because of their marked histidine decarboxylating activity.

(iv) The presence of large amounts of histamine in foodstuffs that have caused food poisoning may justify the classification of such cases as bacterial food poisoning rather than as poisoning by chemicals.

Recommendation

When very large numbers of *Proteus* or *Pseudomonas* are found in food known to have caused illness in one or more persons (an outbreak), examination of the food should include investigation for large quantities of histamine and also for the intensity of histidine-decarboxylase activity of the strains of *Proteus* and *Pseudomonas*.

References

Ackermann, D. (1910). Uber den bakteriellen Abbau des Histidins. *Hoppe-Seyler's Z. physiol. Chem.* **65**, 504.

Ahlmark, A. (1944). *Acta physiol. Scand.* **9**, suppl. 28, cited by Knut A. (1950).

Aiiso, K. (1958). On the role of the *Proteus morganii* in histamine poisoning caused by fish meat. *VII Int. Congr. Microbiol., Uppsala.* Stockholm: Almquist & Wiksells.

Anrep, G. V., Ayadi, M. S., Barsoum, G. S., Smith, J. R. & Talaat, M. M. (1944). The excretion of histamine in urine. *J. Physiol., Lond.* **103**, 155.

Asatoor, A. M., Levi, A. J. & Milne, M. D. (1963). Tranylcypromine and cheese. *Lancet ii,* 733.

Barsoum, G. S. & Gaddum, J. H. (1935). Pharmacological estimation of adenosine and histamine in blood. *J. Physiol., Lond.* **85**, 1.

Baumgarten, U. (1970). *In vitro* Versuche zum Abbau von Histidin zu Histamin durch enzymatische Aktivität (Histidindecarboxylase) von Mikroorganismen. *Vet. Dissertation,* Hannover.

Bayard, H. (1916). *Chicago Med. Recorder,* **38**, 60 cited by Jones, H. (1918).

Belitz, H. D. & Schormüller, J. (1965). Aminosäuren, Peptide, Proteine und andere Stickstoffverbindungen. In *Handbuch der Lebensmittelchemie.* Berlin/New York: Springer Verlag.

Bertrand, D. M. & Berthelot, A. (1913). Ptomaine-producing bacteria in the human intestinal flora. *Lancet i,* 523.

Billy, C. (1962). Recherches sur les odeurs dégajées par les anaérobies. V. *Clostridium sporogenes,* souches marines et souches telluriques. Comparaison de leur pouvoir lipidolytique. *Annls Inst. Pasteur, Paris* **103**, 605.

Bisbini, F., Possati, F. & Marinelli, D. (1960). Due episodi di intossicazione alimentare da filetti di sgombro. *Nouvi Annali Ig. Microbiol.* **11**, 377.

Blackwell, B., Mabbitt, L. A. & Marley, E. (1969). Histamine and tyramine content of yeast products. *J. Fd Sci.* **34**, 47.

Blas, J. & Sarraf, A. (1960). *J. Chromat.* **54**, 456 cited by Prévot, A. R. (1967).

Boyer, J., Depierre, F., Tissier, M. & Jacob, J. (1956). Intoxications histaminiques collectives par le thon. *Presse Méd.* **43**, 1003.

Boyer, J. & Jacob, J. (1961). Les aliments vecteurs d'histamine. *Progr. Méd. nr.* **2**, 35.

Bryan, F. L. (1969). Infections due to miscellaneous micro-organisms. In *Food-borne Infections and Intoxications.* Ed. H. Riemann. London: Academic Press.

Cantoni, C., Bianchi, M. A., Cerruti, F. & Golfi, M. L. (1970). Sulla presenza di ammine non volatili e di aminoacidi liberi in prosciutti stagionati normali ed alterati. *Arch. vet. ital.* **21**, 89.

Center for Disease Control (1968). *Food-borne outbreaks,* Annual Summary, p. 30.

Center for Disease Control (1969). *Food-borne outbreaks,* Annual Summary, p. 31.

Center for Disease Control (1970). *Food-borne outbreaks.* Annual Summary, p. 30.

Code, C. F. (1937). Quantitative estimation of histamine in blood. *J. Physiol., Lond.* **89**, 257.

Code, C. F. & McIntire, F. C. (1956). Quantitative determination of histamine. In *Methods of Biochemical Analysis.* Ed. D. Glick. New York: Interscience Publishers Inc.

Colowick, S. P. & Kaplan, N. O. (1955). *Methods in Enzymology.* New York: Academic Press.

Cooper, M., Husman, D. & Condit, P. K. (1964). Scombroid fish poisoning–California. *Morbid. Mortal. Atlanta* **13**, 166.

Cousin, H. & Noyer, B. (1944). *Rev. méd. fr. Extr.-Orient* **4**, 382 cited by Ramel, P. *et al.* (1965).

Dale, H. H. & Laidlaw, P. P. (1910/1911). The physiological action of beta-iminazolyl-ethylamine. *J. Physiol., Lond.* **41**, 318.

de Koning, P. J. (1969). A new method for the fluorometric determination of histamine in cheese. *Ned. Melk-en Zuiveltijdschr.* **22**, 153.

Diémair, W. (1941). *Die Haltbarmachung von Lebensmitteln.* Stuttgart: Ferdinand Enke.

Dolezalék, J. (1956). Volné aminokyseliny v Sýrech. *Prum. Potravin* 7, 175.

Douglas, W. W., Feldberg, W. D., Paton, D. M. & Schachter, M. (1951). *J. Physiol., Lond.* 115, 163 cited by Henry, M. (1960).

Douglas, H. M., Huisman, J. & Nater, J. P. (1967). Histamine intoxication after cheese. *Lancet ii,* 1361.

Dumont, J. G. (1959). Epidémie familiale de maladie de R. Legroux, J. C. Levaditi, G. Boudin & D. Bovet. *Maroc Méd.* 38, 739.

Edlbacker, S., Jucker, P. & Baur, H. (1937). Die Beeinflussung der Darmreaktion des Histamins durch Aminosäuren. *Hoppe-Seyler's Z. Physiol. Chem.* 247, 63.

Eggerth, A. H. (1939). The production of histamine in bacterial cultures. *J. Bact.* 37, 205.

Eggerth, A. H., Littwin, R. J. & Deutsch, J. V. (1939). The determination of histamine in bacterial cultures. *J. Bact.* 37, 187.

Ehrismann, O. & Werle, O. (1947/1948). Histidine decarboxylase, histidine dehydrogenase and histaminase. *Biochem. Z.* 318, 560.

Feigen, G. A. & Campbell, O. H. (1946). The relative sensitivity of the mucosal and peritoneal surfaces of the guinea pig ileum to histamine, acetylcholine and specific antigens. *Am. J. Physiol.* 145, 676.

Feldberg, W. & Schilf, E. (1930). *Histamin.* Berlin: Springer Verlag.

Ferencik, M., Krcméry, Vl. & Kriska, J. (1961). Fish poisoning caused by histamins. *J. Hyg. Epidem. Microbiol. Immun. (Praha)* 5, 341.

Gale, E. F. (1940). The production of amines by bacteria. I. The decarboxylation of amino acids by strains of Bacterium coli. *Biochem. J.* 34, 392.

Gale, E. F. (1941a). The decarboxylation of amino acids by organisms of the groups *Clostridium* and *Proteus. Biochem. J.* 35, 66.

Gale, E. F. (1941b). The oxidation of amines by bacteria. *Biochem. J.* 36, 64.

Geiger, E. (1955). Role of histamine in poisoning with spoiled fish. *Science, N.Y.* 127, 865.

Gleichmann, F. (1934). Über einen biologisch stark aktiven Stoff in der Tomatensaft *(Solanum esculentum)* mit histaminähnlicher Wirkung; zugleich ein Beitrag zur Frage der therapeutischen Wirkung der Frischgemüsesäfte. *Z. klin. Med.* 127, 111.

Gounelle, H. (1960). Les risques d'intoxications alimentaires dans les collectivités à la lumiére de l'expérience française des 40 dernières années. Rôle du diététicon. *Nutritio Dieta* 2, 33.

Gounelle, H. & Pointeau-Pouliquen, M. A. (1961). Les intoxications alimentaires après consommation de poisson. *Rev. Hyg. Méd. soc.* 9, 603.

Guggenheim, H. & Loeffler, W. (1916). Biologischer Nachweis proteinogener Amine in Organextrakten und Körperflüssigkeiten. *Biochem. Z.* 72, 303.

Guirard, B. M. & Snell, E. E. (1954). *J. Am. Chem. Soc.* 76, 745 cited by Schormüller, J. & Hutch, H. (1957).

Halpern, B. N. & Walthert, F. (1945). Détermination biologique de l'histamine en présence de l'histidine. *C.r. Séanc. Soc. Biol.* 139, 365.

Hanke, M. T. & Koessler, K. K. (1922). The production of histamine and other imidazoles from histidine by the action of microorganisms. *J. Biol. Chem.* 50, 131.

Hanke, M. T. & Koessler, K. K. (1924). On the faculty of normal intestinal bacteria to form toxic amines. *J. Biol. Chem.* 59, 835.

Hayashi, M. (1955). Detection of histamine in deteriorated mackerel pickle. *J. Pharm. Soc. Japan* 75, 1.

Henry, M. (1960). Dosage biologique de l'histamine dans les aliments. *Annls Falsif. Expert. chim.* 53, 24.

Hillig, F. (1956). Volatile acids, succinic acid, and histamine, as indices of decomposition in tuna. *J. Ass. off. agric. Chem.* 39, 773.

Hirai, K. (1933). Uber die Bildung von Histamin aus l-Histidin durch Bakterien. *Biochem. Z.* 267, 1.

Hughes, R. B. (1959). The free amino acids of herring flesh and their behaviour during postmortem spoilage. *J. Sci. Fd Agric.* 10, 558.

Huidobro H. (1956). Rôle physiologique et pathologique de l'histamine. *Revue Path. gén. Physiol. clin.* **56**, 489.

Ienistea, C. (1971). Bacterial production and destruction of histamine in foods and food poisoning caused by histamine. *Nahrung* **15**, 109.

Ienistea, C., Dumitriu, S. & Maior, O. (1960). Cercetarea prezentei histidindecarboxilazei la tulpini de *Escherichia coli. Microbiologia Parazit. Epidem.* **5**, 523.

Jones, H. (1918). A determination of the numbers of histidin decarboxylating organisms in the feces in dementia praecox as compared with the numbers in normal feces. *J. Infect. Dis.* **22**, 125.

Jovanović, L. (1957). Trovanje aminskim bazama (histaminom) u hrani. *Prehranb. Ind.* **11**, 122.

Katae, M. & Kawaguchi, H. (1957). Food putrefaction. *Bull. Univ. Osaka Prefect* **7**, 29.

Katae, M. & Kawaguchi, H. (1959). Amine production by putrefactive bacteria. *Bull. Univ. Osaka Prefect.* **9**, 117.

Kawabata, T., Ishizaka, K. & Miura, T. (1955). Studies on allergy-like food poisoning associated with putrefaction of marine products. *Jap. J. med. Sci. Biol.* **8**, 487.

Keil, W. & Kritter, B. (1935). Zur Chemie und Pharmakologie vergorener Nahrungsmittel. *Biochem. Z.* **276**, 61.

Kendall, J. A. (1926). Intestinal intolerance for carbohydrate associated with overgrowth of the gas Bacillus *(Bacillus welchii). J. Am. med. Ass.* **86**, 737.

Kendall, J. A. & Gebauer, E. (1930). The production of histamine by certain strains of the gas Bacillus. Studies in bacterial metabolism. *J. Infect. Dis.* **47**, 261.

Kendall, J. A. & Gebauer, E. (1931). Uber Wunden-Gasbrand und sekundären. *Schock. Klin. Wschr.* **10**, 443.

Kendall, J. A. & Schmidt, E. (1926). The occurrence of a physiologically active substance in cultures of the gas Bacillus. *Proc. Soc. exp. Biol. Med.* **24**, 104.

Kendall, J. A. & Schmitt, F. O. (1926). Physiologic action of certain cultures of the gas Bacillus. *J. Infect. Dis.* **39**, 250.

Kendall, J. A. & Varney, P. L. (1927). The physiological action of histamine applied directly to the mucosa of the isolated surviving intestine of guinea pig. *J. Infect. Dis.* **41**, 143.

Kimata, M. & Kawai, A. (1953). Production of histamine by bacteria causing spoilage of fish. *Bull. Res. Inst. Fd Sci. Kyoto Univ.* **12**, 29.

Kimata, M. & Tanaka, M. (1954a). The production of histamine by the action of bacteria causing spoilage of fresh fish. *Bull. Res. Inst. Fd Sci. Kyoto Univ.* **13**, 29.

Kimata, M. & Tanaka, M. (1954b). The production of histamine by the action of bacteria causing spoilage of fresh fish. *Bull. Res. Inst. Fd Sci. Kyoto Univ.* **14**, 30.

Knut, A. (1950). Bacterial production and destruction of histamine *in vitro. Acta allerg.* **3**, 136.

Koessler, K. K. & Hanke, M. T. (1919). The production of histamine from histidine by *Bacillus coli communis. J. Biol. Chem.* **39**, 539.

Koessler, K. K. & Hanke, M. T. (1920). A method for the quantitative colorimetric estimation of histamine in protein and protein-containing matter. *J. Biol. Chem.* **43**, 543.

Koessler, K. K., Hanke, M. T. & Sheppard, M. S. (1928). Production of histamine, tyramine, bronchospastic, and arteriospastic substances in blood broth by pure cultures of microorganisms. *J. Infect. Dis.* **43**, 363.

Koslowski, L., Schneider, H. H. & Heise, C. (1951). Experimentelle Untersuchungen Uber die Histaminbildung des Bac. enterotoxicus im Vergleich mit anderen pathogenen Anaerobiern. *Klin. Wschr.* **29**, 29.

Kriska, J., Ferencik, M., Krcméry, V. & Liska, O. (1960). Experimentálne stúdie ichtyotoxikoz z údeného tunáka. I. Epidemiologické a chemické údaje. *Čslká Hyg.* **5**, 580.

Legroux, R., Bovet, D. & Levaditi, J. C. (1947). Presence d'histamine dans la chair d'un thon responsable d'une intoxication collective. *Annls Inst. Pasteur, Paris* **73**, 101.

Legroux, R., Levaditi, J. C., Boudin, G. & Bovet, D. (1946a). Intoxications histaminiques collectives consécutives à l'ingestion de thon frais. *Presse Méd.* **54**, 545.

Legroux, R., Levaditi, J. C., Boudin, G. & Bovet, D. (1946b). A propos des intoxications histaminiques collectives d'origine alimentaire. *Presse Méd.* **54**, 743.

Legroux, R., Levaditi, J. C. & Second, L. (1946). Méthode de mise en evidence de l'histamine dans les aliments causes d'intoxications collectives à l'aide de l'inoculation au cobaye. *C.r. Séanc. Soc. Biol.* **140**, 863.

Legroux, R., Levaditi, J. C. & Bovet, D. (1946). Présence d'histamine dans du "Nuoc-Mam" conservé au laboratoire. *C.r. Séanc. Soc. Biol.* **140**, 864.

Levine, R. J. & Watts, D. E. (1966). A sensitive and specific assay for histidine decarboxylase activity. *Biochem. Pharmac.* **15**, 841.

Listick, F. A. & Condit, P. K. (1964). Scombroid fish poisoning–California. *Morbid. Mortal. Atlanta* **13**, 30.

Malachowska, I. & Misiorek, I. (1957). Histaminotwóreze szczepy *Escherichia coli* u dzieci chorych na biegunki. *Medycyna dośw. Mikrobiol.* **9**, 387.

Mellanby, E. (1915/1916). *Q. Jl Med.* **9**, 165 cited by Feldberg, W. & Schilf, E. (1930).

Mellanby, E. & Twort, F. W. (1912/1913). On the presence of beta-imidazol-ethylamine in the intestinal wall; with a method of isolating a bacillus from the alimentary canal which converts histidine into this substance. *J. Physiol., Lond.* **45**, 53.

Meyer, V. (1961). Das Auftreten decarboxylierbarer Aminosäuren bei der Herstellung von Marinaden und ihr Nachweis. *Veröff. Inst. Meeresforsch. Bremerh.* **7**, 264.

Meyer, V. (1962). Uber Milchsäurebakterien in Fischmarinaden. *Zentlb. Bakt. I Orig.* **184**, 296.

Meyer, V. (1968). Einflüsse auf die Zusammensetzung der Fische und Fischerzeugnisse. Biochemie und Mikrobiologie. In *Handbuch der Lebensmittelchemie.* Ed. J. Schormüller. Berlin: Springer Verlag.

Myrhman, G. & Tomenius, J. (1939). Uber das Vorkommen von Histamin in menschlichen Faeces. *Arch. exp. Path. Ther.* **193**, 14.

Neumann, Z. C. (1949). Observations on the treatment of infantile gastroenteritis. *Br. Med. J.* **4619**, 132.

Oishi, K. (1953). On histamine in shark meat. *Bull. Fac. Fish. Hokkaido Univ.* **4**, 119.

Ough, C. S. (1971). Measurement of histamine in California wines. *J. agric. Fd Chem.* **19**, 241.

Parrot, J. L., Gabe, M. & Herrault, A. (1947). Intoxication aigue du cobaye par l'administration simultanée d'histamine et de putrescine. *C.r. Séanc. Soc. Biol.* **141**, 186.

Parrot, J. L. & Nicot, G. (1965). Le rôle de l'histamine dans l'intoxication alimentaire par le poisson. *Aliment. Vie.* **53**, 76.

Parrot, J. L. & Rense, J. (1954). Rôle physiologique et physiopathologique de l'histamine. *J. Physiol.* **46**, 99.

Pauly, H. (1904). Uber die Konstitution des Histidins. *Hoppe-Seyler's Z. Physiol. Chem.,* **42**, 508.

Peeters, E. M. E. (1963). La présence d'histamine dans les aliments. *Arch. Belges méd. Soc.* **21**, 451.

Plagnol, H. & Aldrin, J. F. (1963). Dosage de l'histamine chez les thons du Golfe de Guinée. *Rev. Conserve.* **18**, 143.

Pointeau-Pouliquen, M. A. (1958). Les causes des intoxications alimentaires en France depuis 1920. Paris Thése Médicine.

Popović, M., Bognar, I., Magdić, S. & Andal, M. (1960). Masovna intoksikatija histaminom posle upotrebe sardina. *Glasn. hig. Inst. Beogr.* **3/4**, 43.

Prévot, A. R. & Thouvenot, H. (1962). Recherches sur les odeurs dégajées par les anaérobies. VII. *Clostridium botulinum* et *Clostridium sordellii. Annls Inst. Pasteur Paris* **103**, 925.

Prévot, A. R., Turpin, A. & Kaiser, P. (1967). *Les Bactéries Anaérobies.* Paris: Dunot.

Quevauviller, Pr. (1959). La technique physiologique au secours de l'expert chimiste. *Annls Falsif. Fraudes* **52**, 298.

Ramel, P., Girard, P., Lauteaume, M. T. & Guezennec, J. (1965). L'histamine dans les conserves de poissons. *Rev. Hyg. Méd. Soc.* **13**, 73.

Riemann, H. (1969). In: *Food-borne Infections and Intoxications.* New York: Academic Press.

Rocha e Silva, M. (1944). Inhibition of histamine effects by compounds of histamine, histidine and arginine. *J. Pharmacol. Exp. Ther.* **80**, 399.

Rodwell, A. W. (1953). The occurrence and distribution of amino-acid decarboxylases within the genus *Lactobacillus. J. gen. Microbiol.* **8**, 224.

Sager, O. S. & Horwitz, W. (1957). A chemical method for the determination of histamine in canned tuna fish. *J. Ass. agric. Chem.* **40**, 892.

Sapin Jaloustre, H. & Sapin Jaloustre, J. (1957). Une toxi-infection alimentaire peu commune: l'intoxication histaminique par le thon. *Concours Méd.* **79**, 2705.

Savi, P. (1957). Technologie et contrôle vétérinaire des produits frais de la pêche, destinés à l'alimentation de l'homme. *Off. Int. Épizoot.* **48**, 238.

Schales, O. & Schales, S. S. (1955). Amino acid decarboxylases of plants. In Colowick, S. P. & Kaplan, N. O. (1955).

Scheibner, G. (1968). Zum Vorkommen von Histidin-Decarboxylase bei Salmonellenstämmen verschiedenster Serotypen. *Z. ges. Hyg.* **14**, 204.

Schormüller, J. & Hutch, H. (1957). Beiträge zur Biochemie der Käsereifung. Das Vorkommen weiterer Decarboxylasen in reifendem Sauermilchkäse. *Zentlb. Lebensmitt. Unters. Forsch.* **105**, 82.

Shifrine, M., Ousterhout, L. E., Grau, C. R. & Vaughn, R. H. (1959). Toxicity to chicks of histamine formed during microbial spoilage of tuna. *Appl. Microbiol.* **7**, 45.

Spuzić, Vl., Danilović, V., Djordjević, S. & Pujević, S. (1957). Masovno trovanje tunjevinom. *Higijena* **9**, 22.

Suchet, A. (1959). Appareillage nouveau pour orgenes isolés et technique de l'épreuve de l'histaminopexie. *Annls Biol. Clin.* **17**, 5.

Suzuki, U. & Yoshimura, K. (1909). Uber die Extraktivstoffe des Fischfleisches. *Hoppe-Seyler's Z. Physiol. Chem.* **62**, 1.

Swiatek, A. & Kisza, J. (1959). Beitrag zur Kenntnis der freien diazopositiven Aminen und Aminosäuren in einigen Käsesorten. In *International Dairy Congress III,* 1459.

Tabor, H. (1955). Isolation and determination of histidine and related compounds. In *Methods in Enzymology.* Eds S. P. Colowick & N. O. Kaplan. New York: Academic Press.

Uuspaa, V. J. & Torsti, P. (1951). The histamine content of Finnish cheese. *Annls Med. exp. Biol. Fenn.* **29**, 58.

Van Veen, A. G. & Latuasan, H. E. (1950). Fish poisoning caused by histamine in Indonesia. *Documenta neerl. indones. Morb. trop.* **2**, 18.

Verge, J. & Boyer, J. (1958). Les toxiinfections alimentaires collectives. *Rev. Hyg. Méd. Soc.* **6**, 45.

Vidal, J. & Mouret, A. (1957). A propos de cinq cas d'intoxication histaminique par les sardines en conserve. *Soc. Sci. Méd. Biol. Montpellier* p. 96.

Werle, E. (1940). Über die Histaminzerstorende Fähigkeit von Bakterien. *Biochem. Z.* **306**, 264.

Werle, E. (1941). Über das Vorkommen von Diaminoxydase und Histidin-decarboxylase in Mikroorganismen. *Biochem. Z.* **309**, 61.

Werle, E. & Raub, A. (1948). *Biochem. Z.* **318**, 538 cited by Schales O., Schales, S. S., (1955).

Windau & Vogt cited by Knut, A. (1950).

Winkle, S. cited by Meyer, V. (1968).

Yoshimura, N. (1910). Über Fäulnis-Basen (Ptomaine) aus gefaulten Sojabohnen *(Glycine hispida). Biochem. Z.* **28**, 16.

Zunz, E. (1919). Sur la présence d'histamine dans les muscles atteints de gangrène gazeuse. *C.r. Séanc. Soc. Biol.* **82**, 1078.

Discussion

Sinell

The toxic effects of histamine after oral ingestion depend within certain limits not only on the dose administered but also on the amino-oxydase activity in the human gut. It is known that several therapeutic agents acting as amino-oxydase inhibitors may be responsible for the toxic effect of certain foods normally regarded as harmless.

Riemann

The presence of histamine in fish meal can cause symptoms in chickens and there is evidence that chickens exposed to histamine have increased numbers of *Cl. perfringens* in the intestinal tract.

Shewan

In north-east Scotland, mackerel are traditionally not eaten—this is thought to be due to the presence of histamine. Histamine is not produced in large quantities in fish such as mackerel or tunny which are kept at chill temperatures but can be produced in quantity at temperatures above 8°. The organism *Achromobacter histamineum* has recently been shown at Torry to be a *Proteus* spp.

Gibbs

On a point of information it is recommended by General Practitioners in the medical profession that patients taking tranquillizers should not eat certain foods, especially blue cheese.

Woodbine

Is there not a link between amino acids which give rise to histamine and tyramine (the latter being linked with migraine induction)? If organisms such as those having a histidine decarboxylase, as mentioned by Dr Ienistea, are added to those with a tyrosine decarboxylase (e.g. streptococci and lactobacilli) is there not a basis for postulating an amine toxicity syndrome in relation to foods?

Ingram

What is the connection between Prof. Sinell's statement regarding the effect of histamine on the amine oxydase of the gut and that of Dr Ienestea that its effect depends on a high level of histamine circulating in the blood?

Ienistea

Protecting mechanisms in the body include processes of acetylation and histaminopexis. The proteins of blood serum are considered to exert a histaminolytic action and the adrenals play an important role. For obvious symptoms of poisoning to appear, comparatively large amounts of histamine must overcome the intestinal barrier and gain access to the blood. This may occur after the ingestion of a very large amount of histamine.

Session 4

Legislation and Non-Legal Specifications

Chairman: Professor M. Ingram

Legislation and Non-Legal Specification in the Field of Food Microbiology and Hygiene

N. SKOVGAARD

Royal Veterinary and Agricultural University,
Copenhagen,
Denmark

The problems of chemical contaminants in food are outlined briefly. It is emphasized that the microbiological safety of food will for many years be the main problem when considering health hazards arising from consumption of foods. The establishment of microbiological specifications is one of the safeguards to secure wholesome food. Some of the difficulties in setting up specifications, related to food as a biological system, are mentioned. There is a need to renew the microbiological indicator principles so far used in food microbiology to cover a wider variety of pathogens which may be transmitted to man by foods.

The automation of microbial counts is mentioned as a new approach to setting up specifications based upon a large number of simple investigations instead of few and elaborate ones.

ALL OVER THE world there is an increasing interest in environmental pollution, contaminants, the presence of trace elements, residues in animal tissues, foods and other materials. Problems like PCB compounds in foods, formation of nitrosamines as the result of nitrate/nitrite curing, pesticide residues, carcinogenic effects of artificial colouring matters, mercury, cadmium and lead poisoning due to various pathways of accumulation of such compounds in the food chain are serious and in urgent need of solution.

However pertinent these problems may be when considering health hazards arising from the consumption of foods, the predominant problem in the Western world is still the microbiological safety of foods. Incidence of food-borne disease, is grossly under-reported in most countries, partly because of lack of information and knowledge in the field of food epidemiology and partly because of inadequate reporting systems. It is estimated that food-borne microbiological diseases in some countries afflict roughly 5–10% of the population annually, high-lighting the importance of safeguards at all points in the food chain.

Although the public, led by the daily press, has taken a keen interest in recent years in all aspects of chemical pollution of food, the predominant problems in food hygiene at large will continue for many years to come to be microbial. These facts justify any approach based on scientific research towards improvements in the prevention of food-borne microbial diseases.

The prevention of food-borne diseases and control of microbiological health hazards require a sequence of operations and due emphasis should be placed on the separate steps. Part of this process is the setting-up of specifications for the food in question.

The following remarks will be confined to microbiological specifications and I will try to outline some of the specific problems and questions with which we are faced.

Are Specifications Needed?

The first question to be raised when speaking of microbiological specifications is: "Do we need them?"

As long as they are used as either *official* or *unofficial guidelines* for the industry and also as they are to a certain extent already incorporated in many food laws and regulations, the answer must be "Yes". But legal authorities, private firms and individuals must on the one hand be aware of the advantages and help that properly applied specifications may give in providing safe food and on the other hand be briefed on all their shortcomings and limitations. The specifications finally set up should be a balanced figure, or set of figures, in which *the consumers' risk as well as the producers' risk must be considered.*

Limitations in the Use of Specifications

The microbial world in a given foodstuff is an *ecological unit,* a biological system, with complicated laws of its own, which we have to understand when trying to interpret our findings. Microbiological specifications are easily misused through lack of understanding of these biological laws. Some examples of this will be given.

Synergism and antagonism

Because research is lacking even the food hygienist is not always aware of the synergistic and antagonistic influences which the different groups or species of microbes in a foodstuff may exert on each other. An example will outline the magnitude of the problem.

Staphylococci are known to be bad competitors. In cooked sterile foods, or foods with low viable counts, staphylococci may propagate very rapidly to critical levels when the temperature permits. Even a few staphylococci/g may for that reason be dangerous. If one accepts staphylococci at all in such food the limit should be low. The internationally accepted magic figure is 100/g. On the other hand I would be prepared to accept much higher figures in raw foods, liked minced meat, and also in cured products with a natural microflora, because the staphylococci will never be able to propagate to any significant extent in competition with a normal spoilage microflora.

Shortcomings in total viable count

When speaking of the so-called total viable count, too much stress is often placed on such values without taking note of the analysis of the composition of the microflora. It is important to know whether the demonstrated flora is the characteristic one which to a large extent must be tolerated or whether it is uncharacteristic of the product and hence justifies further action.

The total viable count has other, usually not recognized, limitations. The temperature at which a certain microflora develops, as well as the effect of the various inhibitory principles present, determines to a marked extent the degree of decomposition which the microflora is able to cause in the food. The count is made at $30°$ but the microflora may have developed at $+2°$.

A microflora of about 5×10^7 psychrophilic bacteria/g developing in, for example, raw meat at $+2°$ has only a slight influence on the odour of the product because the proteolytic enzymes as well as many other enzyme systems work with limited effect at low temperatures. On the other hand if a microflora of the same size in the same product results from propagation at room temperature, the result will often be a marked off-odour, necessitating condemnation.

Salt is one of the inhibitory principles in foods which has a most pronounced narrowing effect on the enzymatic activity of a given microflora especially at low temperatures. These factors alone compromise the value of a total count and others could be added.

Indicator organisms

Not much development has taken place in this field in recent decades. The methodology most usually applied when investigating food is to determine the following:

(1) Total viable count at $25–30°$.

(2) Coliforms at $37°$ as an indicator of unsanitary conditions during processing, including recontamination of heat-treated foods and possible propagation at higher temperatures, for example, above $+10°$.

(3) Faecal coli at $37°$ or $44°$ as a specific indicator of contamination of faecal origin.

(4) Faecal streptococci at $37°$. These are good indicator organisms of many kinds of faults in food processing depending on the nature of the product, but they are seldom indicators of faecal contamination.

(5) Staphylococci at $37°$. Different selective and indicative media are in use but most of these will miss some, or even all, of the staphylococci and no medium so far in use will show whether the staphylococci are enterotoxin producers.

(6) Salmonella if considered necessary.

There are increasing numbers of reports of human disease in which food is claimed to be the causative agent without the food microbiologist being able to confirm or deny this.

Enormous efforts are expended upon detection of trichinae in meat. One must admit that this search is performed in vain under many conditions simply because the parasite is not present in the animal kingdom in that area.

It would be well justified to divert these efforts to the development of methods for the detection in foods of organisms like *Toxoplasma* and *Yersinia* so that appropriate specifications could be set up. There is a high rate of positive serum reactions in humans showing that these organisms are common causes of disease. Raw meat is considered to be an important vector for both toxoplasmosis and yersiniosis.

These examples are given only to indicate that we need to modernize our indicator principles in order to establish microbiological specifications.

I do not underestimate the value of faecal indicators. I must mention that one of the most important reservoirs of pathogenic organisms in both man and animals is the pharyngeal-tonsillar microflora. Rapid direct or indirect methods are needed to reveal contamination from such sources. One cannot rely on the faecal organisms as indicators in such cases.

Modern Approach to Specifications

Although we are still using rather old-fashioned and conventional methods in food hygiene, new fields are developing.

The fluorescent antibody technique for detection of salmonellae is now successfully applied as a rapid and sensitive method. This principle can certainly be applied to a much wider range of pathogens.

There is a growing interest in the automation of microbial methods, and it is a well-established fact that it is preferable to perform many simple investigations than a few elaborate ones. The Food and Drug Administration in the U.S.A. reports that an automatic device for plating and counting aerobic bacteria which deposits a sample at a decreasing rate in the form of an Archimedes spiral on the surface of a prepared nutrient-agar plate is being developed satisfactorily and results in a saving in dilution bottles, Petri dishes, etc.

In spite of the draw-backs and difficulties in the correct interpretation of total counts, it is a fact that for many foods such counts do show good correlation with the sanitary condition of the production line. It is obvious that easy and rapid access to total and other counts as the results of automation will be of the greatest value in setting up specifications and in checking a given foodstuff against such specifications. Large-scale investigations will further facilitate a statistical approach to the problem. It should also be mentioned that electronic cell-counting has been used for several years to screen pathological milk from normal milk.

These comments may seem a rather negative approach to the evaluation of the importance of microbiological specifications. However, it is my experience that the more one learns of the ecology of food microbiology, the more reluctant one becomes to discriminate against or condemn food unless the data indicate an obvious health risk to the consumer. I hope I shall not be misunderstood when I say that bacteria are, after all, very rich sources of protein and fat.

Unofficial standards for some foods such as frozen minced meat exist in Denmark with counts as low as 500,000 bacteria/g. There may be some justification for using such low figures, but I consider them to be well below the level which calls for the attention of the food hygienist. Too many results are often accumulated without being properly utilized.

Specifications should be used to the fullest possible extent in conjunction with data obtained on the food in question when problems of microbiological hygiene arise. For this reason our policy is more in favour of specifications as guidelines than as legal standards. We are also advocating the vertical line in the field of food hygiene, stressing the importance of feed-back of information to the production line and, if possible, even further back to the raw materials.

Recommendations

(1) Analysis of the influence of the micro-ecological situation in foods on microbiological counts.

(2) Influence on specifications of the application of automation to food microbiology.

Discussion

Kampelmacher
    Although I understand the general intention of Skovgaard's paper I think some remarks need elaboration. Skovgaard said "Don't look for things which are not there". What does that mean? It is not there when a certain method is applied? Skovgaard mentioned *Trichinella spiralis*. Using trichinoscopy, evidence for trichinosis is not usually found in countries where the incidence of infection is low but it is still possible to detect larvae if immunofluorescence or digestion methods are used. These larvae may be insignificant for human health but we should hold the finger on the probe in order to know what is going on.
Skovgaard
    My reference to *Trichinella spiralis* was only meant as an example of a specification which, when the incidence of the organism in question has been decreased to zero level, needs re-evaluation in order to channel the total available effort towards the most pertinent problems. You may in this way prevent 100 persons from suffering, e.g. botulism at a cost of 1 person suffering from trichinosis.

Method of Sanitary Evaluation of Food Factories

HALINA SADOWSKA

*Sanit-Epidemiological Department,
Ministry of Health and Social Welfare,
Warsaw,
Poland*

The Polish Food Act and a number of Health Regulations determine the sanitary requirements for food processing in Poland. According to those requirements there has been developed a method of sanitary evaluation/Evaluation Scheme/ based on the punch card system. Some of the items to be considered on a score-card are common for all branches of food factories, while the others vary with the type of food processing establishments. The items in a score card have been classified into 2 groups: (1) technical conditions; and (2) cleanliness and housekeeping—they are valued separately in every group. One demerit point is given for each item which is considered sub-standard. The total points received by a factory determines its score and criteria have been established for the maximum number of points acceptable. It is obligatory for all local health authorities in Poland to make an evaluation of all food premises routinely every 12 months in accordance with this Evaluation Scheme. This evaluation is completed every year with microbiological examinations of final products, raw food, surfaces of equipment and personnel. It must be emphasized that this rating system, based on the periodical evaluation of all food premises in Poland, can show chronological evidence of progress or lack of it. Since the manager of a food premises is provided with a copy of the score-card in every evaluation, it is the fundamental index of sanitary progress in every factory.

OF THE MANY factors involved, the sanitary requirements in food premises can have the greatest influence on the quality of the final product.

These requirements outline the practices during transportation, storage and processing of foods which prevent their contamination by micro-organisms, insects, rodents or other pests, and by foreign chemical substances. The need for sanitary practice begins with the raw material and does not end until the food is served to the consumer.

Sanitary practice depends on 2 groups of factors:

(1) Technical conditions, i.e. sanitary facilities, equipment, construction and furnishing of premises.

(2) Cleanliness, housekeeping and personal hygiene.

The first group demands proper investment and the second depends on compliance by the staff with sanitary requirements. Therefore, adequate education and periodical training of the staff in hygienic practices is of great importance.

According to Polish Food Law—The Food Act of 1970 and a number of regulations—which defines the sanitary requirements, management is responsible for establishing and maintaining hygienic practices in food plants. It is obliged to ensure daily supervision by the management in every food factory and to organize the sanitary training of all food handlers.

The Polish Public Health Authority is responsible for food legislation, periodical supervision of food premises and food control. There is a special organization for the purpose, the State Sanitary Inspection directed by the Ministry of Health and Social Welfare. The Vice-Minister is a Chief State Sanitary Inspector to whom the regional and district sanitary epidemiological stations are responsible.

All the 23 regional and some district epidemiological stations have micro-biological and chemical laboratories for food examination, which is made according to legal standards that lay down the sampling rules, methods of analysis and hygienic quality criteria, both microbiological and chemical, for various foodstuffs. When any food is certified not to meet the requirements of a standard it may be treated as unfit for human consumption. The person responsible for the quality of the product is liable to a fine.

The sanitary supervision of food factories, shops, warehouses, restaurants and canteens is carried out by sanitary instructors (Polish equivalents of England's Public Health Inspectors). A person responsible for existing faults in premises pays a fine not exceeding 200 zlotys immediately after the inspection. If more serious faults have been found or the results of laboratory examination indicate that food produced in the plant may be harmful, the Sanitary Inspector orders a stop on production and a prosecution ensues.

Besides some routine inspections 3–4 times a year, all food premises are evaluated annually by sanitary instructors using special score cards. The items to be considered on the score card are divided into 2 groups which are evaluated separately: (1) technical conditions; (2) cleanliness and housekeeping.

One demerit point is given for each item which is considered sub-standard. Total points received by food premises in each group of items determines the score, the technical conditions and cleanliness being taken separately. Criteria have been established for the maximum number of demerit points which constitutes a satisfactory mark.

Figure 1 shows a part of the score card. The first column lists the items describing technical conditions. In the second column the instructors mark the demerits or faults relating to the items in the first column. The third column contains items contributing to cleanliness and housekeeping and the last one is for demerit marks connected with these items.

Explanations have been prepared of a number of the items to ensure that interpretations are uniform, for example:

(1) Location—the food premises are protected against objectionable odours,

Items-technical condition	Demerits	Items-cleanliness and housekeeping	Demerits
1. Location		No dust, smoke from neighbouring areas	
2. Surroundings		Clean, in good repair	
3. Separation from living premises		No personal articles in production rooms	
4. Design, construction of building		Food premises in good repair	
5. Sewage disposal			
6. Solid waste disposal		Containers covered, disinfected	
7. Water supply		Laboratory examination	
8. Hot running water supply		Adequate temperature	
9. Adequate lighting, ventilation		Humidity, temperature, electric bulbs	
10. Satisfactory floors, walls, windows, doors—construction		Clean, in good repair	
11. Ratproof construction		No visual evidence of rodents	
12. Locker room, double lockers		Clean, in good repair	
13. Shower facilities		Used by the staff	
14. Hand washing facilities		Soap, nailbrush, single-use towels	
15. Toilet facilities		Toilets clean, disinfected	
16. Protective clothing		Clean	
17. Package and container storage		Clean, in good repair	
18. Detergents, disinfectants, cleaning equipment storage		Separated from food	
19. Raw material and final product storage		Temperature, humidity, protection against contamination	

Fig. 1. Portion of a score-card.

smoke, dust or other contamination. Sufficient land is available for future expansion of the plant.

(3) Separation from the living premises—there are no joint entrances or toilets.

(10) Satisfactory construction of floors, walls, windows and doors—floors must be sloped towards an adequate drainage system and be constructed of impervious material which is easily cleaned;

(15) Toilet facilities—adequate numbers of toilets are provided with good lighting and ventilation and not opening directly into a food preparation area; adequate hand-washing facilities.

In carrying out an evaluation, the instructor must make an assessment of every item in an objective way, deciding whether or not it is satisfactory with regard to the minimum requirements specifically mentioned in the appropriate regulations. Therefore, all the instructors have been trained in the evaluation procedure. From time to time the evaluations made by them are repeated by another person, the scores are compared and any differences discussed in order to improve the evaluation technique.

The method of sanitary evaluation has been tested statistically to indicate to what extent it is objective. The testing of 150 evaluations of food factories, made in parallel, but separately, by 2 persons (1 of them always the same, the second different in every case) showed very small variations between the scores, which did not detract from the general validity of the assessment. The errors which did occur were statistically unimportant.

Three copies of the score cards are filled in—for the manager of the food premises, for the district office and for the sanitary epidemiological station. According to the nature of the faults recorded, a date is indicated by which all faults should be eliminated.

If after this date the faults are still found, the person responsible pays the fine not exceeding 10,000 zlotys.

Every year sanitary epidemiological stations summarize, from the score cards, the reports of the sanitary conditions in all food premises in their districts. They evaluate not only the total number of plants with unsatisfactory marks but also the numbers in each group of food factories (dairies, bakeries, ice cream factories and others) with particular demerits, for example "x" dairies have inadequate ventilation, "y" have unsatisfactory hand-washing facilities.

Marginal punched cards have been introduced to simplify the collation of demerits in the various types of food premises. Local authorities as well as the food industry organization or co-operatives responsible for the whole group of food premises, are provided with the reports each year. These data are then used for planning the future sanitary programme and to provide an assessment of chronological changes in sanitary conditions.

Discussion

Bostock

Dr Sadowska indicated that each item on the score card scored one point yet some aspects, e.g. the quality of the water supply, would be of much greater significance than others. Perhaps Dr Sadowska could comment on this and indicate at what level legal action would be taken.

Sadowska

Defects in certain important matters such as the water supply would be dealt with separately and action taken to remedy them.

Griffith

Does the Polish Food Act or the Health Regulations require the Local Health Authorities to give their approval before a food establishment can operate?

Sadowska

In Poland every new food premises must have sanitary evaluation before commencing production. The procedures at the plants must also be considered by the State Sanitary Inspectors.

Griffith

Mention was made in your paper of a "Sanitary Evaluation Scheme" based on a punch card system and reference was made to "hot water supply". What is the acceptable temperature of the water for (a) hand washing; and (b) equipment washing? Would the provision of a "shower" warrant the same score as the provision of "adequate refrigeration" in the evaluation scheme or does the scheme incorporate a scale of importance?

Sadowska

The reference to "hot water supply" was in relation to hand washing, and the temperature is not controlled. Another item is concerned with facilities for equipment washing; the temperature of water for washing and rinsing and also the concentration of detergents and disinfectants are defined whether the equipment is washed by hand or in machines. Some items have been treated in a special way and are of greater importance. When it is found that the special requirements have not been fulfilled, the food premises cannot get the necessary satisfactory mark even though it may achieve sufficient points. Special requirements apply to water supply and to the provision of adequate refrigeration.

Linderholm

We have sometimes used check lists or score cards in the inspection of food premises but only on a trial basis. Whilst we have found many advantages with the system there is always the danger that the inspectors will adhere blindly to the list and miss other things. What is your experience of such a risk? If I remember correctly your example of a score card, I could not see points for control of food handling or processing methods nor for the contamination risk between foods of different bacteriological status. Neither was mention made of the risks dependent on communication between areas or rooms with different hygienic standards or different temperatures.

Sadowska

According to our observations, the score card is very helpful for the sanitary inspectors. The supervision made by them using the score card is much more precise and they notice more sanitary points. It depends of course on the level of this group of workers.

Ingram

The opposite halves of the score card correspond in high degree, as Dr Sadowska herself noted. In that case, the scores must largely be duplicated and it would appear that much of one half of the form could be dispensed with, without much loss of information.

Sadowska

Such division between 2 groups of items is of a great importance because group I "Technical Condition" demands proper investment in plant and facilities and group II "Cleanliness and Housekeeping" depends above all on the compliance with regulations by the staff.

Monty

A point of general interest in connection with the score card system as advocated by Dr Sadowska for the hygienic evaluation of food processing premises in Poland is that the practice has been applied in the U.S.A. by certain food organizations to assess their supermarkets. A trophy is awarded to the most hygienic store and evaluation is carried out by hygiene officers making periodic visits. I understand that the score card system for improving hygiene in food premises is being considered by several food firms in the U.K.

Barrow

The food industry sanitation system as used in Poland is clearly useful. You stated however that planning consent with insistence on certain minimum sanitary standards is necessary before a "new" food manufacturer is "licenced". In Britain, many food producers started as "home" or "cottage industries" (for example, farm cheese and clotted cream) and "new" ones may still start in this way. Whether this is right or wrong, it presumably could not now happen in Poland?

Sadowska

In Poland such production is small. The products may be sold only in the market and not in the shops. Samples are taken only in the market and home production is not controlled.

The Value of Agreed Non-Legal Specifications

N. Goldenberg and D. W. Elliott

*Marks & Spencer Ltd.,
Michael House,
Baker Street,
London W1A 1DN
England*

The present paper describes the experience gained over the years 1955–1972 in
developing a range of "perishable foods" for retail sale, bought to agreed
specifications of quality and safety. The bacteriological specifications are
described in the present paper. Nowadays, it is fairly common practice for food
manufacturers to buy 'perishable" raw materials to specifications, which often
include bacteriological safety specifications. It is more rare for a retail firm to buy
"perishable" products for retail sale to such standards. The development and use
of agreed non-legal bacteriological specifications is described in the present paper.

THIS PAPER describes a system whereby the technologists at Marks & Spencer
have been able to co-operate with suppliers of "perishable foods"* to ensure
their microbiological safety at the point of sale. This has been achieved by close
co-operation with the suppliers at all levels, and, above all, by direct contact
with their production management and technologists. In this operation, much
was done to ensure that suppliers understood the need for the various
procedures and processes involved. This is described in a subsequent paper
(Goldenberg & Edmonds, 1973).

The present paper deals with the establishment of bacteriological standards
agreed between our suppliers and our technologists. Their establishment and
implementation enabled objective analytical checks to be made which revealed
to both our suppliers and ourselves that the various measures designed to ensure
microbiological safety at the point of sale were being adequately implemented
not only in the factory but also in distribution and in the retail stores.

At this stage it would be appropriate to quote a view on bacteriological
standards expressed by Sir Graham Wilson in 1969:

"Bacteriologists are better employed in devising means to prevent or
overcome contamination than in examining more and more samples. Their

---

* By "perishable" foods we mean foods which would allow the growth of either
pathogenic and/or spoilage micro-organisms, if contaminated and if held under conditions
suitable for microbial growth.

chief contribution to the supervision of our food supply should be to ensure that the preparation and processing are properly carried out. Control of processing, using this term in its broadest sense, is of far greater importance than examination of the finished article. Processing, that is to say, should be controlled primarily by technique and only secondarily by results" (Wilson, 1970).

We very much agree with this wise and eminently sensible view.

The use of microbiological standards has a long history. We have drawn on established standards where available and have received much valuable co-operation from the many organizations which we have consulted at various times and to whose valuable advice, freely and generously given, we owe a great deal. These include the British Food Manufacturing Industries Research Association; the (then) British Baking Industries Research Association; the Campden Food Preservation Research Association; the Food Hygiene Laboratory of the Public Health Laboratory Service; and the (then) Low Temperature Research Station of the Agricultural Research Council. The reader is referred here to a number of recent publications which summarize much of this knowledge (Bashford & Herbert, 1966; Hobbs, 1969; Symposium on microbiological standards for foods, 1970).

The purpose of bacteriological standards for perishable food products may be considered under 5 headings:

(a) There is an obvious need to control the risk to the consumer arising from the possible presence of pathogenic organisms and/or spoilage organisms. The development and implementation of bacteriological standards ensures the screening of food products for these organisms.

(b) The successful production of perishable foods calls for the following controls: (i) the use of clean, good quality raw materials to reduce the risk of introducing potentially harmful bacteria into the factory. Where this is not possible or is difficult to ensure, adequate steps should be taken to re-process the material, for example, the re-pasteurization of coconut to eliminate any salmonella bacteria that may be present; (ii) production should take place in clean, well-equipped and well-managed factories, with particular reference to the cleaning of equipment and to minimizing the risk of contamination of the product by factory personnel; (iii) correct packaging of the finished product; (iv) adequate processing including the identification of danger points and proper temperature control at all stages of manufacture. The continual bacteriological monitoring of a process allows a pattern to be established. A breakdown in any of the above controls manifests itself in the bacteriological results.

(c) Control over the distribution of the product including retail sale is achieved through the evaluation, by bacteriological analysis, of the shelf life of the product under various temperature conditions. This shelf life is then implemented by means of a date-coding system enabling stock to be correctly rotated in retail stores and ensuring that the product is fresh when purchased by the customer.

(d) Distribution through a "cold chain" system, involving refrigerated storage in the manufacturer's premises, refrigerated vehicles, refrigerated depots, and sale from refrigerated counters.

(e) The packaging carries full instructions concerning cooking times and temperatures (if relevant) together with the "customer" shelf life of the product under stated storage conditions (e.g. "Eat on day of purchase or 3 days if kept in a refrigerator") previously determined by bacteriological analysis. This should include a "built-in" tolerance to allow for some mishandling by the consumer. Tables 1 and 2 summarize these factors.

Table 1

Marks & Spencer's approach to bacteriological standards

Factors for successful production

(1) Use of good quality, clean raw materials
(2) Clean factories
(3) Clean equipment
(4) Correct processing techniques
(5) No contamination by personnel
(6) High standards of personal hygiene
(7) Adequate use of refrigeration

Table 2

Control over distribution

(1) Cold chain in storage distribution and sale
(2) Implementation of date coding system
(3) Full instructions to the consumer concerning
    (a) Cooking time and temperature
    (b) Keeping life
    (c) Storage conditions

The use of bacteriological standards enables the controls over distribution to be implemented.

Choice of Standards

The choice of bacteriological standards must take into account the point of sampling and the type of food product. There are a number of key sampling points in the production and distribution process, they are listed in Table 3.

Standards for samples from each of these stages are agreed with the food manufacturers and suppliers of raw materials.

The bacteriological standards implemented must take into account the type of product. The same bacteriological standards cannot be applied to different classes of food. When we consider the so-called perishable foods sold by the "cold chain" system, we can see that 5 classes of product exist, identified by the method of production and subsequent treatment by the consumer. These are discussed in more detail below.

Table 3

Bacteriological standards*

(a) Raw materials, especially highly perishable ingredients
(b) In-line process samples
(c) Finished product in the factory
(d) End of shelf life in retail stores
(e) When eaten by customer

* Must take into account the point at which sampling occurs.

Group A

This includes unprocessed items of food to be fully cooked by the consumer before eating; for example, poultry, fresh meat.

Bacteriological standards for these products have only limited value. Both poultry and meat contain potentially pathogenic bacteria from which they could not reasonably be required to be completely free, for example, *Salmonella* spp. or *Staphylococcus aureus* in raw poultry. Emphasis is therefore placed on the other controls outlined above.

Group B

This group includes raw products, involving some preparation in the factory, which must be fully cooked by the customer before eating—for example, sausages, beefburgers.

It is considered reasonable to expect the consumer to keep sausages overnight without refrigeration. Standards are therefore based on the examination of samples held at $65°-70°$ F for 24 h; they are given in Table 4.

This type of product may not always be adequately cooked and it is therefore considered important to ensure the absence of both salmonellae and coagulase-positive staphylococci.

Table 4
Bacteriological standards: group B: sausages *

Test	Satisfactory	Acceptable but investigate	Stop production
Viable count	Less than 500,000/g	More than 500,000/g and less than 2 million/g	50% of repeat samples greater than 2 million/g
Coliforms	Less than 500/g	Over 500/g and under 5000/g	50% of repeat samples greater than 5000/g
Salmonella Shigella	Absent in 25 g	–	Present in 25 g
Coagulase positive *Staph aureus*	Less than 100/g	Greater than 100/g and less than 1000/g	Greater than 1000/g in any repeat samples

* Raw products involving some preparation in the factory to be fully cooked by the consumer before eating.
Note: Samples examined after 24 h at room temperature.

Group C

This class includes fully cooked products to be eaten cold by the consumer or to be re-heated and it is an extensive and important group. Here again, samples are incubated before examination (24 h at $65°-70°$ F). The criteria are absence of pathogens and low total counts. These foods are packed immediately after baking and subsequent cooling and, providing that these processes are adequate, the standards are realistic (Table 5).

A standard for *Streptococcus faecalis* is included because of its heat tolerance, which provides a useful indication of failure in process hygiene or of inadequate baking.

Group D

This includes highly perishable products to be eaten without further processing, for example, fresh cream cakes.

Although a period of storage at room temperature is "built into" the life of these products, it is not considered realistic to apply bacteriological standards after incubation of the product at room temperature. Standards are applied to the in-going perishable raw materials, for example, cream and milk, and to the finished product at the point of packing. A further examination is carried out on samples at the end of the product's shelf life. Special precautions are taken in bakery creameries and an agreed Code of Practice for the handling of fresh

Table 5

Bacteriological standards: group C: cooked meat pies*

Test	Satisfactory	Acceptable but investigate	Stop production
Viable count	Less than 200/g	Over 200/g and less than 10,000/g	50% of repeat samples above 10,000/g
Coliforms	Less than 10/g	Greater than 10/g	50% of repeat samples greater than 10/g
Strep. faecalis	Less than 10/g	Greater than 10/g	50% of repeat samples greater than 10/g
Salmonella/ Shigella	Absent in 50 g	–	Present in 50 g
Coagulase positive *Staph. aureus*	Less than 10/g	Greater than 10/g	Any repeat samples greater than 10/g
Clostridium welchii	Less than 10/g	Greater than 10/g	Any repeat samples greater than 10/g

* Fully cooked products to be eaten cold by the consumer or to be re-heated.
Note: Samples examined after 24 h at room temperature.

cream products is operated by our suppliers (Goldenberg & Edmonds, 1973). Details are given in Table 6.

"Melbase" Products

This refers to cakes made with a synthetic cream ("Melbase") which, if contaminated with *Staphylococcus aureus,* will support slow growth; salmonellae, *Escherichia coli* and typhoid organisms do not grow and eventually die as a consequence of plasmolysis (Goldenberg & Edmonds, 1973).

Emphasis is placed on the observance of strict hygiene precautions during manufacture, on temperature control throughout production and on rapid distribution and sale thereafter (Goldenberg & Edmonds, 1973). Standards for these products are shown in Table 7.

Method of Operation

It should be made clear at the outset that the results obtained by microbiological analysis in relation to standards such as those described above are of use only as a *post-factum* operation. They cannot generally be carried out in time to deal with current production, except in a minority of instances, because of frequent deliveries to retail stores, coupled with relatively short values for "shelf-life". It

Table 6

Bacteriological standards: group D: cream cakes*

Test	Satisfactory	Acceptable but investigate	Stop production
Viable count	Less than 10,000/g	More than 10,000/g and less than 100,000/g	More than 100,000/g
E. coli type I	Absent from 1 g	Present in 1 g and absent from 0·1 g	Present in 0·1 g
Coagulase positive Staph. aureus	Absent from 1 g	Present in 1 g and absent from 0·1 g	Present in 0·1 g
Salmonella/ Shigella	Absent from 25 g	–	Present in 25 g

* Highly perishable products to be eaten without further processing.
Note: Samples examined off the packing line and after 24 h at room temperature.

Table 7

Bacteriological standards for "Melbase"-filled cakes

Test	Satisfactory	Acceptable but investigate	Stop production
Raw material			
Viable count	Less than 500/g	More than 500/g and less than 1000/g	More than 1000/g
E. coli type I	Absent from 1 g	Present in 1 g and absent from 0·1 g	Present in 0·1 g
Finished product			
Viable count	Less than 1000/g	More than 1000/g and less than 10,000/g	More than 10,000/g
E. coli type I	Absent from 1 g	Present in 1 g and absent from 0·1 g	Present in 0·1 g
Staph. aureus (coagulase positive)	Absent from 1 g	Present in 1 g and absent from 0·1 g	Present in 0·1 g
Salmonella/ Shigella	Absent from 25 g	–	Present in 25 g

is, in fact, a method of monitoring production to ensure that the correct procedures are being implemented by all concerned at all stages of the distribution chain.

Bacteriological analysis, coupled with bacteriological standards of the type described, have nevertheless been found to be a very valuable method of control for the following reasons: (a) it ensures correct and reliable process control; (b) it is valuable psychologically, as all concerned in both production and distribution know that such analyses are being done and will therefore take greater care to ensure adequate supervision within the limits of their own responsibility; (c) it gives top management both at Marks & Spencer and at our suppliers a greater degree of confidence in merchandising perishable foods, and, indeed, in developing and extending their range of foods.

Reportage

We have an agreed system in which the manager or bacteriologist, at each supplier producing perishable foods, sends to a specific technologist at Marks & Spencer a weekly summary of bacteriological results. This is accompanied by an interpretation of results using the agreed standards as a reference basis. If the bacteriological results obtained on any 1 line show a sudden deterioration, the telephone is used and agreed action taken.

Resident Technologists at Suppliers

All manufacturers of perishable foods should have a resident technologist or bacteriologist of some experience. It is part of his duties to organize the sampling of products for bacteriological analysis, to supervise bacteriological analyses and to issue reports both to his management and to the Marks & Spencer technological team as described above. It is our view that the safe production of perishable foods cannot be undertaken without having a resident technologist or bacteriologist in the factory.

Analytical Methods

The methods of bacteriological analysis used by our suppliers and by ourselves have been largely standardized to facilitate the comparison of results and to make it meaningful.

Assessment

The bacteriological standards outlined above are considered to be practical working standards. Experience over the years has shown this to be the case.

They have been found both by our suppliers and by Marks & Spencer to be useful and important factors for ensuring microbiological safety standards in foods, many of which have in the past been responsible for food poisoning and where spoilage is costly. They do not represent levels beyond which we consider the food to be dangerous, but they do represent levels beyond which action is called for.

Acknowledgement

The authors wish to thank the Directors of Marks & Spencer for permission to publish.

References

Bashford, T. & Herbert, D. (1966). Safeguards against microbial infection of canned foods. In *The Safety of Canned Foods*, p. 75. London: The Royal Society of Health.

Goldenberg, N. & Edmonds, M. S. (1973). Education in microbiological safety standards in food factories (A review of experience 1955–1971). In *Proceedings of the 8th International Symposium. The Microbiological Safety of Food*, p. 435. London: Academic Press.

Hobbs, B. C. (1969). Staphylococcal and *Clostridium welchii* food poisoning. In *Bacterial Food Poisoning*, p. 67. London: The Royal Society of Health.

Symposium on microbiological standards for foods. (1970). *Chemy Ind.* pp. 186–109, 215–229, 271–274.

Wilson, G. S. (1970). Concluding remarks. In *Symposium on Microbiological Standards for Foods. Chemy Ind.* p. 273.

Discussion

Insalata

In my opinion the primary specification of total count of 200/g for pre-cooked meat pies—and the subsequent analysis after 24 h at "room temperature"—is unrealistic and misleading, for these reasons:

(a) Cooking probably produces a "commercially sterile" product for which a count of 200/g is excessive.

(b) If 200/g is a reality, it cannot be expected that these 200 viable organisms will remain static at room temperature ($20-30°$). Within 24 h the product would not be fit for consumption.

(c) This arbitrary design of "consumer misuse" is contrary to the speaker's definition of the recognized conditions for product stability, i.e. controlled refrigeration and consumer use of the product. The burden is placed wholely on the manufacturer. What of consumer education and responsibility?

Goldenberg

All pre-cooked meat pies are baked to an internal temperature of $185°$F and cooled rapidly. Providing this process is carried out correctly the pies will indeed be "commercially sterile". The 24 h incubation period is a control measure to ensure that the above process has been carried out correctly. The figure of 200/g total count seems to us to be a practical level akin to commercial sterility and one which is consistently achievable. Despite storage instruction on the packaging materials not all customers follow the instructions and not all customers have refrigerators. The manufacturer and retailer therefore have a responsibility to produce a safe product which includes a period of stability at room temperature.

Foster

Did you say that you consider it important for each food processing plant to have its own microbiology laboratory?

Goldenberg

I consider it essential.

Kampelmacher

What are your measures to prevent contamination of food by handlers? What about toilet hygiene in your establishments and how do you control carriers?

Goldenberg

This is dealt with in my paper this afternoon.

Olson

Would you please indicate the mechanism by which your company assures itself that the specifications you impose on your suppliers are actually complied with?

Goldenberg

The specifications are not "imposed". They are jointly agreed between our suppliers and ourselves, and are based on mutual confidence and co-operation, together with random sampling in our stores.

Skovgaard

My question is simply: are some of your standards, e.g. for cream cakes, achieved by the use of preservatives?

Goldenberg

No.

Malkki

I should like to mention work which has been done in our country to overcome hazards from salmonellae in bakery products. Nurmi has found that the pH value of creams and custards can be lowered below pH 5 by the use of glucono-delta-lactone without impairing the taste or whipping properties. This has been practiced successfully in one of our leading bakeries during the past year. It therefore seems possible tentatively to set a pH value limit of 4·8 for cream and custard-filled bakery products to control salmonellae.

Brachman

Can you discuss the methods by which you establish the various bacteriological standards you outlined for the different categories of foods sold by your company?

Goldenberg

By using available information published in the literature, by discussions with established authorities, e.g. the Food Research Associations; the ARC Research Institutes, the Public Health Laboratory, etc.; and finally by bacteriological analysis of specific foods made under appropriate conditions under our technical supervision.

Food Microbiological Specifications in Australia

J. H. B. CHRISTIAN

*Commonwealth Scientific & Industrial Research Organization,
Division of Food Research,
North Ryde,
New South Wales,
Australia*

Domestic food standards in Australia are laid down and enforced by the Departments of Health in the 6 States of the Commonwealth. The 6 Pure Food Acts, which include microbiological specifications, are not uniform, although work towards uniformity is in progress. All 6 Acts contain microbiological standards for water, milk and other dairy products; the other foods covered vary between States. At the Federal level, a much more comprehensive set of regulations, the Commonwealth Food Specifications, is applied to foods purchased for the armed services. The National Health and Medical Research Council convenes a number of committees which suggest new or revised standards that the Council may recommend to State governments for uniform adoption. Other groups are currently charged with proposing standard microbiological methods for the examination of meat, egg and dairy products and pre-cooked frozen foods. As a major food exporter, the Australian food industry is influenced by the microbiological standards imposed by importers. Most important here are those for meat, dairy products, egg products and crustaceans.

Introduction

The Commonwealth of Australia is a federation of 6 sovereign States, each of which exercises its prerogative of setting and enforcing food regulations. For some special purposes, the Commonwealth Government has standards, the enforcement of which does not conflict with State regulations.

State Regulations

Each of the 6 States, through its Department of Health, lays down and enforces its Food Regulations. These relate to foods on sale within the particular State, whether imported or produced locally. The standards within these Regulations are concerned, in the main, with the composition of foods (including such additives as preservatives, colourings, flavourings and antioxidants), and packaging and labelling. Relatively little emphasis is placed at present on microbiological standards, but these exist in all States for water, milk and some other dairy products. As will be discussed below, the 6 sets of Regulations are not uniform. Only some of the States have microbiological standards for oysters, prawns and tomato products.

MSF—13

369

Federal Regulations

During World War II, Commonwealth Food Specifications (C.F.S.) were formulated to cover all food items required by the Armed Services. These specifications are still in use for this purpose and are continually revised and updated by a committee convened by the Commonwealth Department of Primary Industry. C.F.S. are much more comprehensive than the State regulations and include microbiological standards for a much wider range of foods, covering dairy products, egg products, dehydrated foods, freeze-dried products and miscellaneous items. In addition, incubation tests are prescribed for canned foods.

The quality and composition of foods exported from Australia are controlled under Commonwealth Export Regulations, promulgated under Customs Acts but administered by the Commonwealth Department of Primary Industry.

The Commonwealth Department of Health, although conducting research into food-borne disease, does not recommend or introduce food standards. However, it provides the secretariat for the Food Standards Committee (see below) and is represented on many other related committees.

Establishment and Co-Ordination of Specifications

States have traditionally established their own food specifications, so that 6 almost independent sets of regulations have developed. Concern at this situation led the Council of Australian Food Technology Associations (CAFTA), a body composed of representatives of the food industry, to work actively for the co-ordination and standardization of legislation. Under the urging of CAFTA, the National Health and Medical Research Council has established a Food Standards Committee which draws up model standards that are recommended to the States for adoption. This Committee is advised by a number of expert groups, including a Food Microbiology (Reference) Sub-committee.

Uniformity has to some extent been achieved in respect of non-microbiological standards. As the number of purely microbiological specifications in the various Regulations is relatively small, CAFTA's task, at least in this respect, would seem relatively simple. However, it is symptomatic of the difficulties inherent in a Federal system that the greatest divergencies exist in some of the oldest and most important standards, namely those for milk. There is in this field an active Australian Dairy Products Standards Organization already proposing uniform regulations, but as yet these lack unanimous State support. It should be noted that few States use specifications as guidelines. In nearly every case, the only specifications acceptable to Government are those which can be enforced by law.

In the last 5 years there has been increasing pressure from health authorities for the establishment of microbiological standards for a much greater range of foods. There has been, at the same time, an admirable reluctance by microbiologists to recommend new specifications until such time as standard microbiological methods for both old and new specifications have been accepted by enforcement agencies and by the food industry To this end a number of specialist committees is presently considering methods which might be accepted as standard for the examination of meat and meat products, eggs and egg products, dairy products and pre-cooked frozen foods. This involves repeated consultation between the committees and both industry and government, but in most cases is proceeding reasonbly rapidly.

At the same time, attempts are being made by the Food Microbiology Sub-committee of NHMRC to introduce into food legislation a multiple sampling procedure similar to that favoured by the IAMS International Commission on Microbiological Specifications for Foods. Progress here may be slow, as in some cases the Acts themselves would have to be changed to accept the type of specifications that results.

It is much easier and less expensive to promulgate a microbiological standard than to enforce it. Some countries with quite elaborate standards appear to employ insufficient staff for effective enforcement. This is doubly unfortunate, as an unenforced law tends to fall into disrepute and as it is often only by enforcement that the adequacy or otherwise of a standard becomes apparent.

Australia has been a member of the Codex Alimentarius Commission since its inception and is taking seriously its obligation to adhere to Codex requirements. However, the practical situation is complicated by the fact that Codex specifications will have to be incorporated into all 6 sets of State regulations.

Non-Government Specifications

The major non-government specifications in Australia are those written into commercial contracts. These are common to most countries and are set by food processors for the materials they purchase. They apply increasingly to imported as well as locally produced raw materials. In these days of multi-national corporations, it is not uncommon for the same standards to be applied to raw materials produced in different countries under very different conditions, and it should be no surprise that such standards cannot always be met.

However, among the microbiological specifications which most affect the Australian food industry and indirectly the consumer are those set by overseas buyers of Australian exports. These apply particularly to meat, egg products, dairy products and crustaceans. These standards may be enforced by the government of the importing country, by an industry association in such a country, or by the importer himself. In some cases, e.g egg products, compliance of exported material with the buyer's specification may be monitored and certified by a Federal Government laboratory, in most cases this is the exporter's responsibility.

Where a large export market exists, there are 2 possible consequences of such standards to the local consumer.

First, it is clear that importers' specifications may raise to some extent the product quality throughout the segment of the industry concerned. Second, it is equally likely that, in the absence of an appropriate domestic standard, there will be in the short term a tendency for the highest quality product to be exported and the unexportable to be released on the home market. Whether the first consequence compensates rapidly for the second is debatable Either way, a good case can be made that local standards comparable with externally imposed standards should always be enforced in such situations.

Education

The chances of foods meeting reasonable microbiological standards are poor if the industry lacks experienced microbiologists and workers conscious of the principles of hygiene and sanitation. The Australian situation is still far from ideal in this respect. What food microbiologists there are in industry and government employment have in the past been trained mainly in medical or agricultural bacteriology. There are now, however, several tertiary institutions producing professional food technologists and post-graduate work in food microbiology is being undertaken on a rapidly expanding scale. The Australian Institute of Food Science and Technology (AIFST) is about to run its first course on techniques of particular relevance to microbiologists already in the food industry. Courses are given annually by the Commonwealth Department of Health for Government Medical Officers of Health, and there is emphasis here on food sanitation, food inspection, and the epidemiology of food-borne disease.

It is at the non-professional level that an appreciation of hygiene and sanitation and hence of the significance of microbiological standards is most difficult to foster. This problem may well be exacerbated by the burgeoning of the fast-food industry with its own peculiar problems in food handling. Instruction at this level is already well advanced in the meat industry, with the CSIRO Division of Food Research conducting practical courses in major meatworks throughout the country. The AIFST plans to reach supervisory staff in other branches of the food industry with similar courses to commence this year. The food trade courses in the technical colleges already include extensive segments on hygiene and sanitation in their curricula for butchers, chefs, pastrycooks and similar trades.

It is expected that these activities in education will accelerate and that, as a consequence, the Australian food industry will be much better equipped to meet the increasing range of microbiological specifications likely to be introduced in the near future.

Discussion

Goldenberg

Do the Australian regulations for the export of top-quality cheese demand that the milk used be pasteurized? What view do you take of the possible presence of faecal coli and coagulase positive *Staph. aureus* in such cheeses?

Christian

The milk should be pasteurized but Miss Dick is probably better qualified to answer this question.

Dick

For top-quality cheese, coliforms and staphylococci should not be detected in 0·1 g cheese. However, sampling is a problem since the organisms grow in colonies in the cheeses.

Mocquot

Post-pasteurization contamination of cheese milk is an important point one has to look at as far as food bacteriology is concerned. But the activity of the lactic acid bacteria starter culture is of major importance. A milk of very good bacteriological quality may give a cheese which is heavily contaminated if the starter is not active (or even not very active). On the contrary a milk with a low to moderate level of contamination may give a cheese of excellent bacteriological quality if the starter is very active because the drop in pH value will prevent the growth of or even destroy the contaminating bacteria likely to be present.

Elliott

Contaminating organisms may not grow in the cheese itself but if they survive and the cheese is used in a food mixture the organisms may grow.

Olson

The problem with cheese is the wide variety of types from Cheddar to soft cheeses. If one is buying cheese, one should ensure: (a) that the milk was pasteurized from 15–20 sec at 150° F; (b) that the vat record shows the acidity to have reached at least 0·4%–the cheese should otherwise be rejected; and (c) that the staphylococcal count 48–72 h after manufacture is not in excess of 100/g.

Control of Contamination with *Vibrio parahaemolyticus* in Seafoods and Isolation and Identification of the Vibrio

R. SAKAZAKI

National Institute of Health,
Tokyo,
Japan

Vibrio parahaemolyticus is one of the inhabitants of coastal sea water. It is not possible to protect raw seafoods from contamination with this vibrio. Control of infection in man depends mainly on limiting multiplication in seafoods and in preventing secondary contamination of cooked seafood from raw materials. Methods for the isolation and identification of *V. parahaemolyticus* from human stool specimens and from seafoods are discussed.

VIBRIO PARAHAEMOLYTICUS, a marine bacterium, is widely distributed in coastal seawater, sediment and plankton, and in the summer season in Japan, it can readily be isolated from seawater. Coastal sea fish are always contaminated with *V. parahaemolyticus.* Akiyama *et al.* (1964) studied the occurrence of *V. parahaemolyticus* in horse mackerel, which are often eaten in summer, and reported that it was present in 96% or more of fish in August, although none was found in May. The vibrio is usually present not only on the surface of the fish but also in the gills, stomach and intestines. *V. parahaemolyticus* is not found, as a rule, in the ocean or in ocean fish. Such fish are, however, usually contaminated with the vibrio in the fish market.

For the reasons mentioned above, it is impossible to protect raw seafoods against contamination with *V. parahaemolyticus.* The most important method for the control of infection in man lies in simple hygienic measures to reduce the multiplication of the vibrio in seafoods and the secondary contamination of foods from raw sea materials. These measures apply equally to market, transport, shop and home.

Multiplication of *V. parahaemolyticus* is quicker than that of *Escherichia coli.* Aiiso (1964) calculated that the generation time of the vibrio at 37° was 12–15 min. Although *V. parahaemolyticus* is present on sea fish, it is usually found only in small numbers when the fish are fresh. However, a few hours of multiplication on fish at a favorable temperature easily converts a small number of vibrios into many millions.

The size of the infecting dose necessary to produce clinical symptoms may vary with the strain, but it is probably about $10^6–10^9$ viable cells according to the data of Takikawa (1958), Aiiso & Fujiwara (1963), and Sakazaki *et al.* (1968).

Refrigeration or freezing is the most important method for preventing multiplication of *V. parahaemolyticus* in seafoods. The growth of *V. parahaemolyticus* on the surface of sea fish may be arrested between 5°–8°, although the organism may survive for long periods at these temperatures. Slow growth occurs at 10° (Tenmei & Yanagisawa, 1962). Freezing not only arrests multiplication of *V. parahaemolyticus* but also causes a rapid initial decrease in the number of viable organisms. Tenmei & Yanagisawa (1962) showed that *V. parahaemolyticus* was more sensitive to refrigeration than *E. coli* and other organisms.

Takeuchi *et al.* (1957) reported that distilled water killed *V. parahaemolyticus* within a minute, probably due to osmotic destruction of the bacterial cells. For this reason, the washing of fish and of equipment such as dishes, containers, chopping board, and kitchen knives with tap water may effect some decrease in the numbers of viable vibrios, although it is also known that traces of salts and organic substances in freshwater allow their survival (Yanagisawa & Takeuchi, 1957).

V. parahaemolyticus is more sensitive to acidity than *E. coli* and growth is completely inhibited at pH 4·5-5·0. Kondo *et al.* (1960) observed that the vibrio was killed in vinegar within 1 h and in 0·5% acetic acid solution within several minutes. However, raw fish acidified with vinegar, which is widely used in Japan, has frequently caused food poisoning due to *V. parahaemolyticus*. There seems to be little information about the effect of curing and smoking on *V. parahaemolyticus*. Tenmei, Osemura *et al.* (1961) and Tenmei, Yanagisawa *et al.* (1961) reported that *V. parahaemolyticus* is very sensitive to drying, and hot smoking of fish may kill it. Nakanishi (1968) and Tamura (1970) examined curing brines for the presence of *V. parahaemolyticus,* but they could not detect it in any samples. It should be noted, however, that *V. parahaemolyticus* food poisoning from cured vegetables has often occurred in Japan.

Heating is the most effective method of eliminating *V. parahaemolyticus* from seafoods. If all seafoods were heated to 100° shortly before consumption, food poisoning due to *V. parahaemolyticus* would never occur. *V. parahaemolyticus* shows the same degree of heat resistance as *E. coli* and is killed by heating at 60° for 15 min. It is emphasized that raw seafoods and cured vegetables are mostly implicated in food poisoning outbreaks. Boiled or roast fish has sometimes been incriminated when the food was contaminated with the vibrio from raw fish either directly or indirectly. Oysters seldom cause vibrio food poisoning in Japan because they are eaten in the cold season.

Although most cases due to *V. parahaemolyticus* are connected with the consumption of raw fish in Japan, infection from this source is unlikely to occur in South-east Asia and western countries where people are not in the habit of eating raw fish meat. *V. parahaemolyticus* infection in those countries may be caused by other foods which have been contaminated from raw materials or water, or by kitchen utensils.

Little work has been undertaken on the numbers of *V. parahaemolyticus* in seafoods. Asakawa (1968) described a membrane filter method using TCBS agar, and Vanderzant & Nickelson (1972) used a fluorescent antibody technique involving the "A" substance of *V. parahaemolyticus* (Miwatani *et al.*, 1969, 1970). However, these methods are still unsatisfactory for the purpose, because many other marine organisms may be confused with *V. parahaemolyticus*. Indeed, related halophilic Gram-negative rods are troublesome when seafoods are examined for *V. parahaemolyticus*. A modification of the most probable number technique has also been used for the enumeration of *V. parahaemolyticus*. This method involves enrichment culture from a known quantity of food using Salt-Polymyxin broth at 37° for 8 h and subsequent subculture on TCBS agar (Nakanishi *et al.*, 1972).

No relation has yet been established between the numbers of *V. parahaemolyticus* and the hygienic quality or safety of seafoods. For hygienic control of seafoods it will be necessary to establish suitable criteria as soon as possible.

Isolation and Identification of *V. parahaemolyticus*

The diagnosis of gastroenteritis due to *V. parahaemolyticus* depends upon isolation of the organism from patients. Stool specimens for culture should be collected in the acute stage of illness. Detection of the vibrio later in illness may be difficult, because it decreases rapidly in number with recovery from diarrhoea. Specimens should therefore be cultured as soon as possible. If delay is inevitable, Cary-Blair transport medium (Cary & Blair, 1964) is suitable for the preservation and transportation of stool specimens containing *V. parahaemolyticus*.

Isolation of the vibrio from stool specimens is not difficult, since *V. parahaemolyticus* may be the only Gram-negative, halophilic organism present in diarrhoea. Although several selective agar media have been devised for the isolation of *V. parahaemolyticus*, TCBS agar (Kobayashi *et al.*, 1963) and BTB-Salt-Teepol agar (Akiyama *et al.*, 1963) modified by Sakazaki (1969, unpublished data) are suitable. Modified BTB-Salt-Teepol agar contains 1% peptone, 0·3% beef extract, 4% sodium chloride, 2·0% sucrose, 1·5% agar, 0·02% sodium heptadecyl sulphate (Tergitol 7, Union Carbide Co., New York), 0·0004% bromthymol blue and 0·0004% thymol blue (final pH 9·0). TCBS (Thiosulphate-Citrate-Bile Salt-Sucrose) agar is much more selective than BTB-Salt-Teepol agar and is satisfactory for the isolation not only of *V. parahaemolyticus* but also of *Vibrio cholerae*. For 1 l, the dehydrated medium contains 5 g yeast extract, 10 g each of peptone, sodium citrate, sodium thiosulphate, and sodium chloride, 5 g ox gall, 3 g sodium cholate, 20 g sucrose, 1 g ferric citrate, 15 g agar, and 0·04 g each of bromthymol blue and thymol blue (final pH 8·6).

After 24 h incubation, isolated colonies of *V. parahaemolyticus* on these media are round, 2–3 mm in diameter with centres stained green or blue with alkaline bromthymol blue and thymol blue. In contrast, *Vibrio alginolyticus* which is present in the sea and is sometimes isolated from human stools, forms large yellow sucrose-fermenting colonies on these agar media. Some marine organisms related to *V. parahaemolyticus* may grow on BTB-Salt-Teepol agar, but the majority are inhibited on TCBS agar. Occasional strains of *Proteus*, enterococci and coliform organisms may grow on these media, but their colonial morphology, colour and opacity differentiate them from *V. parahaemolyticus*.

Isolation from stool specimens

Enrichment culture is not necessary for isolation of the vibrio in the acute stage of diarrhoea, but it may be necessary to detect *V. parahaemolyticus* in specimens from convalescent patients (Sakazaki *et al.,* 1971). Alkaline peptone water, which is commonly used for cholera, is also satisfactory for the selective enrichment of *V. parahaemolyticus*. Cultures of alkaline peptone water containing 1% peptone and 1% sodium chloride (pH 8·6–9·0) should be incubated for 8 h at 37°. Longer incubation should be avoided, otherwise many intestinal inhabitants may grow. Secondary enrichment may permit isolation of *V. parahaemolyticus* from patients who have received antimicrobial treatment. For this one loopful of an 8 h peptone water culture is transferred to a second alkaline peptone water for incubation at 37° for 8 h. Overnight culture in Monsur's Tellurite-Bile salt broth (Monsur, 1963) at 37° is also effective for secondary enrichment (Sakazaki *et al.,* 1971).

Isolation from marine sources

Compared with faeces, the isolation of *V. parahaemolyticus* from seawater, sea fish, and seafoods may be hindered by other marine organisms, and it is necessary to use more selective media than those for stools. A selective enrichment broth which has been widely used is Polymyxin-Salt broth (Sakazaki, 1965, unpublished data) which contains 0·3% yeast extract, 1% peptone, 2% sodium chloride, and 250 µg/ml of polymyxin B or 500 µg/ml of colistin methansulphonate (final pH 7·4). Glucose-Salt-Teepol broth of Akiyama *et al.* (1964) which contains 3% beef extract, 1% peptone, 3% sodium chloride, 0·5% glucose, 0·0002% methyl violet and 0·4% Teepol (final pH 9·4), may also be employed. Subculture on to TCBS agar is recommended in order to avoid confusion of *V. parahaemolyticus* with other related marine organisms.

Identification and characterization

The minimum characteristics for the identification of *V. parahaemolyticus* are shown in Table 1 (Hugh & Sakazaki, 1972). During the examination of over

Table 1

Minimal number of characters for the identification of Vibrio parahaemolyticus

		+%
Gram-negative, asporogenous rod	+	100
Oxidase	+	100
Glucose, acid under anaerobic conditions	+	100
Glucose, gas	−	0
D-mannitol, acid	+	99·6
Sucrose (1%), acid	−	5·3
Acetylmethylcarbinol	−	0
Hydrogen sulfide (TSI)	−	0
L-lysine decarboxylase	+	100
L-arginine dihydrolase	−	0
L-ornithine decarboxylase	+	97·3
Growth in 1% tryptone broth	−	0
Growth in 1% tryptone broth with 8% NaCl	+	100
Growth in 1% tryptone broth with 10% NaCl	−	0·6
Growth at 43°	+	100

Media should contain 2% sodium chloride, except for tryptone broth.

100,000 specimens of faeces, marine organisms other than *V. parahaemolyticus,* with few exceptions, were not isolated with TCBS and BTB-Salt-Teepol media. *V. alginolyticus* may sometimes be found in stools in association with *V. parahaemolyticus,* but it is readily distinguished by its sucrose fermentation and positive V. P. reaction.

Gram-negative rods found in human stools, including *Aeromonas, Plesiomonas,* members of the *Enterobacteriaceae,* and *Pseudomonas,* sometimes require to be differentiated from *V. parahaemolyticus,* although they usually fail to grow on plating media for *V. parahaemolyticus.* Colonies thought to be *V. parahaemolyticus* are subcultured into TSI agar, and into MR-VP broth and lysine decarboxylase broth, each containing 2% NaCl, and into peptone water with 8% sodium chloride. The oxidase test should be done with growth from TSI or nutrient agar slants. Some additional tests are necessary for strains resembling *V. parahaemolyticus* isolated from marine sources, even when TCBS agar is used. These include the ability to grow in peptone water containing 10% NaCl, growth at 43° (±0·05) in peptone water with 2% salt, production of arginine dihydrolase, and fermentation of some sugars.

The procedures for the isolation and identification of *V. parahaemolyticus* are given in Figs 1 and 2, and its characteristics and differentiation from other organisms are shown in Tables 2 and 3.

For lysine decarboxylase and arginine dihydrolase tests, Taylor's broth (Taylor, 1961) containing 1% salt is recommended. MOF medium (Leifson, 1963) is appropriate for sugar fermentation tests. The shaking culture method

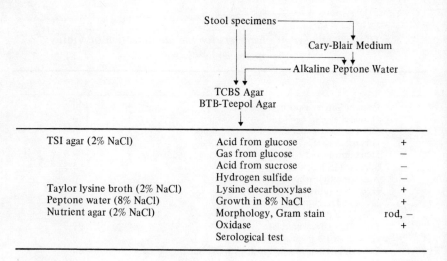

Fig. 1. Isolation and identification of *Vibrio parahaemolyticus* from human stool specimens.

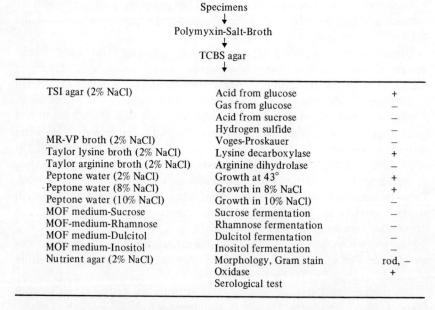

Fig. 2. Isolation and identification of *Vibrio parahaemolyticus* from seawater, sea fish and foods.

Table 2

Physiological and biochemical reaction of Vibrio parahaemolyticus *and related vibrios*

	V. cholerae	*V. parahaemolyticus*	*Vibrio* sp. 6330	*V. alginolyticus*	*V. anguillarum*	*Vibrio* sp. 6267
Growth at 25°	+	+	+	+	+++	+
Growth at 35°	+++	+++	+++	+++	d	+++
Growth at 40°	++	++	++	++	−	++
Growth at 43°	d	+	d	+	−	−
Growth in 0% NaCl	+	−	−	−	−	−
Growth in 1% NaCl	+++	+	++	+	+++	++
Growth in 2% NaCl	+++	+++	+++	+++	+++	+++
Growth in 4% NaCl	+	+++	++	+++	++	++
Growth in 6% NaCl	d	+++	++	+++	d	++
Growth in 8% NaCl	−	++	−	+++	−	−
Growth in 10% NaCl	−	−	−	++	−	−
Indole	+	+	+	+	+	+
Voges-Proskauer	d	−	−	+	d	−
Ammonium citrate	d	+	+	+	−	+
Lysine decarboxylase	+	+	+	+	−	−
Arginine dihydrolase	−	−	−	−	+	+
Ornithine decarboxylase	+	+	d	+	−	−
Sucrose (1%)	+	−	−	+	+	d
Rhamnose	−	−	−	−	−	d
Dulcitol	−	−	−	−	−	d

d = variable

Table 3

Differentiation of Vibrio parahaemolyticus *isolated from human stools*

	TSI agar Slant/Butt	Oxidase	Lysine Decarboxylase	Growth in 8% NaCl	VP	Vibrio 0-1 serum
V. parahaemolyticus	−/A	+	+	+	−	−
Enterobacteriaceae	d/A or AG	−	d	−	d	−
Aeromonas	+/AG or A	+	−	−	d	−
Plesiomonas	−/A	+	+	−	−	−
Pseudomonas and related rods	−/−	+	−	−	−	−
V. cholerae, 1	+/A	+	+	−	d	+
V. cholerae, NAG	+/A	+	+	−	+	−
V. alginolyticus	+/A	+	+	+	+	−

d = variable

with 1% trypticase peptone water without adjustment of pH is recommended for the ability to grow in 8 and 10% salt. In this test, 0·01 ml of an overnight broth culture is inoculated into 10 ml of the peptone water and the result read after overnight incubation at 37^c with shaking in a water bath. Only heavy growth is interpreted as positive. In laboratories where shaking culture apparatus is not available, Cybalski's gradient plate method is useful. Nutrient agar containing 15% of salt, 0·001% of triphenyl tetrazolium chloride, and 0·001% of Tergitol 7 is first solidified in a sloping position in a Petri dish. Molten nutrient agar ($55°$) containing no salt is then poured on and solidified in a level position. Overnight broth cultures of test organisms are inoculated with cotton swabs across the plate, and the results read after overnight incubation. A known strain of *V. parahaemolyticus* is inoculated at both sides, and unless growth of the test strains is similar to that of the control strain, *V. parahaemolyticus* cannot be identified.

Kanagawa test

The Kanagawa test should be performed with Wagatsuma agar (Wagatsuma, 1968). The basal medium contains 0·5% yeast extract, 1% Bactopeptone, 7% sodium chloride, 0·5% D-mannitol, 1·5% agar and 0·0001% crystal violet (final pH 7·5). Other peptones should not be substituted for Bactopeptone. Over-heating or autoclaving may cause equivocal results in the test. For use, 10 ml of a 20% suspension of washed fresh human red cells are added to 100 ml of the basal medium at $50°$, mixed thoroughly and poured into Petri dishes. Only human red cells should be used in this medium. Loopfuls from 15–20 overnight broth cultures of test strains of *V. parahaemolyticus* are spot-inoculated on to 1 plate, and the results read after incubation at $37°$ for 18–24 h. A positive Kanagawa reaction is shown by a clear zone of beta haemolysis formed around the growth. Alpha haemolysis or discolouration of red cells under the growth should be interpreted as negative. Incubation for more than 24 h may result in misinterpretation of negative reactions as positive.

Serological typing

Serological identification of the O and K antigens of *V. parahaemolyticus* is performed by slide agglutination tests. Diagnostic O and K antisera are now available from commercial sources. *V. parahaemolyticus* is usually O-inagglutinable in the living state because of the presence of the surface K antigen, so that cultures for O agglutination should be heated to $100°$ for 1 h. Washing the packed cells with 1% saline twice after centrifugation may increase O-agglutinability of the culture. If the heated suspension is still not agglutinable with O antisera, O form (bright and/or transparent) colonies should be sought on nutrient agar. If O form colonies are not found, O agglutination may be obtained after autoclaving OK colonies for 2 h at $121°$.

For K agglutination, dense suspensions of live cultures of *V. parahaemolyticus* in 1% saline may be used. Suspensions prepared from cultures with poorly developed K antigens (O form) may give weak agglutination with K antisera. On the other hand, with cultures containing excess K antigen (mucoid form), agglutination is rather weak. In both these cases, colonies suitable for agglutination should be reselected. Determination of the K types of *V. parahaemolytics* strains is time-consuming because of the number of different antigens, and the use of pooled antisera in which equal amounts of 5 or more monovalent K antisera are combined, is essential for practical purposes. The procedure and interpretation of the results of agglutination tests are similar to those for enteric bacteria. Serological tests are not diagnostic as many strains of other marine organisms may be agglutinated with *V. parahaemolyticus* antisera.

The antigenic schema established by Sakazaki *et al.* (1968) was based on strains of *V. parahaemolyticus* isolated from man. Although most isolates could be typed with known O and K antisera, many cultures from marine sources were not identifiable with these antisera. For simplification, the Committee of Serotyping of *V. parahaemolyticus* has therefore decided that the antigenic schema should include only serotypes from human sources.

References

Akiyama, S., Takizawa, K., Ichinohe, H., Enomoto, S., Kobayashi, T. & Sakazaki, R. (1963). Application of Teepol in a medium for the isolation of the enteropathogenic halophilic bacteria. *Annual Rept. of the Kanagawa Prefect. Inst. Publ. Hlth.* **12**, 7. (Text in Japanese.)

Akiyama, S., Takizawa, K. & Obara, Y. (1964). Study on enrichment broth for *Vibrio parahaemolyticus. Annual Rept. of the Kanagawa Prefect. Inst. Publ. Hlth* **13**, 7. (Text in Japanese.)

Aiiso, K. & Fujiwara, K. (1963). Feeding test of the pathogenic halophilic bacteria. *Ann. Res. Inst. Fd Microbiol., Chiba Univ.* **15**, 34.

Aiiso, K. (1964). Multiplication and generation time of *Vibrio parahaemolyticus.* In *Vibrio Parahaemolyticus* Eds T. Fujino & H. Fukumi, p. 205. Tokyo: Naya Publ. (Text in Japanese.)

Asakawa, Y. (1968). Personal communication.

Cary, S. G. & Blair, E. B. (1964). New transport medium for shipment of clinical specimens. I. Fecal specimens. *J. Bact.* **88**, 96.

Hugh, R. & Sakazaki, R. (1972). Minimal number of characters for the identification of *Vibrio* species, *Vibrio cholerae,* and *Vibrio parahaemolyticus. Publ. Hlth Lab.* (In press.)

Kobayashi, T., Enomoto, S., Sakazaki, R. & Kuwahara, S. (1963). A new selective isolation medium for pathogenic vibrios: TCBS agar. *Jap. J. Bact.* **18**, 367. (Text in Japanese.)

Kondo, R., Yoshimura, Y., Yamaguchi, M., Tanaka, K. & Yanagisawa, F. (1960). Multiplication of the pathogenic halophilic bacteria on vegetables and dressing materials. *Jap. J. Publ. Hlth* **7**, 752. (Text in Japanese.)

Leifson, E. (1963). Determination of carbohydrate metabolism of marine bacteria. *J. Bact.* **85**, 1183.

Miwatani, T., Shinoda, S. & Fujino, T. (1969). A common antigenic substance of *Vibrio parahaemolyticus.* I. Isolation and purification. *Biken J.* **12**, 97.

Miwatani, T., Shinoda, S. & Fujino, T. (1970). Purification of monotrichous flagella of *Vibrio parahaemolyticus. Biken J.* **13**, 149.
Monsur, K. A. (1963). Bacteriological diagnosis of cholera under field conditions. *Bull. Wld Hlth Org.* **28**, 387.
Nakanishi, H. (1968). Personal communication.
Nakanishi, H., Teramoto, T., Murase, M. & Maejima, K. (1972). Personal communication.
Sakazaki, R., Tamura, K., Kato, T., Obara, Y., Yamai, S. & Hobo, K. (1968). Studies on the enteropathogenic, facultatively halophilic bacteria, *Vibrio parahaemolyticus.* III. Enteropathogenicity. *Jap. J. Med. Sci. Biol.* **21**, 325.
Sakazaki, R., Tamura, K., Prescott, L. M., Bencic, Z., Sanyal, S. C. & Sinha, R. (1971). Bacteriological examination of diarrheal stools in Calcutta. *Ind. J. med. Res.* **59**, 1025.
Takeuchi, T., Hirose, S., Tanaka, K., Yamawaki, M., Dei, S. & Sekiguchi, S. (1957). Studies on halophilic bacteria. Part II. The effects of distilled water on halophilic bacteria. *Bull. Tokyo med. Dent. Univ.* **4**, 359. (Text in Japanese.)
Takikawa, I. (1958). Studies on pathogenic halophilic bacteria. *Yokohama med. Bull.* **2**, 313.
Tamura, K. (1970). Personal communication.
Taylor, W. I. (1961). Isolation of salmonellae from food supplies. V. Determination of the method of choice for enumeration of Salmonella. *Appl. Microbiol.* **9**, 487.
Tenmei, R., Yanagisawa, F., Yamaguchi, M., Tanaka, K. & Hirose, S. (1961). Experimental study on prevention of food poisoning due to *Pseudomonas enteritis* Takikawa. *Publ. Hlth (Lond.)* **25**, 340. (Text in Japanese.)
Tenmei, R. & Yanagisawa, F. (1962). Resistance of the pathogenic halophilic bacteria to low temperature. *Jap. J. Publ. Hlth* **9**, 447. (Text in Japanese.)
Tenmei, R., Osemura, S., Funabashi, E. & Arie, K. (1961). The pathogenic halophilic bacteria and dryness. *Shokuhin Eisei Kenkyu* **11**, 31.
Vanderzant, C. & Nickelson, R. (1972). Procedure for isolation and enumeration of *Vibro parahaemolyticus. Appl. Microbiol.* **23**, 26.
Wagatsuma, S. (1968). A medium for the test of hemolytic reaction of *Vibrio parahaemolyticus. Media Circle* **13**, 159. (Text in Japanese.)
Yanagisawa, F. & Takeuchi, T. (1957). Halophilic bacteria. *Shokuhin Eisei Kenkyu* **7**, 11.

Discussion

Hobbs

What is the significance of the small buds or daughter colonies which surround the large colonies of *V. parahaemolyticus* growing on TCBS agar? Is there any serological significance?

Sakazaki

I am sure the small buds may not be those of *V. parahaemolyticus*. Some other marine vibrio and enterococci may grow on TCBS agar forming small colonies.

Kampelmacher

Just for Dr Sakazaki's information fish is consumed raw in the Western Countries also, for example in the Netherlands, especially the so-called "green herring". This has caused the disease "anisakis" which is known also in Japan. So far, however, we have not seen *V. parahaemolyticus* infections due to the consumption of green herring.

Skovgaard

What is the salt concentration in the polymyxin salt broth?

Sakazaki

Two per cent.

Shewan

Did I understand Dr Sakazaki to say that some marine vibrios other than *V. parahaemolyticus* can be agglutinated by the K antiserum?

Sakazaki

Yes.

Beuchat

Dr Sakazaki, you mention that temperatures of $5°-8°$ are low enough to arrest growth of *V. parahaemolyticus* in fish. I have read in the literature that the organism will grow at $2°$. Was this growth in artificial medium?

Sakazaki

In my experience *V. parahaemolyticus* does not grow at $2°$ either in laboratory media or in foods.

Pye

Is it necessary to use enrichment techniques (e.g. alkaline peptone water) to increase the isolation of vibrios from stools? Or is direct inoculation on to selective media sufficient?

Sakazaki

Enrichment is not necessary for stool specimen examination in the acute stage of diarrhoea.

General Consideration on Microbiological Standards

W. J. HAUSLER, JR.

State Hygienic Laboratory,
University of Iowa,
Iowa,
U.S.A.

The American Public Health Association has been actively involved in the publication of laboratory standard methods since the turn of the century. Most notable of these publications are Standard Methods for the Examination of Water and Wastewater and Standard Methods for the Examination of Dairy Products. The Thirteenth Edition of Standard Methods for Dairy Products which became available to the public in May 1972 was the first to be prepared under the guidance of an Intersociety Council and with funds from the Food and Drug Administration. This process of achieving a publishable document has been regarded as quite successful and has fostered development of a correlated publication. A Technical Committee (soon to be called an Intersociety Council) has been formed, a chairman appointed, and funds are being sought to achieve eventually publication of a manual on Methods for the Microbiological Examination of Foods. The establishment of a mechanism for an Intersociety Council approach of the selection, validation and preparation of a compendium or manual of microbiological methods presents the greatest likelihood of success in obtaining wide acceptance of such procedures.

I AM PLEASED to have this opportunity to present a discussion of the activities of the American Public Health Association in the area of laboratory methods for the examination of foods.

The American Public Health Association, which will be referred to as APHA throughout the remainder of this presentation, is celebrating its Centennial year. Its history contains a large number of accomplishments which have been made by its many members who represent every aspect of professional public health performance.

As with many organizations of this type the membership grew and started collecting into groups of professional similarities. In 1899 the first sub-organization of APHA, the Laboratory Section, was established. At the present moment there are 19 sections covering all aspects of public health activities.

The Laboratory Section of APHA has been the most active in the collection and presentation of standard laboratory methods. In 1905 Professor S. C. Prescott of the Massachusetts Institute of Technology reported on "The Need for Uniform Methods in the Sanitary Examination of Milk" at the annual meeting of the APHA in Boston. In his presentation Professor Prescott mentioned that differences in composition of the culture medium employed, variations in methods, dilutions and temperature and duration of incubation, as well as other minor differences, all tended to produce results which were valueless for comparison. As a result of these comments, a committee was formed and in 1910 the first edition on "Standard Methods of Bacterial Milk Analysis" was published. In subsequent editions of this publication there has been co-operation with other agencies and associations, but they were primarily under the auspices of the Co-ordinating Committee on Laboratory Methods of the APHA and not a product of the Laboratory Section. The Co-ordinating Committee on Laboratory Methods was concerned not only with milk analysis but also with water, air and diagnostic microbiological methods of importance to public health.

Milk and milk products available to the American consumer are among the safest foods which can be obtained. No single factor has been responsible for this development in the dairy industry. Instead, a number of factors have, over the years, contributed to this progress. Among the more significant are: the eradication of diseases from dairy cattle, compulsory pasteurization, improvements in sanitation on the farm and in the processing plant, improvements in refrigeration and distribution, and adequate and uniform laboratory control of raw and finished products and of the environments in which they are produced and processed. Achievement of adequate laboratory practices is to a large degree the result of 12 editions of *Standard Methods for the Examination of Dairy Products.*

The procedure used to prepare the first 12 editions involved: (a) appointment of a chairman who was responsible to APHA; (b) appointment of committee members selected by the chairman; and (c) consultation between chairman and committee members to develop an edition of standard methods for dairy products.

The committee responsible for recent editions was divided into a series of subcommittees each of which was responsible for updating and otherwise revising one or several chapters in Standard Methods. For example, 49 persons contributed to the 12th edition of Standard Methods for Dairy Products.

The 12th edition under Dr William G. Walter was the first to be developed under an enlarged scope. The basic philosophy of this edition was that "No new method or modification of an old method should be introduced unless it has undergone careful comparative testing in several laboratories, with the data available to the committee and to any other interested parties, preferably by publication in a recognized scientific journal. Notice of intention to include or modify should appear in print in several places, with enough time to present evidence for or against to be submitted by any interested party with recommendations".

This basic philosophy was fairly successful but it met with the age-old problem that when methods are required to be evaluated and new ones developed there must be a source of money readily available to conduct the necessary studies.

When the 12th edition was completed, the editor, Dr Walter, asked the question, should there be a 13th edition of Standard Methods for the Examination of Dairy Products? Dr Walter indicated the need for: (a) support of applied research on methodology; (b) a mechanism to keep Standard Methods up-to-date; (c) meetings of committees when parts of the manual are being revised; (d) a mechanism for collaborative testing of methods; (e) an advisory committee made up of representatives from industry, government, and universities; and (f) financial support to do the things needed for further improvement of the publication.

In 1968 a contract was negotiated between the American Public Health Association and the U.S. Public Health Service which provided for the establishment of an Intersociety Council on Standard Methods for the Examination of Dairy Products. The basic purposes of the contract were to provide funds for regular meetings of the Intersociety Council, necessary collaborative testing, initial money for research and development of new methods, review of chapter manuscripts and publication of a 13th edition. After actual signing, responsibility for management of the contract within the U.S. Department of Health, Education and Welfare was assigned to the Food and Drug Administration.

An editor was chosen by APHA and was designated to organize an Intersociety Council which would be responsible for the development of the 13th edition. The editor also served as chairman of the Intersociety Council; the author had served as the chairman and editor. The Council was made up of 9 persons who represented professional societies, regulatory agencies, the dairy industry, and the academic community. The Council further enlisted the help of 58 other persons who were experts in various fields and jointly prepared the 13th edition of Standard Methods for the Examination of Dairy Products which was ready for sale in May 1972.

The success of an Intersociety Council for the development of standard methods for the examination of a particular food must await not only the acceptance and critical review of this new edition but also the development of several more editions on the same basis.

The American Public Health Association believes that utilization of Intersociety Councils for the preparation of manuscripts of manuals of laboratory methods is the most logical step, so that 1 organization is no longer solely responsible for the development of methods of analysis that cross many scientific disciplines, governmental agencies and professional groups. In other words, the APHA recognizes that its role should be reduced to one of co-ordination and that it should not act as the sole responsible organization. In recent years, in the preparation of Standard Methods for the Examination of

Water and Wastewater and in Methods of Air Sampling and Analysis, the APHA has served as a secretariat and fiscal office for publications but the actual work has been carried out by an Intersociety Council.

Efforts to bring food-borne illness under control and to improve sanitary practices in the processing and handling of food have emphasized the need for microbiological standards of quality for many foods. Because of this concern, several attempts have been made to set standards of quality. Enforcement of microbial standards for foods is ineffective at present because, with few exceptions, there are no nationally recognized standards.

In 1958, under the chairmanship of Dr Harry E. Goresline, the APHA published the first edition of "Recommended Methods for the Microbiological Examination of Foods". Because the committee realized that there may be other methods of equal value they refrained from designating the methods as "Standard Methods". The 2nd edition in 1966 under the chairmanship of Dr John M. Sharf continued this primary philosophy. Methods were recommended which, in the opinion of the editorial committee, continued to be basic and reliable. It was expected that they would give reproducible results under the normal variety of conditions encountered in laboratories in widely separated locales and without unusual or extremely sophisticated apparatus and routines.

Although these recommended methods have been useful they have also been responsible for stressing the need for standard methods to provide a sound basis for setting quality standards and in order to assess whether the standards have been met. In addition, it is often difficult for regulatory agencies and industry to reach a common understanding on a particular problem because of the variability of the methods used. Though several attempts have been made to develop standard methods, none has brought the necessary groups together nor has there been a systematic approach to method evaluation. An Intersociety Council approach to the selection, evaluation and preparation of a compendium or manual of microbiological methods presents the greatest likelihood of the wide acceptance of such procedures.

Continuing in its role of co-ordinator, the APHA has initiated the formation of a Technical Committee on Methods for Microbiological Examination of Foods. It is called a Technical Committee rather than an Intersociety Council according to the new APHA administrative rules of procedure. However, the term Intersociety is still likely to appear in the working title.

The major impetus for the formation of this Technical Committee and its goals arose from recommendations made at the conclusion of the 1971 National Conference on Food Protection in Denver, Colorado. It was stated at the conclusion of that meeting that "The Application of National Microbiological Standards for Foods will demand extensive laboratory support, and the success of the program will depend, in part, on the uniformity of laboratory methods used together with the precision with which these methods are used".

Dr Marvin L. Speck of the University of North Carolina is Chairman of the new Technical Committee and its members represent the Food and Drug Administration, Association of Official Analytical Chemists, Institute of Food Technology, National Environmental Health Association, Association of State and Territorial Public Health Laboratory Directors, Association of Food and Drug Officials of the United States, Center for Disease Control, American Council of Independent Laboratories, National Marine Fisheries Service, U.S. Department of Defence, National Canners Association and the U.S. Department of Agriculture.

Currently the staff of the APHA is negotiating a contract proposal to the Food and Drug Administration to provide financial support to the Committee for its meetings, for the development of methods, for collaborative study of methods and for research.

Although no official action will be taken by the Technical Committee until funds are available, the group has tentatively adopted a classification of methods and outlined the content of its proposed publication on Methods for the Microbiological Examination of Foods.

The classification of methods considered has been proposed by the Council on Laboratory Standards and Practices of the APHA for use in future methods publications. The proposed classifications are as follows.

1. Official method

Procedures known to be applicable for specific purposes on the basis of extensive use, and evaluation by collaborative studies that demonstrate acceptable reliability.

2. Candidate method

Widely used methods that have demonstrated their utility by extensive applications but which have not been evaluated by acceptable collaborative studies.

3. Provisional method

Published methods that have been devised or modified explicitly for routine examination of specimens, or which have been successfully used in research and evaluated by the originator or other investigators to demonstrate superiority to other methods for the same purpose.

4. Presented for information

Methods of unknown value but which have been recognized by laboratory workers as useful.

In addition to these considerations, the Technical Committee has expressed concern about the International acceptability of any published methods and will seek to establish liaison with the International Commission on Microbiological Specifications for Foods and the International Standards Organization.

In summary, it can be seen that the American Public Health Association has progressed from being solely responsible for the publication of methods for the examination of foods to its current role of initiating and co-ordinating the activities of specially designed Intersociety Committees. It is the hope of APHA as it enters its second century that this new role will be far more beneficial than its past accomplishments. The prestige and acceptance of Standard Methods for the Examination of Dairy Products and Standard Methods for the Examination of Water and Wastewater as well as other publications on laboratory methods in air and diagnostic procedures should be sufficient to encourage the development of an acceptable and influential publication on laboratory methods for foods.

Recommendation

The importance of food methods publications by the American Public Health Association is recognized as significant to programs within the United States of America. However, greater acceptance must be achieved because of international trade agreements and product marketabilities. Liaison should be established immediately by the American Public Health Association with the International Commission on Microbiological Specifications for Foods and the International Standards Organization.

Session 5

Education

Chairman: Professor F. A. Aylward

Education and Training in Food Hygiene in the U.K.

F. AYLWARD AND R. K. ROBINSON

Department of Food Science,
The University of Reading,
Berkshire, England

Aspects of food hygiene and microbiology are dealt with in formal courses in 2 main groups of educational establishments in the U.K. Firstly, in those Universities which have faculties in science, medicine, veterinary science or agriculture, and secondly, in educational centres such as polytechnics, technical or agricultural colleges. In the Universities the teaching will include aspects of microbiology and its application in different fields. In the other centres the courses range from the fundamental approach of degree work through to instruction on "codes of practice" associated with "craft" courses.

As far as the University sector is concerned there are some 78 faculties, schools or departments (Medicine, Veterinary Science, Agriculture, Microbiology and Food Science) whose courses have some content of food microbiology. In some cases, such as where microbiology is taught within Departments of "Biology", the section of the course dealing with food hygiene may be very limited, while in Schools of Veterinary Science, for example, the teaching of microbiology will have a definite bias towards animal welfare and slaughterhouse practice.

The recent development of Departments of Food Science goes some way towards fulfilling the need for microbiologists in the food industries. For obvious reasons the handling of pathogenic micro-organisms is omitted from most of these courses. Nevertheless, with appropriate post-graduate training, food science graduates, with their background knowledge of food chemistry, microbiology and processing methods should be able to deal with problems of food hygiene.

In the non-University sector of education, courses related to food stuffs are many and various. Thus, a number of polytechnics now offer degree courses in food-orientated topics (including a consideration of food hygiene), while a large number of technical centres teach diploma and certificate courses related to catering and food management, or special commodities. Many of these courses do not deal with food hygiene or microbiology in any depth, but their great value is in creating an awareness of the need for hygienic practices in relation to food at all levels within the catering and food industries.

Education and Training Programmes

IN RELATION TO food hygiene these can be considered broadly under 4 headings.

(1) Formal courses in universities including technological universities, and in associated Faculties, Colleges or Institutes concerned with subjects such as Medicine and Public Health, Veterinary Science and Agriculture.

(2) Formal courses in other educational centres (such as polytechnics, technical and agricultural colleges).

(3) Training provided through *ad hoc* courses and seminars arranged by scientific and technical bodies, and/or by teaching and research centres.

(4) Training provided for personnel on an in-service basis within the food industries.

Activities in the United Kingdom under this last heading are treated in detail in this book by Goldenberg & Edmonds (1973), so this paper is limited largely to the first 3 areas. Within teaching centres, food hygiene is normally taught in conjunction with other courses such as food science, microbiology or public health. One of the noteworthy developments in the past 12 years in the U.K. universities has been the growth of degree courses in food science (Aylward, 1971; Aylward *et al.*, 1971), and parallel developments in food processing and other courses within the non-university sector (Anon., 1971*a*). There has also been a significant increase in university departments and/or courses in microbiology.

As this paper has been prepared primarily for an audience from abroad, certain points about the educational system in the U.K. are included; these aspects may be taken for granted by the British reader.

University Sector—General

There are now 46 universities in the United Kingdom, some of these (notably Oxford, Cambridge and London) are organized on a federal basis so that they include a number of constituent or affiliated colleges or institutes. Although many universities have grown out of technological institutions (and may still retain a strong technological bias) there is no distinction in Britain between "classical" and "technological" universities. The typical university may include Faculties (or other groupings) covering broad areas such as: Arts, Letters and Philosophy; Economic and Social Sciences; Natural Science; Applied Science, Technology or Engineering; Medicine and Public Health; Agriculture and/or Veterinary Science. Within these *Faculties* are *Schools* or *Departments,* and thus, in 1 university there may be several sections concerned (or potentially concerned) with food hygiene.

A distinguishing feature of the British University pattern is its traditional freedom from *direct* state control. The major part of university funds come from the national exchequer through the Department (Ministry) of Education and Science, but the funds are allocated through an intermediate or *buffer* body—the University Grants Committee. The universities have, therefore, considerable freedom to plan and carry out their activities, and there are wide variations in the academic arrangements from 1 centre to another.

The Commonwealth Universities Year book (Springer & Craig, 1972) provides the following statistical picture of courses directly related to *microbiology,* or likely to include at least some relevant material.

Table 1

Faculty, School or Department	No. of University Centres
Medicine	19
Veterinary studies	10
Agriculture	12
Microbiology (including sub-sections of Departments of Botany or Biological Sciences)	30
Food Science	7

In the first 3 groups of subjects, microbiology is introduced at some stage in the curriculum. The emphasis will vary with the type of course and also with the university, and, *a fortiori,* the proportion of time devoted specifically to food hygiene will vary considerably. Thus, in agricultural courses, food hygiene may be treated in connection with animal production (and in particular dairy production courses); in veterinary science there will be a deeper treatment of some aspects of hygiene in relation to animal disease and to meat inspection. In the medical and public health fields, undergraduate courses will include sections on food-borne diseases; these problems will receive much deeper treatment in centres, such as the London School of Hygiene and Tropical Medicine and the Liverpool School of Tropical Medicine, which provide post-graduate courses for medical practitioners and others concerned with public health at home and abroad.

Over the past 15 years there has been a rapid increase in microbiology courses in British Universities; the 30 centres listed include many where microbiology has been introduced as an optional or subsidiary subject in relation to degrees in botany or general biology, and in some of these, courses in relation to food hygiene may be very limited. In other universities, independent professional departments of microbiology have been established with provision for courses leading to the B.Sc. Honours (or higher degrees) in Microbiology. These departments are important in that they can provide academic leadership in teaching and research, and at the same time train specialists for posts in industry and in governmental or other research centres.

University Departments of Food Science

There are in the U.K. 4 universities with well-established B.Sc. Honours courses in Food Science, namely Reading, Leeds, Nottingham and Strathclyde. Reading offers in addition a course leading to the B.Sc. Honours in Food Technology (Weybridge) on a "sandwich" basis, that is alternate periods in the College and in industry. In addition degree courses have recently been initiated at Queen's

University, Belfast (for the B.Sc. Food Science) and at Queen Elizabeth College, London (Food and Management Science).

Undergraduate courses in Food Science

The curricula for the B.Sc. Honours courses vary from one department to another both in the subject matter and in the titles used for the various courses, but all include subjects which can be grouped under the following headings:

(i) Chemistry, Biochemistry and Microbiology.

(ii) Applied Biochemistry and Applied Microbiology related directly to the composition of foods and changes during storage, processing and distribution.

(iii) Courses in specialized topics such as Nutrition, Statistics, Quality Appraisal by Physical, Chemical and Organoleptic methods, Food Hygiene, Food Standards and Legislation.

(iv) Courses in the principles or practice of food preservation and processing.

The B.Sc. Honours course in *Food Technology* (Weybridge/Reading) follows a somewhat similar pattern but with a greater emphasis on section (iv), that is with more detailed studies of food processing and engineering techniques.

Undergraduates (normally aged about 18 years) are admitted to the courses on the basis of good passes in the terminal secondary school examinations, that is the General Certificate of Education—Advanced level, or the equivalent in Scotland. Candidates are expected to possess a satisfactory grade in Advanced level Chemistry, and preferably in 2 other subjects as well; the latter being chosen from Biology (or Botany and Zoology), Physics or Mathematics. The actual requirements vary from University to University, but all departments attach special importance to the personal qualities of the applicant, his interest in the course, and to his general education.

M.Sc. "taught" courses in Food Science

M.Sc. "taught" courses have been known in the British Universities for many decades, but it is only in the past 20 years that they have become part of the normal university pattern. The M.Sc. courses in Food Science as provided at Leeds, Nottingham, Reading and Strathclyde are *conversion* courses; they are designed so that a graduate with a good Honours degree or equivalent (in Chemistry, Biochemistry, Microbiology or some appropriate discipline), is able over a period of 1 academic year to cover aspects of Food Science with some degree of specialization or "options". Thus, Reading offers an M.Sc. in General Food Science and a course with options in Dairy Science. Other options such as Food Analysis, Toxicology and Nutrition and Public Health are planned. At Weybridge, the M.Sc. taught course in Food Technology offers the specialization in 1 of 3 fields—Food Processing and Engineering, Quality Control and Microbiology.

Provision is made in some departments for M.Sc. students to receive, supplementary to the course, some external industrial training. An M.Sc. course is normally of 1 year's duration, but provision may be made for students who have not fulfilled all entry requirements to follow a "qualifying year"; this includes courses in basic Chemistry, Biochemistry and Microbiology. At Reading where the 2-year programme has been formalized the entrants to the *qualifying* year are mainly overseas graduates with agricultural or biological backgrounds, and who are being prepared for teaching, research or governmental posts in their own countries.

The relative merits of undergraduate courses leading to a first degree in Food Science in comparison with post-graduate courses are still being debated in many countries. For the man or woman who has chosen to take a first degree in Biochemistry, Microbiology or some other scientific subject there may be great personal advantages in spending a year studying Food science, before entering the food industry or doing research. There are also considerable potential advantages to the firm or institution in which the M.Sc. student takes his first post, in that he can make a contribution based both on his first degree honours subjects and on his inter-disciplinary studies in Food Science. On the other hand, it is much easier to give a properly balanced Food Science education by a carefully planned 3- or 4-year course than is possible in the 9–11 months formal tuition of the M.Sc. Course.

Higher degrees by research

The Food Science (and Technology) departments provide, in addition to the "taught" courses for M.Sc., research studies leading to the M.Phil. or M.Sc. (Research) or the Ph.D. Entry for higher degrees is normally based on performance at the first degree, with a first class Honours or at least an upper second being a common requirement. A minimum residence period of 2 years is needed to obtain the M.Phil. (or M.Sc. by research), and 3 years for the Ph.D. Although research training is primarily based on individual research under the supervision of a university teacher, there is an increasing tendency on the part of universities to insist that research students take a number of formal courses outside their main subject in order to broaden and to deepen their knowledge. This is becoming the normal procedure for graduates in 1 of the basic sciences, Engineering, or Agriculture who are commencing research in Food Science.

Food Hygiene in Relation to Food Science

The 1970/71 statistics (see Aylward *et al.,* 1971) showed that, in the 5 University centres reviewed, 187 undergraduates were following B.Sc. Honours courses, 42 graduates were preparing for the M.Sc. "taught" courses and 58 graduates registered for research degrees. When one considers that none of these degree courses were available in their present form 15 years ago, it will be seen that rapid progress has been made. In all the universities concerned with Food Science (and Technology) there is formal teaching in both pure and applied Microbiology, including Food Hygiene. In addition, questions of Food Hygiene are reviewed in many of the general courses concerned with preservation and processing of foods.

The actual number of students, undergraduate or post-graduate, from the Departments of Food Science who will finally specialize in Food Hygiene or Microbiology is difficult to estimate. There will clearly be variations in the type of post taken up by the new graduates, but all food science graduates should be aware not only of the principles of Food Hygiene but also of the need to apply these in practice.

The dairy industry, in the post-Pasteur days of the last century, was one of the first sections of the food industry to recognize the importance of hygienic control. Microbiology and hygiene has, therefore, occupied throughout this century a central place in any courses associated with dairy science and milk processing. The University of Reading, Department of Food Science developed, in fact, from a Department of Dairying, which had been established as a professorial department in 1938; this latter department had its origins in the British Dairy Institute (from 1888).

Food Courses in Polytechnics and Other Centres

Types of colleges

Parallel to the universities, there is in the U.K. a network of institutions which, although forming part of the tertiary sector of education, have their finances administered by local or regional authorities. This sector includes the Polytechnics, together with a large number of technical centres of different types. They offer courses on a wide variety of subjects at different levels, and provide part-time evening or *day-release* courses as well as full time courses. Most of the leading technical colleges are established in industrial centres, and in the past there was a fairly sharp distinction between such centres and the agricultural colleges situated, normally, in rural surroundings. Agricultural colleges, up until recently, limited their courses to those concerned directly with agricultural production and management, but several agricultural colleges have, in the past few years, broadened their courses to include food preservation and processing.

Types of courses

The courses related to foodstuffs at technical (including agricultural) colleges and polytechnics are directed towards qualifications in topics such as the following:

General food science and/or technology and/or processing.
Specific commodities, such as bakery, dairying.
Catering and food distribution.
Home economics and domestic subjects.
Institutional management: hotel management.

Qualifications

The courses lead to many different types of qualifications, including B.Sc. pass and Honours degrees, Higher National Diplomas (H.N.D.) and Certificates (H.N.C.), Ordinary National Diplomas (O.N.D.) and Certificates (O.N.C.).

The B.Sc. is awarded by the National Council for Academic Awards (N.C.A.A.) in a number of technical centres which have had their courses approved by the Council. Such Degree courses include Food Science (Polytechnic of the South Bank, London), Catering Studies (Huddersfield Polytechnic) and Dietetics (Leeds Polytechnic) (Anon., 1972*a*).

Ordinary National and Higher National Diplomas (and Certificates) are awarded through a *Joint Committee* of the Department of Education and Science and an appropriate professional and/or industrial body (Anon. 1972*b*). Certificates with a more "applied" bias are awarded through national examination schemes sponsored by the City and Guilds of London Institute (Anon., 1971*b*).

Number of centres

In the United Kingdom as a whole there is a very large number of centres offering diploma courses (usually full-time or "sandwich") or certificate courses (usually part-time) in general food processing or in more specialized fields. Thus, there are over 100 centres for catering education and for bakery education. In the national pattern, the diploma and certificate courses play a very important role and for several different reasons: (i) they attract many hundreds of students from all sectors of the food production and distribution industries; (ii) many of these students are on "sandwich" or part-time day release courses, so that they have the opportunity of applying their "learning" to their day-to-day work; (iii) many of the courses are organized in close co-operation with industry, that is through industrial or trade associations; (iv) part-time lecturers from industry are often directly concerned with the teaching; (v) the colleges, because of the flexibility, can provide *ad hoc* courses for special groups, or in special subjects.

Food hygiene training

It can be assumed that food courses in technical colleges and associated institutions include some instruction in food hygiene. The content and standard of the courses varies from the fundamental approach associated with degree work through to instruction on "codes of practice" as found in "craft" courses. Nevertheless, the value of these courses should not be underestimated, for in those workers directly concerned with food products, this type of technical training may encourage an awareness of the need for hygiene in a food context.

This problem of creating an interest in the hygienic quality of food is one of the major activities of the Royal Society of Health (Anon., 1970). This Society, among its other functions, serves as an examining body and offers a Certificate in the Hygiene of Food Retailing and Catering. Courses for the Certificate may be given at some 97 colleges or centres approved by the Society. In addition some of these centres run courses leading to a Certificate of Meat Inspection and/or a Diploma for inspectors of meat and other foods. The role of the Royal Society of Health in the field of Food Hygiene is an extremely valuable one, the more so as their influence (through courses and examinations) extends to many areas overseas.

Scientific and technical bodies

There is in the U.K. a variety of scientific and technical bodies concerned with different aspects of microbiology and hygiene. The list includes the Society of General Microbiology, the Society of Applied Bacteriology, The Society of Chemical Industry (with a Food Group and a Microbiology Group), the Institute of Food Science and Technology, the Society of Dairy Technology and the Royal Society of Health. All these bodies through their meetings, symposia and publications make some contribution to the education and training of their members, and in some cases, of a wider public. As already noted the Royal Society of Health is an examining body; its meetings and publications are designed to reach different sectors of the general public, including the officers of Local Authorities concerned with the implementation of food hygiene regulations.

The universities and other colleges already discussed are to an increasing extent organizing summer schools and other types of vacation courses and symposia on selected topics, and these have included, in recent years, occasional meetings devoted to Food Hygiene.

The food research centres in Britain include several supported by industry, but receiving also public subsidies through the Ministry of Agriculture, Fisheries and Food; three of these centres, namely the Campden Food Preservation Research Association (RA), the British Food Manufacturing Industries R.A. and the Flour Milling and Baking R.A., deal with, among other matters, problems of Food Hygiene. Through bulletins issued to members, through symposia and/or training courses, these bodies assist in the dissemination of information. Other research centres, financed largely out of public funds, include the Food Research

Institute (Norwich), the Meat Research Institute (Langford, Bristol), the Torry Research Station (Aberdeen), the National Institute for Dairy Research (Reading) and the Hannah Research Institute, Ayr. In a different category are the laboratories sponsored by the Department (Ministry) of Health and Social Security, and in particular the Central Public Health Laboratory, Colindale, serves a valuable function in relation to Food Hygiene.

The above centres are not usually concerned with education in the formal sense, but usually provide facilities for the supplementary training of research workers and technicians, and make available members of staff as lecturers at scientific and other meetings.

References

Anon. (1970). *Certificate in the Hygiene of Food Retailing and Catering; Regulations and Syllabus.* Royal Society of Health, 13 Grosvenor Place London S.W.1.
Anon. (1971*a*). *Careers and Courses in Food Science and Technology.* Institute of Food Science & Technology, 41 Queen's Gate, London S.W.7.
Anon. (1971*b*). *Renumbering of City and Guilds Examination Schemes.* City and Guilds of London Institute, 76 Portland Road, London W.1.
Anon. (1972*a*). *Compendium of Degree Courses–1972.* Council for National Academic Awards, 3 Devonshire Street, London W.1.
Anon. (1972*b*). *Aims and Activities.* Institute of Biology, 41 Queen's Gate, London S.W.7.
Aylward, F. (1971). Food Science and Technology in Europe, Canada and Africa. In *Third International Congress/Food Science and Technology.* Eds. G. F. Stewart & C. L. Willey. Illinois, U.S.A.: Institute of Food Technology.
Aylward, F., Hawthorn, E. J., Lawrie, R. A., Rolfe, E. J. & Ward, A. G. (1971). Food Science and Technology in the Universities of the U.K. *Chemy Ind.* p. 1030.
Goldenberg, N. & Edmonds, M. S. Education in microbiological safety standards. In *Proceedings of the 8th International Symposium. The Microbiological Safety of Food.* p. 435. London: Academic Press.
Springer, Sir Hugh, W. & Craig, T. (1972). *Commonwealth Universities Handbook.* London: Association of Commonwealth Universities.

Training Public Health Workers and Food Service Managers

F. L. BRYAN*

*Health Agencies Branch,
Center for Disease Control,
Atlanta, Georgia 30333,
U.S.A.*

In the United States, vehicles responsible for reported food-borne disease outbreaks are more frequently prepared in food service establishments than in the home or in food processing plants. Faulty operational procedures contribute to these outbreaks. For a national health agency to reach the core of this problem in a nation where the responsibility for health rests with the individual States, training is given to the State and local health department personnel who directly influence managers of, and workers in, these establishments. This laboratory training and field training, as well as the homestudy courses gives health workers specific information about the microbiology of food-borne pathogens, the epidemiology and prevention of food-borne diseases, and administrative approaches. Also, guidelines, slide series and other training aids (examples are given) are prepared to enhance training programmes of State and local health departments, universities, and community colleges. Prevention of food-borne diseases will not be achieved until food service supervisors become cognizant of the factors that contribute to food-borne disease outbreaks and are motivated to require that appropriate preventive measures be routinely practised in the operations they supervise. Training is an effective means to these ends and perhaps the only way to substantially reduce the incidence of food-borne illness.

FOOD-BORNE DISEASES are continuing problems in the United States as in most countries. Even with tremendous strides in the development of sanitary food service and processing, and with an increased level of education of the populace, the reported food-borne illness rate/year is about the same now as in the past (Bryan, 1973*a*). Vehicles responsible for reported food-borne disease outbreaks are more frequently prepared in food service establishments than in homes or in food processing plants (Table 1). Thus, if priorities for improvement are to be set, the aspects of the food chain that would reduce most significantly the incidence of food-borne disease are operations within food service establishments. Foods may be contaminated when they reach food service establishments, and faulty operational procedures (which contribute to further contamination, allow pathogens to survive, and/or promote multiplication of

---

* Chief, Food-borne Disease Activity, Health Agencies Branch, Training Program, Center for Disease Control, Health Services and Mental Health Administration, Public Health Service, Department of Health, Education, and Welfare, Atlanta, Georgia.

Table 1

Place where foods were mishandled in food-borne disease outbreaks (in the United States, 1968–1971)

Place	1968	1969	1970	1971	Total
			Year		
Food service establishments	114	114	115	114	457
Homes	24	48	42	56	170
Food processing plants	16	31	21	27	95
Unknown or unspecified	106	178	185	123	592
Total	260	371	363	320	1314

pathogens) often occur in these establishments. Some* of the more significant of these faulty procedures are:

Failure to refrigerate cooked foods properly
Holding foods at warm (bacterial incubating) temperatures
Preparing foods a day or more ahead of serving
Failure to cook foods thoroughly or to reheat leftover foods adequately
Infected employees who practice poor personal hygiene
Cross contamination of cooked foods from raw foods of animal origin by
    equipment or workers' hands
Inadequate cleaning and disinfection of kitchen equipment
Obtaining foods from unsafe sources

These faulty procedures are not unique in food service establishments. Many of them occur in homes, at picnics, and at camps. They also occur in food processing plants. In the latter case they are usually referred to as raw product contamination, process failure, and post-process contamination. After such contamination, additional mishandling must also occur in transit, in storage, or in homes or food service establishments before outbreaks will ensue.

Food preparation in homes is less likely to occasion outbreaks of food-borne disease than food preparation in food service establishments. Smaller amounts of food are prepared; the food is prepared just before serving; relatively small portions of food are leftover to be stored and handled further. In food service establishments, however, foods are often prepared in large amounts; they are prepared several hours, sometimes a day or more, before serving. During the interval between preparing and serving, the foods must be held at temperatures that inhibit bacterial growth, but do not significantly affect the organoleptic properties of the food. Often large volumes of foods must be cooled and,

* Other faulty procedures and the relative importance of these procedures in causing outbreaks are cited by Bryan (1973b).

perhaps, once again reheated and held at warm temperatures. Other dishes must be made from leftovers so the food may have to be cut, ground, sliced, chopped, mixed, or otherwise handled. This introduces many opportunities for contamination.

The core of the food-borne disease problem in the United States is the operational procedures in food service establishments. To achieve a significant reduction in the incidence of such diseases, food service managers must be trained, as well as supervised, in safe food handling practices by knowledgeable public health workers.

Training Public Health Workers

For a public health worker to cope effectively with the food-borne disease problem, he must know about the agents*—such as *Salmonella* spp., *Staphylococcus aureus,* and *Clostridium perfringens* —that cause most food-borne disease. Essential information about the agents and the epidemiology of the diseases they cause include: reservoirs and sources; modes of transmission; conditions necessary for outbreaks to occur; resistance of the organisms or their toxins to environmental factors such as heat; temperatures and other conditions that permit, and are optimal for, growth of the organism; and principles and applications of prevention and control. With a thorough understanding of these features, public health workers can evaluate the steps of a food operation to identify hazardous situations and critical control points. Preventive measures can then be prescribed and taught to managers and key workers. Members of the public health team should also have skill in investigating food-borne outbreaks, in examining specimens and foods for pathogens or toxins, and in consulting with and training food industry managers. Also, they should be motivated to incorporate aspects of food-borne disease control into their routine inspectional and educational activities and to develop new approaches in food-borne disease control that are workable in their communities. Training can provide this information, develop these skills, and stimulate the public health professional to initiate action.

Several training approaches—classroom courses, homestudy courses, laboratory bench training, field training, seminars, workshops, programmed instruction, training manuals and outlines, articles in professional journals and texts, epidemic problem situations, and teaching kits—have been used. The courses and seminars listed in Table 2 are directed at public health workers—administrators, laboratory workers, and inspectors. Upon request, state health departments, food industries, universities, community colleges, and professional organizations are assisted by agencies of the Federal Government. This assistance consists of

---

* The agents vary in different countries. Those cited are the most important in the United States.

Table 2

Course taught by the Public Health Service that are related to epidemiology, microbiology,
and control of food-borne diseases

Subject	Course	Duration	Primary audience	Agency
Epidemiology	Principles of Epidemiology	5 days	Professional public health workers	CDC[a]
	Epidemiology and Control of Food-borne Disease	5 days	Supervisors of food hygiene and surveillance programmes	CDC
	Epidemiology and Control of Salmonellosis	5 days	Supervisors of food hygiene and surveillance programmes	CDC
	Investigation and Surveillance of Food-borne Outbreaks	1 to 3 days	Sanitarians, other public health workers	CDC
	Seminar on Salmonellosis	$\frac{1}{2}$ to 1 day	Professional public health workers	CDC
	Seminar on Infectious Hepatitis	$\frac{1}{2}$ to 1 day	Professional public health workers	CDC
Microbiology	Food Microbiology	10 days	Laboratory workers	FDA[b]
	Isolation of *Salmonella* from Food Products and Animal Feeds	10 days	Laboratory workers	CDC
	Laboratory Methods in Anaerobic Bacteriology	10 days	Laboratory workers	CDC
	Isolation and Preliminary Identification of Enteric Bacteria	5 days	Laboratory workers	CDC
	Laboratory Methods in Enteric Bacteriology	10 days	Laboratory workers	CDC
	Laboratory Methods in Medical Parasitology, Part I. Intestinal Parasites	20 days	Laboratory workers	CDC
	Basic Laboratory Methods in Virology	10 days	Laboratory workers	CDC
	Laboratory Methods in Enterovirus Infections	10 days	Laboratory workers	CDC
	Pesticide Residual Analysis of Foods	5 days	Laboratory workers	FDA
	Laboratory Methods in General Medical Bacteriology	10 days	Laboratory workers	CDC
	Laboratory Examination of Dairy Products	5 days	Laboratory workers	FDA
	Current Concepts in Milk Analysis	4 days	Supervisors	FDA
Control	Applied Procedures for the Control of Food-borne Diseases	5 days	Sanitarians	CDC
	Current Concepts in Food Protection	5 days	Sanitarians	FDA
	Administration of Food-borne Disease Control Programmes	5 days	Supervisors of food hygiene programmes	CDC
	Food-borne Disease Control Homestudy Course	12 weeks	Sanitarians	CDC

Table 2–*continued*

Subject	Courses	Duration	Primary audience	Agency
	Milk Pasteurization–Tests and Controls	3 days	Sanitarians	FDA
	State Milk Laboratory Survey Officers Workshop	5 days	Sanitarians	FDA
	Sanitary Food Service	2 to 5 days	Dietitians and food service managers	FDA
	Community Hygiene Homestudy Course	18 weeks	Sanitarians	CDC
	Communicable Disease Control Homestudy Course	12 weeks	Professional public health workers	CDC
	Vector-borne Disease Control Homestudy Course	12 weeks	Sanitarians	CDC
	Epidemiology and Control of Vector-borne Diseases (Part I. Basic Vector-borne Disease Control)	10 days	Sanitarians	CDC

[a] CDC = Center for Disease Control, Health Services and Mental Health Administration, Public Health Service, Department of Health, Education, and Welfare, Atlanta, Georgia.
[b] FDA = Cincinnati Training Facility, Food and Drug Administration, Public Health Service, U.S. Department of Health, Education, and Welfare, Cincinnati, Ohio.

consultation on course planning, providing technical staff for training presentations, and supplying literature and training aids. Examples of some of the materials that are loaned—epidemic food-borne disease problems—are listed in Table 3.

Training Food Service Managers and Supervisors

It is difficult for a national health agency to reach the core of the food-borne disease problem—faulty operational procedures in food service establishments—in a nation where the responsibility for health rests with the individual states. In an effort to get round this problem guidelines and visual aids (as well as the

Table 3

Epidemiological problems dealing with food-borne disease outbreaks

Setting	Disease
Bonair Mental Hospital	Salmonellosis
Petersboro (small city)	Typhoid fever
State Prison	Chemical poisoning
Jones County (rural county)	Infectious hepatitis

training just described) are made available to public health workers. This enables them to teach food service managers the principles of, and specific applications of, food-borne disease prevention. Because the food service manager is responsible for day-to-day operations within his establishment and for training his staff, he is very important in preventing food-borne outbreaks.

Several years ago, many health departments conducted training courses for food service personnel. This activity is continuing but waning. Much of it dealt with personal hygiene and dishwashing and was directed primarily at waitresses and dishwashers. Unfortunately it dealt neither with the most important aspect of the food-borne disease problem nor with the people in food service establishments who could be important in preventing outbreaks. It is the managers and food handlers who prepare, cook, and store foods, who should have top priority for training.

Training the food service worker has been frustrating because of the high rate of turnover of personnel in the food service industry. Every person in a food service job will be replaced $1\frac{1}{4}$ times within a year. This high rate is atrributed to employee turnover, part-time employees, and creation of new jobs in a growing industry (U.S. Department of Labor, 1969). Food service managers are somewhat more stable in employment than the workers they supervise. It is more economical, both in use of funds and manpower, for health department personnel to train food service managers instead of the workers they supervise (Table 4). Thus, if time, personnel, or money limit programme activities, health agencies should give priority to the training of managers and supervisors. They should also continue to assist the food service industry by providing consultation, visual aids, and spot teaching assistance in industry-operated food handler training courses.

Training can develop the capability of managers and supervisors to solve food-borne disease problems inherent in their operation or, at least, make them aware of the problems and encourage them to seek professional assistance in finding solutions. Training can also stimulate these persons to initiate needed change. The overwhelming purpose of training managers and supervisors, however, is because they are the people who can effect change in the food service establishments they supervise. Such training will also become a necessity if licences are only granted to restaurant operators who have passed a food hygiene examination; this move was suggested in a National Conference on Food Protection (American Public Health Association, 1972).

Industrial associations or managers of individual establishments or chains should take the lead in training food service workers. Some industrial groups have already taken the initiative in training workers in the retail food industries. Significant training activity has been undertaken by the National Restaurant Association, the Supermarket Institute, the Single Service Institute, and the National Sanitation Foundation. Universities in some states have also made a significant contribution to the training of food service personnel.

Table 4

Effort and cost of food sanitation training courses for food service employees and managers[a]

		Total number, rate, or cost	
Situational factor	Constant or multiples	Food service workers' course	Food service managers' course
1. Number of establishments	431		
2. Number of employees/ establishment (based on national average)	9		
3. Number of estimated employees	$(1 \times 2)^b$	3879	
4. Accession rate of employees/year[c]	125%		
5. Estimated percent who stay in food service work	20%		
6. Yearly turnover rate	(4–5)	1·05	
7. Number of employees needing training first year of programme	$[(3 \times 6) + 3^d]^b$	7952	
8. Number of managers/employee	20%		
9. Estimated number of managers	(3×8)		776
10. Estimated accession rate of managers/year	75%		
11. Percent estimated who stay in food service work	20%		
12. Yearly turnover rate	(10–11)		0·55
13. Number of managers needing training first year	$[(9 \times 12) + 9]$		1203
14. Time for presentation of food handler courses (h/course)	4^e		
15. Estimated time for each preparation (h/course)	8^e		
16. Total time spent (h/courses)	(14 + 15)	12^e	12^e
17. Estimated initial preparation time (h)		40^e	80^e
18. Number of persons/course[f]		25^f	25^f
19. Estimated cost for supplies, handouts, films, etc.		$\$19{\cdot}00^e$	$\$35{\cdot}00^e$
20. Sanitarian's (instructors) workdays/year	220		
21. Sanitarian's work-hours/day	7		
22. Sanitarian's annual salary[g]	$11,000		
23. Cost/sanitarian	(22 ÷ 20) = $50.00		
24. Number of courses required for first year	(7 ÷ 18)	318	
	(13 ÷ 18)		48
25. Total hours required for first year	$[(24 \times 16) + 17]$	3856	656
26. Sanitarian man-days required for training for first year	(25 ÷ 21)	551	94
27. Sanitarians or sanitarian man-years required for training for first year	(26 ÷ 20)	2·5	0·43

Table 4–*continued*

		Total number, rate or cost	
Situational factor	Constant or multiples	Food service workers' course	Food service managers' course
28. Personnel cost of training programme for first year	(26 × 23)	$27,550	$4700
29. Total cost of training programme for first year	[(24 × 19) + 28]	$33,592	$6380
30. Number of courses required for each subsequent year[h]	[(3 × 6) ÷ 18] [(9 × 12) ÷ 18]	163	17
31. Total hours for each subsequent year	[(30 × 16) + 17]	1996	284
32. Sanitarian man-days required for training in each subsequent year	(31 ÷ 21)	285	41
33. Sanitarian man-years required for training programme for each subsequent year	(32 ÷ 20)	1·3	0·19
34. Personnel cost of training programme for each subsequent year	(32 × 23)	$14,250	$2050
35. Total cost of training programme for each subsequent year	[(30 × 19) + 34]	$17,347	$2645
36. Man-years for 10-year programme	[(33 × 9^i) + 27]	14·2	2·14
37. Total cost for 10-year programme	[35 × 9^i) + 29]	$189,715	$30,185

[a] Data based on a real United States community used as a teaching reference community—Dixon-Tiller County, U.S.A. (Source: Center for Disease Control, Atlanta, Georgia).

[b] Numbers in parenthesis or brackets refer to performance or situational factor number to show how calculations were made.

[c] Reference: U.S. Department of Labor (1969).

[d] Omitted the number of employees leaving employment before they were trained. This could conceivably reduce the number (3879) by half.

[e] Number would vary depending on scope of course.

[f] Recommended size for maximum learning but the situation and facilities may change this number. Reference: U.S. Department of Health, Education and Welfare (1969).

[g] Salary would vary with location and with qualifications desired for sanitarian or instructor.

[h] Omitted increased number of establishments and employees caused by new establishments built in the community. In the United States this number could be estimated as 1 to 5% per year.

[i] The 9 refers to number of years and not to performances or situational factor number.

Besides short courses, several approaches have been used to train food handlers. These include programmed instruction using teaching machines (Carter *et al.,* 1964; Moore, 1964; Moore & Klachko, 1967), correspondence courses (Konhauser, 1970), telecommunication programmes (Conklin, 1972), articles in food service journals (Bryan & McKinley, 1972), texts (Longree, 1967; Hobbs, 1968; Longree & Blaker, 1971), and booklets and leaflets (Zottola, 1967, 1968,

1971; U.S. Department Health, Education and Welfare, 1961, 1963, 1964, 1966).

To assist in preventing food-borne outbreaks, the Center for Disease Control has developed a kit for the training of food service managers and supervisors. The kit is available for short term loan to health departments, universities, community colleges, and the food industry. It consists of 56 slides, narrative, quiz, and resource materials to supplement the lecturers technical background. It has been developed with flexibility in mind so that teachers can supplement the standard material with slides of local problems or their solutions. The slide series may be used as a short presentation on the food-borne disease problem and its control in food service establishments or it can be used as a public health base for other presentations dealing with food hygiene. [A more complete programme for training food service managers in several aspects of food hygiene is discussed by Bryan (1969).]

Future Direction

Surveillance, laboratory methods, bacteriological standards, and laws and ordinances are important in effective food hygiene programmes, but before the food-borne disease problem can be significantly reduced, supervisors at all levels of the food chain must become cognizant of the factors that contribute to food-borne disease outbreaks. In addition they must insist that appropriate preventive measures be routinely practised in the operations they supervise. Training is an effective means to this end and, perhaps, the only way to substantially reduce the incidence of food-borne illness.

References

American Public Health Association. (1972). *Proceedings of the 1971 National Conference on Food Protection*. Washington, D.C.: U.S. Government Printing Office.

Bryan, F. L. (1969). Use of visual aids in effective training of food service managers in food-borne disease control. *J. Milk Fd Technol.* **32**, 245.

Bryan, F. L. (1973a). Emerging food-borne disease problems. Part I. Their surveillance and epidemiology. *J. Milk Fd Technol.* (In press.)

Bryan, F. L. (1973b). Emerging food-borne disease problems. Part II. Factors that contribute to outbreaks and their control. *J. Milk Fd Technol.* (In press.)

Bryan, F. L. & McKinley, T. W. (1972). Turkeys: The bad guy of school lunch. *School Foodservice J.* **10** (Nov/Dec), 83.

Carter, E. J., Moore, A. N. & Gregory, C. L. (1964). Can teaching machines help in training employees? *J. Am. diet. Ass.* **44**, 271.

Conklin, M. (1972). *What to do if food-borne illness strikes your institution*. Columbia, Missouri: Columbia Medical Center, Univ. Missouri.

Hobbs, B. C. (1968). *Food Poisoning and Food Hygiene*. 2nd ed. London: Edward Arnold.

Konhauser, A. H. (1970). A correspondence study program. *Hospitals* **44**, 40.

Longree, K. (1967). *Quantity Food Sanitation*. New York: Interscience.

Longree, K. & Blaker, G. G. (1971). *Sanitary Techniques in Food Service*. New York: J. Wiley.

Moore, A. N. (1964). Teaching machines: A new training aid for the dietitian. *Hospitals* **38**, 64.

Moore, A. N. & Klachko, H. W. (1967). Problems in producing programs for auto instruction. *J. Am. diet. Ass.* **51**, 420.

U.S. Department Health, Education and Welfare. (1961). *From Hand to Mouth.* Public Health Service Publ. no. 281.

U.S. Department Health, Education, and Welfare. (1963). *You can Prevent Foodborne Illness.* Public Health Service Publ. no. 1105.

U.S. Department Health, Education and Welfare. (1964). *Cold Facts about Home Food Protection.* Public Health Service Publ. no. 1247.

U.S. Department Health, Education and Welfare. (1966). *Hot Tips on Food Protection.* Public Health Service Publ. no. 1404.

U.S. Department Health, Education and Welfare. (1969). *Sanitary Foodservice* (Instructor's guide). Public Health Service Publ. no. 90.

U.S. Department of Labor. (1969). *Eating and Drinking Places Industry.* Industry Manpower Surveys no. 115.

Zottola, E. A. (1967). *Salmonellosis.* Extension Bull. 339. St. Paul, Minnesota: Agric. Ext. Sta., Univ. of Minnesota.

Zottola, E. A. (1968). *Staphylococcus Food Poisoning.* Extension Bull. 354. St. Paul, Minnesota: Agric. Ext. Sta. Univ. of Minnesota.

Zottola, E. A. (1971). *Clostridium perfringens Food Poisoning.* Extension Bull. 365. St. Paul, Minnesota: Agric. Ext. Sta., Univ. of Minnesota.

Discussion

Monty

Having heard a most interesting and comprehensive programme for the education of food handlers and food management, I noticed in his table on food-borne diseases that mishandling of food in the home was responsible for a comparatively high percentagge of cases. I should like to ask Dr Bryan whether any attempt is made in the U.S.A. to educate the food consumer in the fundamental principles of hygiene and handling to reduce the incidence of food poisoning from this cause?

Bryan

At the recent National Conference on Food Protection (mentioned several times at this meeting) development of a plan to educate the consumer was a major task force assignment. The suggested approach emphasized consumer participation in planning and evaluating food hygiene programs, educating the public through mass media, and training school-aged children. Only a small portion of this plan has been implemented at the national level but action has been taken in some States and localities by various health, agricultural, or educational agencies.

In the past, hygiene in food service and processing plants has been emphasized, with only sporadic efforts to educate the consumer. However, various federal, state, and local health and agricultural agencies, university extension activities, and some industrial groups have developed leaflets and booklets on food hygiene. These publications are given to individuals who request them, to groups planning mass feeding, to community groups during training sessions, and to individuals who are contacted by home economists acting on behalf of university extension activities.

In the health curriculum of elementary education, community health is taught to children in the 4th, 5th or 6th grade. Food hygiene is usually included briefly in the texts on community health. Emphasis on this subject, however, often depends upon the teacher's knowledge of or interest in the subject. Most health workers who have an interest in food hygiene would like to see more emphasis on food hygiene training in grade and high schools.

Education in the Food Processing Industry

H. RIEMANN

University of California,
2046 Haring Hall,
Davis, California 95616
U.S.A.

The available U.S. statistics for food-borne disease outbreaks show that the major problems occur in schools, canteens, hospitals and restaurants, suggesting that the need for education is most acute in the food service sector. Nevertheless, workers from all sections of the food-chain from producer to consumer should have some knowledge of food hygiene; food processors are no exception. Yet in 1965 the food industry employed some 1·6 million persons and most of those directly involved with handling food received no training in food hygiene prior to employment. Some in-service training is available in the larger firms, and some trade associations (American Institute of Bakers, American Meat Institute, National Canners Association) operate training schemes oriented towards the correct handling of specific commodities. However, in recent years, efforts have been made to provide courses in food hygiene and microbiology in recognized educational establishments. Some 250 Technical Institutes in the U.S. now have curricula dealing with aspects of food hygiene, but it is difficult to estimate how many students with a training of this type enter the food industry. In the University sector, 44 Universities or Colleges provide undergraduate and post-graduate education in food science and technology (including food microbiology); the number entering industry from these courses is not known. It is proposed that the number of formal courses in food hygiene and microbiology provided by both Universities and Technical Institutes should be increased in the future; such a move would ensure that the food industry received an adequate supply of trained personnel at all levels.

Introduction

PROTECTION OF FOOD from microbial contamination is mainly a people problem. Inadequate processing and contamination is due to the failure of people to recognize processing requirements or to apply the necessary safety procedures; both of which may be the results of insufficient education.

The educational needs with respect to food safety are not unique to the processing industry but are equally important in food production, agriculture and fisheries, in food distribution and food service. The available U.S. statistics of food-borne disease outbreaks show that the major outbreaks occur in schools, canteens, hospitals and restaurants, suggesting that the education need is most acute in the food service sector. Education must obviously deal with all industries concerned with foods and indeed with all levels of the population. The following account is limited to activities with respect to personnel in the U.S.

417

food industry, and is to a large extent based on the Proceedings of the 1971 National Conference on Food Protection (American Public Health Association, 1972).

Food Production

Three categories of people are involved in the production of foods.

(1) Persons directly involved in production.
(2) Persons whose duty is surveillance of production.
(3) Professional people concerned with human and animal health.

Workers, many of whom are migrant workers, directly involved in the production of field crops receive little or no training. The schools, primary and high school, offer little formal training in foods and there is little organized adult education in connection with the hygiene of crop production.

Information about microbial contamination has so far been considered of minor importance in the education of many categories of agricultural product inspectors and is practically non-existent in courses provided for supervisory production personnel. Animal health inspectors of the United States Department of Agriculture (USDA) and the various State agencies have a high degree of training, but the microbial contamination of food is not the primary focus. The Agricultural Extension Service deals mainly with production problems and the prevention of spread of animal diseases, but in recent years their concern has extended to include sanitation and human welfare.

Food Processing Industry

There were 38,000 food processing plants in the U.S. in 1963. More than 20,000 have less than 20 employees. The food industry employed 1·6 million persons in 1965. Most of the employees directly involved in food handling had no training before their employment; they receive on-the-job training, but this is mainly directed towards the skill required for the job.

Existing Channels of Training

High school programmes exist, but are better suited for the general population than for training employees for the food industry. It has been proposed that courses based on techniques should be substituted for the traditional commodity-oriented food courses in vocational high school programmes; one high school in New York has made this change.

Fifty institutions in the U.S. offer academic-type curricula for technician training for the food industry. One hundred and eighty-one institutions offer education in agriculture and natural resources and about 25 have food science curricular. It has been estimated that 41 schools have 2-year curricula in dairy technology and 21 more have courses in laboratory techniques of potential relevance to the food industry. Accurate evaluation of the present technology training programmes in the U.S. is not available, but probably less than 500 students/year complete such programmes and seek employment in the food industry.

Career oriented programmes combined with practical experience in industry have been started by a few institutions. This type of course work is generally not transferable to academic 4-year colleges.

Some trade associations operate training schools (American Institute of Bakers School, American Meat Institute, National Canners Association) and produce manuals, charts and other teaching aids (American Meat Institute Center for Continuing Education, 1972). Training is also provided often on a less organized basis by companies that sell equipment, professional consultants and Inspectors of the Food and Drug Administration (FDA) and USDA. Most large companies have in-house educational programmes which seem to differ considerably in scope and organization.

A number of seminars, conferences and short-courses are organized every year. Some are organized by trade associations and professional societies, others by the university extension service; the land grant colleges are especially active in this field. Three hundred and twelve programmes were listed in 1970, and these had a total enrollment of 4755, most of them in dairy technology (including sanitation).

Forty-four universities and colleges in the U.S. and Canada provide undergraduate and graduate education in food science and technology. The undergraduate curriculum in most of these schools is in compliance with the recommendations of the Institute of Food Technologists (IFT Committee on Education, 1966, 1969) and includes food microbiology. Of the 44 schools, 32 offer bachelors, masters and doctors degrees, 4 offer bachelors and masters only, 2 masters only and 3 masters and doctors only.

In 1967–68, 290 bachelors, 237 masters and 139 doctors degrees were conferred. The total enrollment in the same period was: 787 bachelor, 368 master and 394 doctoral candidates. It is estimated that the capacity will increase to 2500 enrollment in 1974–75 with 1360 graduating/year. Many of the graduates seek employment in the academic world and in government rather than in industry; the accurate proportions are unknown.

If the number of individuals with technical or academic education were distributed evenly among the food processing plants and were willing to change jobs readily, it is estimated that about 1 plant in 5 could have an employee who had received an appropriate education within the last 10 years; 1 plant in 3 would have an employee who had participated in a seminar or short course within the last 3 years. It is, however, well-known that personnel with an education in food science (including sanitation) are very unevenly distributed among the processing plants. There is obviously a need for an expansion of food education, and recommendations made by the 1971 National Conference of Food Production may be summarized as follows:

(1) A broad National Education programme should be initiated.
(2) Educational material should be developed by a central agency.
(3) Managers in food industry should be required to demonstrate that they possess some knowledge of safe food handling practices.
(4) Use of qualified professional personnel should be encouraged. Mandatory regulations dealing with safety should be introduced in the most sensitive areas, i.e. where the risk of disease outbreaks is greatest.
(5) Competence of regulatory personnel should be established through training, supervised experience and/or examinations.
(6) Encouragement should be given to universities to expand graduate curricula in food science microbiology and public health so as to give young scientists the appropriate interdisciplinary competence.

References

American Meat Institute Center for Continuing Education. (1972). *Handbooks in Meat Microbiology and Meat Plant Sanitation* (J. A. Carpenter and E. A. Zottala).
American Public Health Association. (1972). Proceedings of the 1971 National Conference on Food Protection. U.S. Government Printing Office, 1712-0134. Washington, D.C.
IFT Committee on Education. (1969). Recommendations on subject matter outlines for food oriented courses. *Fd Technol. Champaign* **23**, 307.
IFT Committee on Education. (1966). IFT adopts undergraduate curriculum minimum standards. *Fd Technol. Champaign.* **20**, 1567.

Discussion

Hess

I should like to know the exact connection between the irrigation system in California and the high incidence of *Cysticercus bovis*.

Riemann

The lack of field toilets, or the failure to use them, leads to contamination of water in irrigation channels by carriers of *Taenia saginata*. The same water is used for cattle drinking water in feed lots.

Cooper

With regard to training at the "grass roots level", Riemann has said that the U.S. Extension Service is playing an increasing part in the education of farm workers in hygiene in relation to the production of food. Will he please elaborate? For the information of the Conference, the Agricultural Development and Advisory Service in the U.K. runs short courses together with the Agricultural Training Board and sometimes with Agricultural Colleges. Examples of courses are those for dairy herdsmen in milking techniques, dairy hygiene and mastitis control. Other courses are concerned with aspects of animal and crop husbandry.

Riemann

I would refer you to the Proceedings of the 1971 National Conference on Food Protection organized by the American Public Health Association [U.S. Govt. Printing Office, Report No. 1712–0134 (1972) Washington, D.C.U.S.A.].

Food Hygiene Education in The Netherlands

E. H. KAMPELMACHER

The Agricultural University,
Wageningen,
The Netherlands

University education in the field of Food Microbiology and Food Hygiene in The Netherlands is reviewed. Two institutes, *viz.* the Veterinary Faculty University of Utrecht, and the Department of Food Science Agricultural University at Wageningen, play an important part in this training. They are the sole centres for this sort of education, although no specific training in this field was available until 1971, when a Chair of Food Microbiology and Hygiene was established at the Agricultural University. The training at the Veterinary School was and is directed mainly towards meat inspection and slaughterhouse problems and to a certain extent to Dairy Science. However, a curriculum, which would involve a Chair in Food Microbiology, is contemplated. Besides these 2 institutes a certain amount of marginal training in Food Microbiology is offered at the Technical University of Delft and by several chemical, medical and biological Departments of Dutch Universities. Yet most graduates active in the field of Food Microbiology in The Netherlands owe their professional status to postgraduate or extramural training, often obtained abroad. The need for interdisciplinary cooperation in education, as well as in practising Food Microbiology, is stressed, particularly in view of the increasing need for international co-operation in microbiological quality assurance in the food industry. It is imperative to pay proper attention to aspects of food processing in this connection. It is also indicated that such interdisciplinary co-operation will further the level of practising food microbiology, which in turn will create an increasing demand for food microbiologists.

The Status Quo

TWO INSTITUTIONS undertake training in Food Hygiene, namely the Veterinary Faculty of the University of Utrecht, which is the only school of veterinary medicine in the country, and the Agricultural University at Wageningen, which also is the only institution of its sort in The Netherlands.

As far as the Veterinary Faculty is concerned, there has been a Chair for the Science of Food of animal origin from the beginning of the Faculty's existence in 1922. However, the activities of this chair were mainly in the field of meat inspection procedures (including infectious diseases and pathological processes in connection with slaughter animals), and to a lesser degree also in slaughter house and meat hygiene. In addition the Faculty has a special Chair for Dairy Science, although it must be admitted that veterinarians have played only a minor role in the practice of dairy science in The Netherlands. Summarizing, one may say that in the Netherlands the veterinarian traditionally receives an excellent training in meat inspection and slaughter house hygiene, but not yet in food microbiology proper. However, in recent years an intensive review has been carried out with the purpose of updating the curriculum of the Veterinary Faculty, and it is to be expected that in the very near future the study programme will be changed. This may result in much more emphasis being laid on food microbiology, particularly relating to foods of animal origin, and on food hygiene.

At the Agricultural University of Wageningen a group of laboratories dealing with all aspects of food science was established some 20 years ago. This group now deals with food technology, food chemistry and food microbiology. The last 2 parts were one unit until 1971, when it was felt that in view of recent developments in the food industry more emphasis should be placed on food microbiology and hygiene. The Chair for Food Chemistry and Microbiology was therefore divided and a special Chair for Food Microbiology and Hygiene was established. It was the second of its kind on the European Continent, the University of Louvain, Belgium, having blazed the trail in 1969 (Mossel, 1969). Although the independent Laboratory for Food Microbiology has just been established, it is already offering an all-round training in this field.

Besides the 2 institutes mentioned, the Technical University at Delft has traditionally devoted attention to the technological aspects of food microbiology. Such internationally renowned scientists as Beijerinck and Kluyver have had a tremendous influence in our country, particularly as teachers to several generations of University students; notably from Leiden and Groningen. Professor Kingma Boltjes of Amsterdam University functioned similarly until his retirement in 1971.

In reviewing the professional education of the food microbiologists and public health workers who occupy senior positions today, one encounters chemists, veterinarians, biologists, physicians and even a few pharmacists. They have all received a certain amount of basic microbiological training at university, but have become specialists in their field by further study, usually abroad, after completing their initial university education. As pointed out above, this situation has now changed both in Belgium and in The Netherlands and more progress is anticipated in the future.

Desiderata for the Future

Perhaps because we did not have specific education in food microbiology until recently, several disciplines have claimed in the past that their training makes them especially well qualified in the field of food hygiene. This has led to some competition, which has considerably hampered the development of food hygiene. The veterinarian had obviously no profound knowledge of chemistry, physics and statistics, the chemist was not too familiar with the live animal and customary procedures in slaughter houses and the physician was not well trained in food and nutrition in general. The situation has improved considerably since

1969 when an interdisciplinary Food Microbiology Group was founded within The Netherlands' Society for Microbiology; this Group has organized some 30 very useful mini-symposia.

Such co-operation is very much required. During this symposium we have heard about the great significance of infection cycles, the role of healthy animals as carriers of pathogenic microorganisms, and the antimicrobial residue problem, which even today cannot be fully evaluated with regard to its significance. In addition, there is a tremendous increase and liberalization in international trade which, in conjunction with mass production of foods, complicates problems in the field of food hygiene (Bauman, 1972). It is obvious that the problems cannot be approached by one discipline alone, but that a very broad knowledge is required in the fields of microbiology, general biology, biological and analytical chemistry, physics, mathematics, and also sanitary engineering. In my opinion protection of the consumer, the ultimate goal, can be achieved in future only by an interdisciplinary approach (Kampelmacher, 1972). No single part of our task belongs to one discipline in particular. On the contrary, the diversity of educational backgrounds and specialization, if brought together, can help us to achieve the goal which all of us aim at, namely to produce safe food for human and animal consumption. I have substantiated these points in an earlier paper to this symposium, in discussing the so-called Walcheren project (Edel *et al.*, 1973).

There is a second point which again is not specific for one country, but should interest us all, namely, the future possibilities for well-trained food hygienists. Because of the increase in international trade mentioned above, the need for harmonization of safety requirements and methods of examination becomes urgent. So far, importing countries usually require that special provisions and procedures be imposed in food manufacturing plants, and especially slaughter houses, in exporting countries. There is also, however, a trend to develop end-product specifications, which means that the importing country leaves the exporting country freedom to process foods as long as this corresponds to certain international or bilateral agreements. In view of this development the need for food microbiologists and hygienists in factories and government positions will increase.

Until now, food hygiene control (especially with trained graduate supervision) has been regarded as a luxury in many food factories and especially in the numerous small ones. It was believed that a certain amount of common sense, clean clothes and clean hands were sufficient for the sanitary production of foods. Today food production and processing have become extremely complicated. One has to have advanced knowledge of the origin of the raw products and the microbiological and chemical contamination. As these have a direct effect upon the end product, they must be controlled all the way along. It is clear therefore that specialists will have tasks waiting for them. Moreover, experience shows that work emerges when specialists are available. There is,

426 E. H. KAMPELMACHER

hence, good hope that in future an increasing number of scientists will be required in our field and that this will, in turn, lead to further academic development of basic food microbiology.

References

Bauman, H. E. (1972). Food Microbiology. *Am. Soc. Microbiol. News.* **38**, 312.
Edel, W., van Schothorst, M., Guinee, P.A.M. & Kampelmacher, E. H. (1973). Mechanism and prevention of salmonella infections in animals. In *Proceedings of the 8th International Symposium. The Microbiological Safety of Food.* p. 247. London: Academic Press.
Kampelmacher, E. H. (1972). Since Eve ate apples . . . Fleischwirtschaft **52**, 820.
Mossel, D. A. A. (1969). Out of the Orphanage—Food Microbiology becomes a Scientific Discipline. Inaugural Address. University of Louvain.

Discussion

Elliott

Dr Kampelmacher is to be commended for his suggestion that food science in Europe might be improved by additional training and for his realization of the role of food microbiology in food science. The Codex Alimentarius Committee on Meat Hygiene has recently drafted a report that would require not only slaughter operations but also further processing of meats to be under the supervision of veterinarians. Would Dr Kampelmacher also recommend that a degree in veterinary medicine is a necessary prerequisite for inspection and supervision of further processing?

Kampelmacher

Where would you, Dr Elliott, draw the line as far as specialists are concerned? I agree with you that when meat becomes a food commodity other disciplines also have a part to play. However, we must not forget that the veterinarian has valuable information with regard to the original product, namely the live animal. Slaughterhouse inspection of slaughtered animals and poultry is a task which should, I think you would agree, be performed under veterinary supervision.

Olson

Would anyone care to comment briefly on the impact of the presumed requirement that local food inspection services must hire veterinarians in order to conform with EEC requirements?

Baird

The Directive on Intra-Community trade in fresh meat (red meat) requires veterinary certification for the export of meat within the EEC. The Draft Regulations for Intra-Community trade in meat-based products (again red meat) will require veterinary supervision of hygiene and veterinary certification. At present this is only a draft and it has yet to be implemented. The Directive on trade in fresh poultry meat will affect the U.K. and will be enforced from January 1973 for Intra-Community trade and from 1976 for all trade in fresh poultry meat in the EEC, including the supply of fresh poultry meat for the home market. Many other countries outside the EEC, e.g. Sweden and Switzerland require veterinary inspection and certification for meat and meat-based products exported from the U.K. Veterinary supervision of the hygiene of production and veterinary certification is required for heat-treated meat products exported from Britain to the U.S.A.

Problems of the Training of Food Microbiologists

H. J. Rehm

*Institut for Mikrobiologie,
44 Munster (Westf.) Germany*

In Germany 4 types of Institutes give courses related to food hygiene and microbiology; namely those devoted to Medicine, Veterinary Studies, Food Chemistry and General Microbiology. In every case the curriculum will have a different emphasis, and hence persons currently defined as "food microbiologists" may well have little knowledge of certain vital aspects of the subject. Thus, a food chemist will have little training in the handling of pathogenic micro-organisms, while a veterinarian may not be familiar with the microbial contaminants of foods of plant origin. These differences in all-round competence may raise problems not only in relation to the routine quality control of food, but also in connection with research; the exclusion of chemists from work on pathogenic bacteria has almost certainly hindered investigations on the structure of bacterial toxins. However, a taught course to cover all the aspects of chemistry, technology, and microbiology that are relevant to the task of a food microbiologist, would either be very superficial in its treatment, or take an excessive number of years to complete. In Germany, therefore, a compromise is being sought. On the one hand, food control laboratories are being started with an inter-disciplinary staff capable of carrying out a wide range of investigations. On the other hand, bodies, such as the German Chemical Society, are arranging postgraduate courses in food microbiology, so that chemists can learn the fundamentals of the microbiological control of food. Further developments along these lines should go some way towards producing "food microbiologists" who are conversant with the varied facets of their subject.

THE MICROBIOLOGICAL control of food is becoming increasingly important, as are also investigations on the microbial spoilage of food, and the use of micro-organisms in the production of food.

The extensive development of food microbiology has been caused by various factors, including:

(1) the increased range of different foods for large populations
(2) the rapid development of many new techniques of food production
(3) increased trade with movement of food between countries far apart geographically.

In the centrally situated Federal Republic of Germany with its extensive production of different types of food and its world-wide trade, questions of food microbiology are, for the above reasons, of special importance.

To meet these requirements of the food industry training must be provided for the scientists who are engaged in the microbiological examination of food. In

427

Germany over the last hundred years, 4 types of institutions developed which make microbiological control of food, and food research, their concern.

Types of Training

(1) Human medicine and public health; this is concerned with food hygiene in relation to illness, or poisoning from food micro-organisms or their toxins. Medicine gives an essential basis for food hygiene.

(2) Veterinary medicine and hygiene: this is concerned with microbiological aspects of animal products. It is also interested in the hygiene of food and food production.

(3) Food chemistry: the food chemist is responsible for the chemical control of food; in addition he is concerned with some microbial aspects of food, omitting pathogenic micro-organisms.

(4) Scientific microbiology: this is concerned essentially with scientific investigations on food microbiology.

In these 4 areas the training is different and for this reason 4 different types of food microbiologist exist. Scientific microbiology has until recently been taught in Institutes of Botany, or in the few scientific Institutes of Microbiology; this indicates the very heterogeneous training of food microbiologists. Thus, we have, at present, a situation whereby in several regions of West Germany, similar positions may be occupied by food microbiologists with different types of training. For this reason there are variations in competence and also in judgement, depending on the background of the microbiologist.

Difficulties in food control occur, for example, when a food chemist with a microbiological training wants to isolate and to identify pathogenic micro-organisms. He is often not trained for this type of work. Similar difficulties exist when the veterinarian passes judgement on the microbial contamination of food of plant origin, or on the technology of food production. He has not normally been trained in this area.

In addition, deficiencies in the chemical training of the veterinarians and the exclusion of food chemists from work on pathogenic bacteria effect the progress of food research, and have hindered investigations on the structure of most bacterial toxins. Our knowledge of the structure of these important substances is less than that of many less important proteins. The insufficient training of food chemists in microbiological methods and in judging results, and the incomplete knowledge of veterinarians in the technology of food production and in the chemical composition of foods, illustrates the problems of food control and food research in the Federal Republic of Germany. Almost certainly there are very similar problems in other industrialized countries, and in the under-developed countries also.

Essential Requirements

What steps can be taken to alter this difficult position in respect to the training of food microbiologists, and to standardize their training? First let us consider what knowledge is required by food microbiologists. Among others, the following important subjects may be listed:

(1) A basic training in microbiology.
(2) Specialized knowledge of food spoilage micro-organisms, of micro-organisms used in the production of food, and of food preservation.
(3) General knowledge of hygiene, and a specialized knowledge of food hygiene.
(4) Some knowledge of diseases caused by pathogenic micro-organisms occurring in food.
(5) A good knowledge of the chemical composition of food.
(6) A good knowledge of the technology of food production.
(7) General knowledge of food legislation in relation to microbiology; both in the home country and in other countries with which an intensive trade in food products exists.
(8) In addition there should be some knowledge of marginal subjects, such as the phytopathology (for example to judge spoilage in citrus fruit and apples), industrial microbiology and parasitology.

Surveying all these points it is evident that such a broad training—even if it is done in a specialized course—may be very superficial, or take an excessive number of years to complete. It is necessary, therefore to concentrate on certain points. In food control it is essential that microbiologists should judge only those areas of food microbiology in which they are trained.

In Bavaria attempts are being made to avoid these difficulties. It is planned to start a very extensive control institute in which medical, veterinary-hygienic and chemical investigations (including microbiological investigations of food) can be carried out. We look forward to success in promoting inter-disciplinary contacts.

In the German Federal Republic it is proposed to offer postgraduate courses for graduates of different disciplines. The section for food chemistry of the German Chemical Society has arranged a series of courses in Karlsruhe. In these courses food chemists can learn the fundamentals of microbiological control of food, and gain experience in discussing and assessing the results of microbiological investigations. The programme of these courses is given in Table 1.

Following these courses, more than 250 food chemists have been trained in microbiological methods. These food chemists may not be food microbiologists, but they are able at least, to investigate some foods by microbiological methods. A similar course was held for veterinarians in Berlin.

For food chemists with basic microbiological knowledge, courses for

Table 1

Programme for basic courses

(1) Problems of food microbiology	L
(2) Problems of the microbiology of water and drinks	L
(3) Fundamentals of microbial methods	L + P
(4) Taxonomy of some food micro-organisms	L
(5) Methods for the quantitative enumeration of viable micro-organisms	L + P
(6) Importance of food hygiene in plants	L
(7) A short review of food spoiling micro-organisms and their physiology	L
(8) Factors that influence microbial activity	L
(9) Preservatives and disinfectants	L
(10) Special microbiological methods for food products	L + P
(11) Microscopic-bacteriological methods in food microbiology	L + P
(12) Membrane filtration techniques in food microbiology	L + P
(13) Quantitative enumeration of micro-organisms and judging the numbers–microbiological problems of sampling	L + P
(14) Food poisoning and food-borne pathogens	L
(15) Methods for determination of *Escherichia coli* and coliform bacteria in foods	L + P
(16) Equipment of small microbiological laboratories–necessary security rules in maintaining a microbiological laboratory	L + P

L, Lecture; P, practice.

advanced food microbiological training are offered. The programme is given in Tables 2 and 3.

Future Needs

More research institutes are required employing food microbiologists with scientific professional training, and with a good knowledge of food chemistry, food processing and in the handling of food pathogens.

It would be useful to offer courses for veterinarians that would provide an insight into the technology of food production, as well as a knowledge of non-pathogenic food spoilage micro-organisms (and their physiology and biochemistry). It is difficult to decide whether it will be possible to train them also in food chemistry.

Until changes are made, a great part of food microbiological research remains in the hands of scientific microbiologists and veterinary microbiologists. Microbiologists with chemistry and food chemistry as subsidiary subjects can step into this breach, and nowadays there are many such "food microbiologists". It should be noted also that in the German Federal Republic there is an increased number of food chemists who have taken microbiology as part of their degree course, and who prepare a dissertation on a food microbiological theme; in the future greater progress can be expected.

Table 2

Course for advanced scientists in food microbiology

Part I Lectures

    (1) Microbiology of margarine, butter, fats and oils
    (2) Microbiology of salads with mayonnaise
    (3) Microbiology of ice-cream
    (4) Microbiology of nonalcoholic drinks (including soft drinks)
    (5) Microbiology of milk and milk products
    (6) Microbiology of canned foods
    (7) Methods for enumeration of air-borne micro-organisms
    (8) Microbiology of dried foods
    (9) Antimicrobial substances in food production
(10) Problems of the production of vinegar and acid products
(11) Microbiology of fish delicatessen products, anchovies and fish preserves
(12) Problems of control methods of precooked foods
(13) Microbiology of chocolates and sugar products
(14) Microbiology of sugars and molasses
(15) Microbiological problems of frozen and freeze-dried foods
(16) Microbiology of bakery products and potato chips

Table 3

Course for advanced scientists in food microbiology

Part II Methods

    (1) Methods for quantitative enumeration of specific micro-organisms in foods
    (2) Methods for cultivating anaerobic bacteria
    (3) Diagnostics of clostridia
    (4) New general and selective methods for enumeration of microbial cells in foods
    (5) Diagnostics of *Bacillus* species
    (6) Diagnostics of enterobacteria
    (7) Diagnostics of staphylococci
    (8) Diagnostics of enterococci

Furthermore, attempts are being made to provide training for food microbiologists by filling professorships in scientific microbiology with food microbiologists, and by founding new departments of food microbiology related to existing departments of food chemistry.

Many of the above problems of the training of food microbiologists are world-wide, and should be discussed by close contact between people of different countries. Efforts are now being made to establish microbiological standards for the whole world. Consequently, efforts should be made to train food microbiologists to interpret these standards in the same way, or at least in similar ways. This would be especially useful, because the appraisal of biological results by non-specialists is extremely difficult. It is important that food microbiologists in all countries should use the same methods of examination, and also the same criteria to assess results. Difficulties will soon arise if lawyers have to pass a judgement on microbiological results.

I believe that the Committee on Food Microbiology and Hygiene of the IAMS is a very suitable body to make recommendations for training, and perhaps to promote international courses for food microbiologists. We must obtain a sound basis for microbiological food control, to protect the consumer as well as the producer.

Discussion

Reinius

With reference to the papers presented and to the discussion held on educational aspects during this symposium, as the representative of W.H.O. I am glad to express how highly the Organization appreciates the work in the field of microbiology and hygiene undertaken by this Committee. The organization looks forward to continuous fruitful cooperation.

As one of the most important joint activities between your Committee (or members of it) and W.H.O., I would like to point out the efforts to co-ordinate and intensify education and training in food microbiology, particularly on the postgraduate level. Contacts are established and programmes are presently at various stages of development and implementation in Denmark, France, The Netherlands, United Kingdom and Venezuela as well as in some other countries. The need for this kind of training is evident and factual numbers of candidates to attend courses annually have been presented by most of the Regional Officers of W.H.O.

The need for long term planning and co-ordination of activities for postgraduate training of students of various disciplines in food microbiology would call—in accordance with some proposals made during the discussions—for a specific meeting or meetings to deal with this important problem. It is clearly not possible in the frame of the agreed agenda and considering the time available to discuss the wide range of questions involved at this time.

The present situation in the world offers a variety of solutions as to the distribution of responsibilities in food control between various professions. In fact, in many countries the infrastructure in this field is a result of years or decades of development towards one or another of the possible directions.

Logically the responsibilities of each professional group should be concentrated in the field which corresponds to the respective education and training of this particular group. However, successful food control work includes as an essential component, co-operation between various professionals especially in overlapping areas of activity. Basically, food control activities are multi-disciplinary, by their very nature, and thus a certain degree of variability concerning "who does what" is one of the intrinsic characteristics of the work. In each individual country the ideal distribution pattern of responsibilities to be developed is both dependent on the present structure and on the cost-benefit aspects of possible changes in the curricula of training and education. The main problem is to find out which is the most efficient way to satisfy the need for

DISCUSSION 433

highly qualified food microbiologists. This question has particular relevance when the responsibilities of various professional groups are under consideration in food control work at the local level.

The formulation of useful and well founded recommendations concerning the above problem obviously has to be undertaken by a multidisciplinary group of experts, as is the common practice when specific problems in this field are tackled by expert committees, scientific groups and study groups convened by the W.H.O.

Harrigan

You consider that this is perhaps not the time and place to discuss demarcations between the professions. However, it is my impression that veterinarians in the U.K. are perhaps educated in a rather narrower field than are their counterparts in the rest of Europe. If I understood correctly the remark made earlier by Dr Olson, when the U.K. enters the E.E.C. it will be necessary for public health departments and other control agencies concerned with food production and servicing industries to be administered and staffed by veterinarians. Could anyone here perhaps clarify the position with regard to the implications of this change both from the point of view of inspection systems and agencies and, equally important, the impact on the educational programmes within the U.K.?

Kampelmacher

We must be careful not to mix up different problems which have arisen during discussion this morning. Firstly, there is the world situation as outlined by Dr Reinius. At this level, mutual understanding is badly needed between the disciplines. It may be that this group, in which so many disciplines are represented, is a good forum for such discussions. Secondly, there is the specific problem of the E.E.C. and the future E.E.C. countries, where rules have been established and requirements made as to who should perform certain tasks. Thirdly, there is the very specific situation in the U.K. where the veterinary education has been directed mainly to other goals than in the European veterinary colleges. To make it more complicated exports to the U.S.A. require veterinary inspection in slaughterhouses and meat factories.

The E.E.C. situation needs early contacts, discussions and agreements as far as the interdisciplinary nature of food hygiene is concerned. With some exceptions (e.g. slaughter house meat hygiene inspection) it is not true that a certain field belongs to a certain discipline. The best educated man in the field and his willingness to work within an interdisciplinary team will finally establish the future pattern of European food hygiene.

Baird

In the U.K. veterinary training in the past has not placed much emphasis on food microbiology but there is no reason why the Universities' Veterinary Faculties should not include this subject in the curriculum. There is some indication that veterinary educationists appreciate the need to do this. There is no reason why veterinary graduates should not undertake post-graduate training in food hygiene and food microbiology.

Jarvis

In his remarks on education in Food Science and Technology in the U.K., Prof Aylward mentioned the 1 year M.Sc. taught course in Food Microbiology which is offered at the National College of Food Technology, a department of the University of Reading. This course has previously trained veterinarians from

Europe (in addition to microbiology graduates from the U.K. and many other countries of the world) and could provide a suitable course for postgraduate education in food microbiology for veterinarians in the U.K. This course is based upon principles in food microbiology with suitable examples drawn from appropriate commodities. Subsidiary courses in food chemistry, food quality control, food process engineering and statistics provide the student with an integrated course which gives the necessary background for employment in the food industry. Short courses in food microbiology are offered at various levels to the staff of Member Companies at the British Food Manufacturing Industries Research Association. Such courses are oriented towards the needs of industry.

Riemann

Veterinarians who are employed in the U.S.D.A. Meat Inspection Service receive additional education through courses established by the Service. Schools of Public Health in the U.S.A. offer education leading to M.S. or Ph.D. degrees in public health. These schools are frequented by veterinarians, medical doctors, engineers, nurses, etc. who qualify to serve in public health, including food hygiene.

Linderholm

During the past few days we have discussed food microbiology mainly from the human health point of view but we should not forget what Dr Baird has said, that the requirements of the importing countries include certificates about the animal diseases and zoonoses in the country of origin. It has been found very unpractical to have different certificates and therefore the veterinary services have the responsibility since they must be involved. That is one of the reasons why veterinary education in many countries includes a great deal of food hygiene although this is not so in all countries. I should like to add a further point to the four made by Kampelmacher: The increasing interest in environmental problems means that the environment in which food is produced will be more closely scrutinized. The agricultural and veterinary services will and should probably, be more and more involved in food hygiene, in its widest sense, instead of being isolated.

Rehm

For future education, I think we essentially have 2 points: (a) Education in the long term: extensive training is needed to be a real food microbiologist; (b) education for the moment: a series of courses in food microbiology and food technology should be used to enable practicing food microbiologists to interchange ideas on methodology, etc. in order to widen the outlook of all.

Education in Microbiological Safety Standards

N. GOLDENBERG AND GRETA EDMONDS

Marks and Spencer Ltd.,
Michael House,
Baker Street,
London W1A 1DN, England

The authors summarize their experience over the years 1955–1972 in working with a number of Food Manufacturers producing "perishable foods" to agreed specifications on quality and safety. Their experience in this field was first developed in the production of cakes containing fillings of whipped "synthetic cream". This is described in some detail with reference to the education of (i) Top Management; (ii) "Middle-Management; and (iii) the Operatives on the factory floor on the procedures and risks involved. Based on the success of this operation, the principles evolved and the experience gained were then applied to the development of a large range of other "perishable foods", including meat and poultry pies; sausages; sliced meats; sandwiches; fresh creams; trifles; desserts; yoghourts and other dairy products; cooked and roasted poultry products etc. The authors conclude that the education of all concerned, from Top Management down to the Operatives, to appreciate the hazards involved in the production of "perishable foods" is essential; without this, it may not be possible to get the factory "Establishment" to achieve adequate measures to ensure the safety of their products to the consumer.

IT MAY SEEM rather surprising at first glance that we in Marks & Spencer should be speaking of education within the food industry for as I imagine you know, we are large retailers of textiles, footwear and foods. However, our interest in the merchandise we sell does not start when its reaches our stores—quite the contrary.

It is our policy only to sell foods which are of high quality; of good value; are safe to eat, and which are made in food factories with high standards of hygiene. So far as the Food Technologists in Marks & Spencer are concerned, we achieve these aims by working closely, and, indeed, intimately with our suppliers and with the manufacturers of the raw materials we recommend to our suppliers. We do so by working with our suppliers to agreed specifications, which "cover" raw materials; the use of the right type of packaging; methods of process control; inspection of the finished products; methods of storage and distribution; agreed "shelf lives"; and the establishment and maintenance of good standards of hygiene. Many of these aspects of our work have been described elsewhere (Goldenberg, 1960, 1962, 1963, 1964a, b, 1968), and will not be referred to here again except to emphasize that microbiological standards are only a part of our overall approach to the many problems of safeguarding quality and safety (Goldenberg & Tall, 1971).

It is therefore clear that we are not only retailers, but that we also have considerable experience in the production of foods. Our Food Development Department or "Food Technical Services" in Marks & Spencer consists of a team of technologists, technicians, chefs, hygiene officers, chemists and bacteriologists. Many spend a major part of their working week at our food suppliers. A close working relationship has been built up between our suppliers and ourselves.

Prior to 1955, the cakes we sold in our Stores were of the "non-perishable" type, i.e. cakes such as Madeira; fruit cakes; Swiss rolls; chocolate-enrobed lines and cakes containing "buttercream" fillings.

At this time, however, we decided to extend our range of cakes and decided to enter the field of cakes containing "synthetic" cream fillings; we decided against using fresh cream fillings at that time because: (a) our suppliers had had no experience in handling fresh cream; (b) we were concerned about the bacteriological hazards involved; (c) refrigerated counters were then not available in our stores.

Together with one of our suppliers of raw materials, we were able to develop a new type of "synthetic cream", containing a high sugar content in the aqueous phase. Exhaustive tests showed that under normal conditions (15–25°) of all the normal pathogens, only staphylococci would grow, and these relatively slowly. *Salmonella* spp. and faecal coli died off.

We therefore had a "cream" which had good whipping qualities; was stable at normal temperatures; had reasonably good eating qualities and which was bacteriologically "safe", if good standards of personal hygiene and equipment cleaning were observed during the handling at the bakery.

Together with the supplier, we made certain that there was good process control at the factory making the "cream" and that factory hygiene and personal hygiene were of a high standard. Each batch was bacteriologically tested before being sent out to the bakeries. Cans were date-coded to ensure strict rotation of usage. The cans of "cream" were held in refrigeration immediately after processing and transported to the bakeries in refrigerated vehicles. There was very close co-operation between our technologists and the manufacturer of the "cream", and the technologists in charge of production and of quality control, including bacteriological control.

This "cream" was used for filling such lines as eclairs, sponges and fruit tartlets. There were then sold on a "daily delivery—daily sell-out' basis in our stores.

We soon realized, however, that neither our suppliers nor we could achieve the right conditions in the factories for producing these "perishable" foods unless all concerned understood something of the microbiological problems involved; of their causes; of the importance of safety to the ultimate consumer;

and how the risks involved could be either eliminated, or, at least, minimized. We had to educate all concerned to understand the problems and risks involved, and why certain actions and disciplines had to be introduced to deal with them.

Management at Suppliers

Although our cake suppliers were efficient and had good standards of hygiene, they had had no experience of handling "perishable" foods. The "fillings" largely used were jams, and "buttercream" type fillings. There were no serious bacteriological or safety problems associated with their use; we found that some of our suppliers and their staffs were a little frightened of "bugs" and "bug ɔ problems". We therefore had an education problem here and work to do in drawing up acceptable, realistic and agreed working standards.

Medical Authorities

We were at first discouraged by various medical authorities from entering this field. They thought—understandably—that there were so many cases of food poisoning already associated with "cream" cakes that Marks & Spencer should not encourage its use. We thought, however, that as we had a number of "interlocking" factors on the side of safety, that the risks were relatively small. They were: (a) formulation of the "cream", so that only staphylococci out of the normal pathogens could grow; (b) good handling at suppliers and good standards of hygiene to eliminate the possibility of contamination in the bakery; (c) refrigerated storage at suppliers; (d) good stock control at our Stores on the basis of "daily delivery—daily sell-out".

Education of Management

Our venture into this new field was put over as a challenge to our suppliers. They responded excellently and together we discussed the potential hazards of the operation. These were neither minimized nor exaggerated. The special precautions to be taken were jointly discussed and explained. These could be summarized as:

(1) It was essential to have a completely separate area for handling "cream", so that there would be no "cross-contamination" from raw eggs, flour etc.

(2) High standards of hygiene in this area.

(3) Education in personal hygiene.

(4) The understanding of sterilization procedures.

(5) Purchase of necessary equipment, taking particular care that this could be easily dismantled and readily sterilized, and used only for "cream" lines.

(6) Bacteriological control; the need for a resident bacteriologist and for regular analyses.

(7) Adequate refrigeration.

(8) Daily delivery of cakes to our Stores.

Members of our technical team worked very closely with our suppliers in setting up production areas for this project. In fact, the challenge was so well accepted that the idea of a "creamery" within a bakery became a matter of prestige.

A great deal of work went into the emphasis of (1) adequate process control; (2) bacteriological control; (3) quality control in its widest forms.

Strangely enough, we found that although our suppliers had become quite

Table 1
"Imitation cream": no. I

---

1. *Premises*

   (a) Separate "creamery" to be partitioned off from rest of bakery
   (b) "Creamery" must be cool; have good ventilation; preferably temperature controlled at 65°F (18°C)
   (c) Lighting must be good and protected by plastic shades
   (d) Double sink for washing and rinsing; with ample supplies of hot water
   (e) Sufficient hand washing basins near entrance to the "creamery"

2. *Processing equipment*

   (a) Stainless steel mixing equipment
   (b) Stainless steel filling and depositing equipment
   (c) Stainless steel tin openers
   (d) Cold-room at 40°F (4°C). Sufficiently large to hold all stocks of unopened cans and all production; with adequate space between trays
   (e) An autoclave or pressure cooker for sterilization
   (f) Equipment which cannot be steam sterilized, must be sterilized with chemical sterilizing agents, previously agreed with M & S

3. *Sterilization of equipment*

   (a) All equipment coming into contact with "cream" to be cleaned as follows at the end of every day:

      (i) Wash off excess "cream" by rinsing off with hot water
      (ii) Dismantle as completely as possible; wash all parts first in hot water to remove all cream and then in hot detergent-sterilizer solution
      (iii) Rinse in clean hot water
      (iv) Sterilize in the autoclave (suggested conditions 15 lb/in² for 15 min). Ensure that all air is vented from the autoclave before the pressure is raised.

   (b) At lunchtime and when otherwise necessary, hand equipment, such as stencils and palette-knives, washed in hot water, then in detergent-sterilizer solution, followed by a rinse in very hot water
   (c) Conveyor belts and tables to be washed in detergent-sterilizer solution and then rinsed in hot water

---

enthusiastic about this venture and understood the precautions to be taken, there was still a certain amount of reluctance to accept the need to have a bacteriologist in the production team. Understandably, management and production personnel associated a bacteriologist with a laboratory, agar plates and test tubes. We had to explain to management that the technologist could make a valuable contribution to the achievement of quality and safety standards on the factory floor, provided that he was integrated and accepted as part of the "production team".

Eventually we succeeded. Our first Code of Practice on the handling of imitation cream was drawn up and agreed as a working document (see Tables 1–3).

Table 2

"Imitation cream": no. II

1. *Cream handling*
   (a) "Cream" delivered twice weekly in insulated vans
   (b) "Use by" dates stamped on each can and must be adhered to
   (c) "Cream" to be stored at $40°F$ ($4°C$). After delivery, it must be held overnight before use
   (d) Opened cans or whipped cream must not be kept overnight
   (e) Tops of tins to be cleaned before opening. (Wipe with paper towelling dipped in detergent-sterilizer solution, then with another paper towel dipped in clean hot water)
   (f) Tin openers to be immersed in detergent-sterilizer solution after use and rinsed in clean hot water immediately before re-use

2. *Hot weather conditions*
   Where there is no temperature control, production to be stopped when the temperature in the creamery reaches $75°F$ ($24°C$)

3. *Bacteriological analysis*
   (a) Samples of cream from the finished product to be taken under sterile conditions for analysis. At first samples to be taken every other day, and later weekly from each product
   (b) Copies of results obtained to be sent to Marks & Spencer; unsatisfactory results to be discussed immediately with "cream" suppliers and Marks & Spencer

4. *Supervision by resident chemist or bacteriologist*
   Production must be supervised by a resident qualified chemist or bacteriologist, who will visit and inspect production at least twice a day to ensure agreed precautions being observed
   Evening shifts must be regularly inspected. Inspection must include the sterilizing operation. In the initial stages the technologist should be present full-time until staff have been fully trained. Technologist should have power to stop production immediately and reject merchandise at his discretion

5. *Finished products*
   To be wrapped immediately after finishing and at once transferred to the cold room until ready for despatch. They must be held sufficiently long in the cold room to have cooled completely to $40°F$ ($4°C$). This is usually about 7 h

Table 3a

Hygiene rules for personnel handling "perishable" foods

1. All staff handling "perishable" foods to be medically examined to make certain that they are not "carriers" of pathogenic bacteria
2. Each morning, before work begins, supervisor to check if any operatives are suffering from stomach disorders; diarrhoea; boils or sores on hands or faces; heavy colds; sore throats; or have been in contact with infectious disease. If so, they must be transferred at once to other duties until return to full health. An inspection record should be kept and signed by the person responsible for the inspection
3. All persons with cut or burned hands must not handle "perishable foods"
4. Operatives must *always* wash their hands thoroughly on entering or re-entering the production room and apply bactericidal barrier cream after washing. Bactericidal liquid soap to be used, with paper towels or "continuous" linen towels
5. Finger nails to be kept short and well cleaned with a nailbrush every time the hands are washed
6. Paper handkerchiefs to be provided in a dispenser near the hand washing sink. Operatives must use paper handkerchiefs rather than a cloth handkerchief and wash their hands after use
7. Persons handling cooked meats must *not* handle uncooked meats

Table 3b

Prevention of food poisoning: "stool" tests

1. The following recommendations apply to all operatives and supervisory staff working on
   (a) All production units which produce "foods at risk" or "perishable foods"
   (b) All staff canteens

2. *Production units and staff canteens already producing "foods at risk"*

   Stool tests should be carried out
   (a) Whenever there is any question of staff suffering or having suffered from an intestinal illness or been put at risk by close contact with known infected patient, e.g. by eating food prepared by such a patient
   (b) On food handlers involved in an outbreak of food-poisoning or dysentery

3. *New staff*

   Newly-engaged food handlers or supervisory staff require stool tests if the relevant medical doctor judges that these are required in the light of the patient's medical history

4. *Employees returning from abroad*

   Employees who have recently been abroad or recently come from abroad require to have their medical history carefully checked by the relevant doctor before beginning work or returning to work. Any with a history of intestinal disease *must* be submitted to stool checks before starting work.

5. *New production units and new staff canteens*

   All food-handlers and supervisory staff should be given stool checks if the relevant medical doctor judges these to be necessary

Role of the Technologist

One of the conditions of this operation being carried out was that there should be a resident technologist in the bakery. Part of his time would be spent in the "cream" room each day. During this time he would supervise standards and ensure that the agreed Code of Practice was being implemented. It would also be his responsibility to re-visit the bakery on occasion in the evenings to supervise the final cleaning and sterilizing of equipment.

This would be in addition to carrying out the necessary regular bacteriological analyses. It was agreed with management that copies of these analyses would be sent to a specific technologist at Marks & Spencer. If there was any unusual result, we would be notified immediately. The relevant standards have already been discussed in a preceding paper (Goldenberg & Elliott, 1973) and in other previous papers (Stephens, 1970; Goldenberg, 1964a).

The technologist at each bakery, as a member of the management team, was given authority from the Board of Directors to stop production, either for short or long periods, if "key" standards were not satisfactory.

Marks & Spencer organized courses for these technologists. This enabled our suppliers' technologists and ourselves to exchange ideas and discuss any problems encountered at their factories. We were able to agree on both working standards and bacteriological standards. A visit to the manufacturer of the "cream" was always included on these courses. These courses had the effect of building up confidence between the technologists at Marks & Spencer and at the suppliers. It also helped to broaden the approach of the newly-appointed technologists. Sometimes we found that the new technologist was looking for the bacteriological sterility of a hospital operating theatre. This was neither necessary or feasible. He also had to learn how to work with people and communicate to factory personnel. These courses were very helpful in this field of work.

Education of Line Management and Operatives

Having educated management and established the role of the technologist, we had to communicate our standards to the supervisor and his/her team of operatives.

Short talks were given by Marks & Spencer Technologists and hygiene officers and the suppliers' technologists to the supervisors at their bakery. It was explained that this was a new venture into handling a "perishable" material. If personal hygiene standards fell or the hygiene of equipment deteriorated, it could lead to food poisoning, spoilage and possible loss of confidence in their products. Very elementary talks on bacteriology were also given.

The function of the technologist had to be explained to the supervisors and the need for good working co-operation between them emphasized.

The Code of Practice was discussed and explained. We helped the Supervisor to set up a system to keep a daily record book to ensure that all people working in the "creamery" were: (a) "Stool-tested"—see Table 3a; (b) free from coughs, colds, cuts, boils, enteric troubles etc.; (c) carry out daily hand inspection; (d) ensure provision of soap, paper towels, nail brushes and barrier creams.

Some considerable time was spent in explaining the chemical sterilization of equipment, and, in particular, the operation of the autoclave, which was often quite new to the bakery.

Supervisors and operatives responded very well to these talks. There was no bonus given for working in the "creamery"; instead, it became a "prestige area". There was no problem encountered in "stool-testing" operatives—they knew why it was being done and that it was no substitute for good hand-washing practice. They rather enjoyed being involved with the supervisors, the technologists and management in maintaining standards. Our success was due to the fact that it was a new venture; everyone was kept informed of requirements, standards, bacteriological results and the Code of Practice.

It was emphasized, at all levels and at all stages, that what we were asking was not something untoward and extraordinary, but fairly simple and not difficult to achieve. We tried to put over the view that operatives who worked on the production of "perishable" foods had a social duty to the buying public to produce foods which were both good to eat and safe to eat. This was readily accepted once it was fully explained and understood.

Temperature Control

Specific temperatures were agreed for: (a) the storage of the "cream"; (b) the working temperature of the "cream" room; (c) the cold storage room for holding the finished goods prior to despatch; (d) it was agreed that production would be stopped if the temperature in the "cream" room rose above the agreed maximum level.

These cakes were sold from "ambient" counters in our stores. Any cakes not sold at the end of the day of delivery were disposed of.

The prestige of the "creamery" was maintained by management visiting it daily and going through the same hand-washing drill before entering the room, as was asked of the operatives. Visitors from Marks & Spencer always made a point of spending some time in the "cream" section—they too had been "stool tested"; had washed hands on entry and were free from colds etc.

Results

There have been no serious bacteriological problems in the years we have been involved in this operation. When occasional indicative organisms were found, investigations at the bakery usually led to the tightening-up of procedures. These investigations helped to make operatives and supervisors even more conscious of the nature of their work and of the need to take the necessary precautions.

Development of Other "Perishable" Lines

With the experience of handling foods containing "synthetic cream", we felt sufficiently confident to develop other "perishable" products. Generally speaking, the Codes of Practice drawn up for these products were based on our experience in "cream" handling. In addition, we introduced what we call "cold chain handling". That is to say, the product is cooled after production; stored in cold storage, then transported to our Stores in refrigerated vehicles and sold from refrigerated counters. These products are listed below and the details of production and handling given in the relevant Tables. Meat pies, Table 4; sliced

Table 4

Meat pies and sausage rolls

1. *Cooked fillings*

   Cooking time to be specified. After cooking: transferred to shallow sterilized trays: cooled rapidly: kept at less than $40°F$ $(4°C)$ and used within 48 h
   Pies to be filled cold. Maximum resting time for filled pies 3 h at room temperature or overnight at $40°F$ $(4°C)$ followed by 1 h at room temperature

2. *Uncooked fillings*

   Uncooked fillings to be held below $40°F$ $(4°C)$; pies to be baked within 24 h of preparation

3. *Baking*

   (a) To be more than sufficient to destroy vegetative bacteria, with a wide margin of safety
   (b) Pies with pre-cooked fillings to be baked to a minimum centre temperature of $180°F$ $(82°C)$
   (c) Pies with uncooked fillings to be baked to a minimum centre temperature of $200°F$ $(93°C)$
   (d) Centre temperature to be checked frequently and recorded by quality control

4. *Jellying*

   Jelly to be *boiled* for 5 min and then maintained at $170°F$ $(77°C)$ during filling. Total count less than 1000/g.

5. *Cooling*

   Pies to be cooled by forced draught (filtered air) to below $50°F$ $(10°C)$ (centre temperature) without delay

6. *Wrapping*

   Handling to be minimized. All pies to be fully cooled in the centre to $50°F$ $(10°C)$ before wrapping. To be marked with visible sell-out date code. Pies to be placed in refrigerator (temperature $40°F$ $(4°C)$) immediately after wrapping meats, Table 5; dairy products—fresh cream, Table 6; cold chain, Table 7).

The relevant bacteriological standards have already been discussed (Goldenberg & Elliott, 1973).

Having applied these standards to production, we thought it was essential that they should be extended to factory canteens; the standards are given in Table 8. This was done for 3 reasons:

(1) To avoid any incidence of food poisoning in factory canteens and its possible transference to production areas.

(2) To avoid illness due to food poisoning in the staff canteens and consequent problems of staff shortages.

(3) To make it clear to all staff that "perishable foods" should be handled with the same care as in production.

The importance of this approach can perhaps be best illustrated by reference to an incident that took place at one of our suppliers, and, which, because of the various "interlocking" precautions agreed to, was dealt with in time.

There were no reports of any food-poisoning from the consumer public; both the supplier and we were alerted to the problem by analyses showing the

Table 5

Sliced meats and sandwiches

---

1. *Personnel*

   Standards as for all other perishable foods

2. *Processing area*

   To be separate from other manufacturing areas likely to cause "cross-contamination", e.g. flour dust, uncooked fillings etc. it should be temperature controlled (65°F, 18°C)

3. *Equipment*

   Slicing machines
   (a) At the end of the day's production the machine must be dismantled; washed; sterilized using a detergent/sterilizer; and rinsed in clean hot water
   (b) Slicing blades must be cleaned and sterilized as above when changing from slicing one meat to another, e.g. from ham to tongue

   Hand utensils
   All hand utensils to be washed and sterilized as above at break periods and at the end of the day's production

   Canned meat prior to slicing
   Prior to opening, the top of the can to be wiped with paper towelling soaked in detergent/sterilizer

   Dispatch
   After packing, the sliced meats should be put into refrigeration (40°F, 5°C) as soon as possible prior to dispatch through the "cold chain"

Table 6

Fresh cream

---

1. *Receipt of cream from creamery*

    (a) Delivery vehicle temperature to be checked and recorded
    (b) Cream of temperature above 10°C (50°F) to be rejected

2. *Storage*

    To be at 3-4°C (38-40°F)

3. *Ageing*

    Cream to be used within 2 nights of arrival at bakery

4. *Production at bakery*

    (a) Production to conform to practices laid down in Marks & Spencer code of practice on "the handling of perishable bakery products"
    (b) Personnel—all people involved in production to be stool-tested and cleared medically
    (c) Equipment—to be of stainless steel
    (d) Cleaning of equipment—all equipment to be thoroughly washed in warm water, followed by hot detergent to remove *all* food material *before* sterilization
    (e) Sterilization of equipment—all equipment to be autoclaved (121°C/15 min) wherever possible. Where not possible, equipment to be chemically sterilized with quaternary ammonium detergent-sterilizers, or chlorine sterilizers, as agreed with Marks & Spencer
    (f) *The sterilization of savoy bags used for fresh cream*
       (i) At each "production break", savoy bags to be washed thoroughly with warm water, and then turned inside out
      (ii) Then wash in hot detergent solution to remove *all* traces of cream
     (iii) Soak in solution containing 200 p/m available chlorine for a *minimum* of 1 h
     (iv) Rinse in cold water and allow to dry
      (v) Swab counts to be taken from depositing end of bags twice daily, to check sterilization procedure

---

presence of *Salmonella panama* in some of the meat products produced. We discontinued sale of all the products and, shortly afterwards, the factory management, without pressure from us, closed the factory temporarily after consultation with the local Medical Officer of Health.

Subsequent investigation showed that: (a) there was an outbreak of food-poisoning in a nearby village due to *Salmonella panama*; (b) there were no cases among the factory staff; "stool" tests showed, however, that there was one "carrier" in the factory; (c) there was probably "cross-contamination" in the factory, where raw meat and cooked meat were handled in the same room by the same operatives; (d) there may have been some "cross-contamination" in the staff canteen, where the infected products had been sliced and served to the operatives.

Table 7

"Cold chain" temperature control

(a) The finished products to be held in cold storage. All cold stores to have external chart recorder

(b) Products to be transferred to cold storage immediately after production; "bottlenecks" and delays must be avoided

(c) All perishable raw materials and all finished products must be placed in cold storage during "breaks"

(d) To achieve the desired product temperature thermostat settings on both cold storage rooms and refrigerated vehicles will generally be well below this figure depending on the particular installation

(e) Frequent checks must be made to ensure that products are cooled to the desired temperature before dispatch

(f) Refrigerated vehicles must be pre-cooled prior to loading and must be inspected regularly to ensure cleanliness

(g) *Reception at store*
Before opening the door of the delivery vehicle, the temperature shown on the dial thermometer should be checked

---

The lessons learnt were: (a) raw and cooked meats must not be handled in the same room but in different rooms, using separate equipment and employing different teams of operatives; (b) the standards for handling "perishable foods" should be the same in the staff canteens as in production areas.

Cheese

A different and rather interesting aspect of this operation concerns the standards of safety of cheeses, which we began to sell at the same time. We took the view that cheese should be made from pasteurized milk to avoid any danger from pathogenic bacteria which may have been carried by the milk or got in through contamination. We were also concerned about the presence of toxin derived from the growth of coagulase-positive *Staphylococcus aureus,* where the bacteria are destroyed but the heat-tolerant toxin may remain. Such cases have, in fact, been reported (Hobbs, 1969; Hobbs & Gilbert, 1970).

Some of our suppliers were, at the time, unwilling to accept this view. There was apparently a tradition in some parts of the dairy industry that cheese of good quality can only be made from non-pasteurized milk. Finally, we persuaded one supplier to carry out a simple experiment. One lot of milk was divided into 2 lots, only one of which was pasteurized. The same mature cheddar cheese was then made from both and the 2 lots of cheese matured side by side under exactly the same conditions, and examined side by side until fully mature.

We found—and the Supplier agreed—that there was no difference in texture, flavour, bouquet or eating quality at any stage nor at the end when the cheese was ready for sale.

Table 8
Factory canteens—basic hygiene requirements

Essential equipment
1. A double sink unit with stainless steel drainer. One sink marked "wash" and the other "rinse". Supply of hot water
2. Cool storage for raw vegetables and salads outside of main kitchen
3. Adequate refrigerator
4. Kitchen waste to be removed daily

Equipment cleaning
1. All equipment used for raw meat or poultry must not be used for cooked meats unless thoroughly cleaned, either by boiling water or by detergent/sterilizer, with subsequent rinsing
2. All cooking equipment, preparation tables, can openers, cutting boards, etc., to be cleaned thoroughly daily
3. Meat slicing machines to be cleaned and sterilized at least daily as in Table 5

Personnel Hygiene
1. Staff to wear clean overalls and adequate head coverings
2. Hands to be thoroughly washed using hot water, soap and nail brush.
   (a) Before starting work in the morning
   (b) On returning to work after all breaks or using the lavatory.
   (c) After handling raw meats or poultry

Daily health checks
1. Freedom from colds, sore throat, boils, sores or skin infections on hands and face
2. Freedom from stomach disorders or diarrhoea
3. No contact with infectious disease
   Any persons suffering from these disorders to be transferred to another job, where foods are not handled, until restored to health

Food freshness
1. All hot meat dishes, including gravies, soups and stews, to be cooked fresh daily, i.e. not warmed up
2. Meat for serving cold to be cooked thoroughly in small joints, cooled rapidly, and then placed in a refrigerator until required
3. Cold sweets, such as trifles and custard, to be partly cooled and held in a refrigerator until required

---

All the cheese sold in our Stores has been made from pasteurized milk without loss of quality; this even applies to the special French and Italian cheeses.

"Non-Acid" Canned Products

This class of foods comprises canned meats and poultry; canned vegetables; canned dairy products and canned fish. In all cases, the pH is higher than 4·5, so that if contamination occurs after cooking, pathogenic bacteria will be able to multiply easily, and especially so in the absence of any "competing" bacteria.

Like other organizations (Bashford & Herbert, 1966; Hobbs & Gilbert, 1970), we found that bacteria can enter well-made, normal cans if they are handled with dirty hands, particularly whilst the cans are still wet shortly after cooking and cooling. We had a proven case of food-poisoning due to coagulase-positive *Staph. aureus* bacteria in canned peas; the cans were normal and well-made; the "canners end" was well-made; the cans had, however, been transferred by hand from the retort basket (after cooling in chlorinated water) to a travelling belt conveyor. Incubation tests are not normally designed to detect the small number of cans which could have been infected.

Investigations with the help of the Campden Canning Research Association helped to clarify the causes; it was highly probable that the bacteria had come off the operatives' hands on to the wet cans and had been drawn into the cans through the seams as the vacuum had formed.

To overcome this danger, we finally issued a Code of Practice on the handling of non-acid packs. This is summarized in Table 9.

Safety in Retrospect

As stated earlier, we have had no serious bacteriological problems in terms of significant occurrence of food poisoning. However, on occasions we have had to ask our suppliers to discontinue production for a short period whilst investigations were carried out to resolve a particular problem. Because of the attention paid to the regular bacteriological analyses, we have generally been able to stop potential problems in time.

There is some evidence that the standards for handling "perishable" foods of the type described above, as developed and applied jointly with our suppliers, have played a part in raising standards of hygiene generally throughout the food industry.

Suggestions for Further Research

(1) The presence of salmonella in various meats and in poultry to varying degrees has been attributed in the past and also at the present Symposium to imported feedingstuffs. This is a major problem in the U.K. It should be dealt with in 2 parallel ways: (a) the use of official legal bacteriological standards for imported feedingstuffs; (b) visits by qualified technologists to factories of production abroad to pin-point sources of contamination jointly with the producing countries; action for elimination should be agreed.

This should be put to the Ministry of Agriculture, Fisheries and Food by a suitable body.

(2) The skin of poultry seems always to be infected with coagulase-positive

Table 9

"Non-acid" canned goods—vegetables; dairy products; meats and fish

---

Processing

1. Cans should be processed as soon as possible after filling
2. Process times and temperatures used must ensure a commercially sterile product. F_0 values to be determined and agreed with M & S before production commences and re-determined if there is any recipe modification affecting heat penetration
3. Time/temperature recorders must be fitted to all retorts. It is recommended that controllers are also fitted, particularly for temperature
4. Retorts to be checked daily to ensure that the venting cocks are in good working order and that both recorders are registering correctly

Chlorination of cooling water

1. Can cooling water must be chlorinated and the chlorine water allowed a contact time of 20 min before use
2. Water at the retort overflow, in cooling sections of continuous sterilizers or at the exit end of cooling canals should:
   (a) Have 5 p/m free chlorine present (positive reaction with diethyl-para-phenylene diamine method) when tested every 2 h
   (b) Have less than 100 colonies/ml after incubation for 5 days at 22–25°C tested once per day until satisfactory pattern has been established and then once or twice/week

Can handling after processing: "dry-handling"

1. After processing, cans must be dried as quickly as possible to prevent infection through the seams. Measures recommended to accelerate drying are:
   (a) The use of suitable wetting agents for static retorted cans
   (b) The tilting of retort crates to remove most of the residual water
   (c) The use of air driers on continuous cooker exits or busse unloaders
   (d) Allowing cans in retort baskets to stand overnight
2. Cans must be handled mechanically until packed or allowed to dry completely in retort baskets. Cans must not be touched by naked or gloved hands until completely dry
3. Particular attention regarding disinfection must be paid to those sections which are unavoidably in contact with wet can seams, e.g. take-off from continuous sterilizers
4. Can runways of conveyor systems should be designed so that
   (a) They can be easily cleaned by regular treatment with bactericidal detergents
   (b) Cans do not roll on their seams
   (c) Can abuse is avoided

Staph. aureus bacteria on the skin. This is a danger in under-cooked birds and may be a source of cross-contamination in the home of the consumer. It would be interesting to know where these bacteria came from and what, if anything, can be done to eliminate them or greatly reduce their incidence.

The authors wish to take this opportunity of thanking the Directors of Marks & Spencer for permission to publish.

References

Bashford, T. & Herbert, D. (1966). Post-process leakage and its control. In *The Safety of Canned Foods,* pp. 75–91. London: Royal Society of Health.

Goldenberg, N. (1960). Odour problems in the merchandising of foodstuffs. In *Odour in Packaging,* pp. 19–25. London: Institute of Packaging.

Goldenberg, N. (1962). Cake packaging. In *Recent Advances in Packaging of Food.* Proc. Inst. of Packaging Conf. Harrogate, 1962. London: Batiste Publications.

Goldenberg, N. (1963). Quality control in the cake industry. *Proc. 16th Ann. Conf.* British chapter affiliated to Amer. Soc. Bakery Engineers; pp. 40–47, Davis House, High St. Croydon, Surrey.

Goldenberg, N. (1964*a*). Food Hygiene: Standards in manufacture, retailing and catering. *R. Soc. Hlth J.* **84,** 195.

Goldenberg, N. (1964*b*). The selection, use and handling of dried fruit in cake-baking. *Proc. 19th Ann. Conf.* British chapter affiliated to Amer. Soc. Bakery Engineers; pp. 21–28. Davis House, High St. Croydon, Surrey.

Goldenberg, N. (1968). Food quality control. *Fd Mf.* 29.

Goldenberg, N. & Tall, P. D. (1971). The need for consultation, communication and agreed codes of practice–a retailers' experience. Symposium on *Progress in taint-free packaging,* p. 43. Organized by Society Chemical Industry Food Group and Pira. 25 November 1970, Zoological Society, London.

Goldenberg, N. & Elliott, D. W. (1973). The value of agreed non-legal specifications (1955–1971). In *Proceedings of the 8th International Symposium. The Microbiological Safety of Food.* p. 359. London: Academic Press.

Hobbs, B. C. (1969). Staphylococcal and *Clostridium welchii* food poisoning. In *Bacterial Food Poisoning.* Ed. J. Taylor. London: The Royal Society of Health.

Hobbs, B. C. & Gilbert, R. J. (1970). Microbiological standards for food: public health aspects. *Chemy Ind.* 215.

Stephens, R. L. (1970). Bacteriological standards for foods: A retailer's point of view. *Chemy Ind.* 220.

Discussion

Tomkin

The successful programme you have described is built around the involvement of people at all levels in your production plants. This would inevitably require your obtaining the support and co-operation of unionized employees. Have you had any difficulty in obtaining the support of the unions and, if so, how did you deal with the problems which arose?

Goldenberg

The discussions with the unions or union representatives are the responsibility of our suppliers; but I can say that the key to getting their support is to explain carefully and patiently *why* certain procedures are deemed to be essential. It is *not* good enough to tell people *what* to do. It is just as important to explain *why.* For example, the social consequences and implications of poor and careless handling of perishable foods were simply not undertstood by operatives; it was only after it was explained to them what these consequences could be in terms of sickness and food poisoning that they accepted the need for such procedures.

Ingram

From similar experience in different contexts, I agree with Mr Goldenberg on the need to convince the responsible operatives at the working level of the necessity for changes in practice. This means convincing the relevant trades' union representatives. I recently attended a meeting in Australia on possibilities for research and development in the meat industry at which a union representative was present who was obviously interested in technical progress and capable of participation in technical discussion. This seemed to me a most welcome development which might be much assisted by appropriate arrangements for training.

Christian

The conference to which Prof. Ingram referred concerned future research and development in meat science and technology. Recent developments in Australia's meat export markets have led to substantial modifications in abattoir practice to meet importers' specifications. This has of course concerned the relevant trades' union and some have responded by taking a much greater interest in technology. Many of us were, like Prof. Ingram, very impressed by the pertinence of the technical comments made at that conference by the trade union representative.

Insalata

There are inherent hazards in using a "resident chemist" rather than professional microbiologists. Has this produced a real problem?

Goldenberg

No, it has not done so. A "resident chemist" can normally be quickly trained to use the necessary methods of bacteriological analysis and to be able to interpret them adequately in relation to plant practice to ensure safety. There are many businesses where it is simply not possible to justify the employment of a full-time professional microbiologist but where the "resident chemist" can quite easily undertake the essential work that needs to be done.

8th International Symposium on the Microbiological Safety of Food

Summing-Up

Sir Graham Wilson

MR. CHAIRMAN, LADIES AND GENTLEMEN

In attempting to summarize the various topics that have been discussed during the past four days I find myself in a difficult and almost impossible position. So much information has been forthcoming and so many suggestions have been made during the course of the Symposium that any attempt to summarize them and to frame the wide generalizations that should be drawn from them must necessarily prove a failure. I ask you, therefore, to pardon me for the inadequacy of my performance.

As an introduction to this task I propose to give a brief review of the history of food poisoning. And here I am dealing specifically with the gastro-enteric form of the disease, and not with the far wider subject of food-borne infection. Chemists in the middle of the last century were already analysing food and pointing to the rôle played by some of the heavy metals, notably lead and arsenic, in the causation of food poisoning. In numerous outbreaks, however, metallic poisoning could be ruled out, and it was then that Brieger in Germany suggested that food-poisoning was caused by toxic amines formed by bacterial action in the food. This so-called ptomaine hypothesis prevailed for 20 years or more. till it was demonstrated that toxic amines became evident only during the later stages of putrefaction, at a time in fact when both the flavour and the smell of the food rendered it uneatable.

During the second decade of the present century, Savage in this country and Jordan in the United States recorded the finding of salmonellae in many outbreaks of food poisoning. This was 25 years or so after Gärtner in Germany had isolated *Salm. enteriditis* from the meat of an emergency-slaughtered cow that had given rise to a serious outbreak of food poisoning, and de Nobele in Belgium and Durham in this country had similarly incriminated the organism we now know as *Salm. typhimurium.* The work of Savage and Jordan was generally accepted, and for a time, food poisoning was regarded as almost synonymous with salmonella infection.

Then came Dack and his colleagues in Chicago who found that in many of the outbreaks in which salmonellae could not be demonstrated enterotoxin-forming staphylococci were present. Staphylococcal enterotoxin had been described three times previously—in Belgium, the United States and the Philippines—and it

453

was not till it was discovered for the fourth time by Dack in 1930 that its significance in relation to food poisoning was realized.

During the 2nd World War *Cl. welchii* was found by Knox, Hobbs and others in England to be associated with a peculiar form of food poisoning; and *Bacillus cereus* also came under strong suspicion—a suspicion that was fully justified later by Norwegian workers.

Also during the war years we became cognizant of the part played by enteropathogenic forms of *Escherichia coli* in the causation of infantile enteritis. Here again is an example of a long lag phase before the discovery of previous workers was appreciated. In the last century Jensen in Scandinavia had brought evidence to show that these organisms were responsible for certain forms of calf dysentery, and in the late twenties of the present century Adam in Germany had found them to be associated with infantile enteritis. Adam's paper was completely overlooked till I happened to run across it 4 years or so after Bray in this country had rediscovered these organisms.

The latest addition to the group of food-poisoning organisms is *Vibrio parahaemolyticus*, described by Sakazaki in Japan. Numerous workers have become interested in this organism, so there is no fear of its suffering from the neglect that attended three of the other organisms I have described.

Whether enteroviruses can cause food poisoning is not known. Cows' milk has been shown to convey the virus of the Far Eastern form of encephalomyelitis, but so far as I am aware there is no well documented account of a viral form of gastroenteritis. Nevertheless, we cannot rule out such a possibility.

Part of my object in giving you this brief historical summary is to make it clear that we now have a fairly complete list of organisms responsible for food poisoning. Besides those I have mentioned, we have evidence that certain other organisms when present in enormous numbers in the food may prove toxic, and an occasional new one may well be described; but on the whole I think we are justified in assuming that our knowledge of the bacterial causation of food poisoning is fairly complete. Not so, however, the epidemiology of the disease, of which our ignorance is still great. The figures, for example, given of the prevalence, or more strictly apparent prevalence, of the disease in different western european countries, in North America and in Australia revealed so many discrepancies as to furnish a real epidemiological puzzle. Why is it, for instance, that salmonellae are far the commonest cause of food poisoning in Britain but play only a small part in the United States where, as we know, the fowl population is heavily infected? Why should staphylococci be the predominant cause of food poisoning in the United States, and yet account for only 2–3% in British outbreaks? And why should only 300 outbreaks of food poisoning be recorded annually in the United States when in Britain there are 6000—20 times as many in a country with only a quarter of the population? These are questions that demand intensive epidemiological investigation. It is perfectly clear that no country has as yet anything like a complete record of the amount of food poisoning that is occurring. Even in England and Wales where reports are received weekly from all public health and hospital laboratories and analysed in a central epidemiological bureau, we have strong reason to believe that a number of the smaller outbreaks in the community, and still more of the family outbreaks, are never recorded. Food epidemiology has lagged a long way behind food bacteriology; and before we can learn the full extent of food poisoning, we shall have to institute a thorough system of surveillance, and get the co-operation of general practitioners and of the public themselves in the reporting of cases of gastro-enteritis associated with food.

As with all diseases of multifactorial causation it is extremely difficult to assess the relative importance of individual factors. In my opening address three days ago I referred to 5 reasons for the great increase in food poisoning that has occurred in this country during the last 30 years, namely communal feeding, pre-cooked food, bulk distribution, importation of new foods and expanded laboratory services. Of these, chief attention has been paid in our meeting to the importance of contaminated materials for incorporation in animal foodstuffs. Spray-dried egg from the western hemisphere during the war years led directly to infection of the human population with salmonellae; and by distribution of the poorer quality batches to farmers to a widespread dissemination of both old and new types of salmonellae among our cattle, pigs and poultry. Now other countries, as well as our own, are suffering from the effects of imported bone and fish meal often heavily contaminated with these organisms. One of the points that was brought out most clearly was that infection of pigs or other animals from contaminated feeding stuffs might lead through a long chain of causation to a major outbreak of food poisoning in the human population. Even though only 1% or so of animals on the farm might be infected, the proportion increased progressively as the result of stress through transportation, prolonged holding in crowded lairages and cross contamination in the abattoirs till 60 or 70% of samples of the final meat as delivered to the butcher or meat processing plant might be contaminated with salmonellae. Such meat in the provision merchant's shop, if served on the same counter as pre-cooked or preserved articles of food and cut up with the same knife, might lead to contamination of foods that, unlike butcher's meat, were to be eaten raw, and were therefore especially dangerous. Happily it was also shown how by good hygienic measures it was possible to avoid this progress in the building-up of infection, and to provide the butcher with meat of which only 1–2% of samples were infected. In this connection a great deal was said about the need for training veterinarians in the subject of food hygiene, and for the desirability of veterinary inspection not only in the abattoirs but also in meat processing plants.

Bacteriologically, figures were presented to demonstrate the continuous flux in the serotypes of salmonellae met with in the outbreaks of food poisoning; and attention was drawn to the curious anomaly that the introduction of new serotypes into Britain did not seem to have increased the total number of outbreaks observed, suggesting that one serotype displaced another. Stress was laid on the growing importance of poultry in the causation of human salmonella infection; and on the need to watch for instances of poisoning by marine foods eaten raw.

Measures for the prevention of food poisoning were discussed at length. The institution of training courses for all classes of food workers was considered desirable, varying in content and duration according to the educational level and the type of operative concerned. It was agreed that, if education in food hygiene was to be successful, it was necessary to begin at the top and work downwards. Once the managers of a firm were convinced of the importance of food hygiene and were determined to see it carried out in practice in their factories, a great step forward had been taken. With the full co-operation of the management and subsidiary staff it was possible to put on sale perishable food products of surprisingly low bacterial count. Cleanliness, though by no means a guarantee of safety, had a profound psychological effect; and when the reasons for it were explained to the supervisors and the operatives their full co-operation could often be assumed. The quite simple principles underlying good food hygiene could easily be explained, so that the consequent importance of adequate heating and refrigerated storage of food could be understood.

There was virtually no discussion on the contrasting use of legislative regulations and of advisory recommendations for improving the hygienic standard of food production and handling. Practices varied in different countries, and in a matter such as this it seemed wise to refrain from imposing uniform standards and methods of securing them on an international scale. Whatever method was employed, good results could be expected only by means of education. The difficulty of gaining compliance was experienced particularly in small factories, canteens, shops and cafeterias where insufficient capital was available to pay for structural improvements and running expenses. Even here, however, education in the principles of food hygiene could result in substantial improvement, provided the goodwill and co-operation of the management and workers could be gained. It was agreed that more attention should be paid to the education of the housewife, teaching her by every possible means to boycott unhygienic places where food was sold or consumed and to state her reason for doing so. How this should be done was not discussed in detail, though informative pamphlets useful for this purpose were exhibited.

Discussion at this Symposium has been carried on at a high level, and a remarkable degree of unanimity manifested in relation to most of the numerous topics considered. The exchange of different views has proved very stimulating and the large gaps in our knowledge have emphasized the need for further research.

Author Index

Numbers with asterisk refer to pages where References are listed at the end of each paper.

A

Abbott, J. D., 173, 178*
Abd-el-Ghani, S., 81, 86*
Abdussalam, M., 229, 236*
Abedi, Z. H., 296, 302*
Abinanti, F. R., 261, 266*
Ackermann, D., 328, 338*
Adams, M. H., 32, 38*
Adelaar, T. F., 294, 303*
Adler, H. E., 122, 125*
Ahlmark, A., 333, 338*
Aicher, J., 233, 236*
Aiiso, K., 23, 27*, 336, 338*, 375, 383*
Ajmal, M., 93, 100*
Akimov, A. M., 91, 92, 104*
Akiyama, S., 22, 28*, 375, 377, 378, 383*
Albertsen, V. E., 14, 15*
Albrecht, P., 260, 267*
Alder, V. G., 82, 85, 87*
Aldrin, J. F., 328, 329, 341*
Alexander, J., 59, 65*
Alford, J. A., 94, 102*
Allan, B. C., 26, 27*, 138, 140*, 160, 161*, 190, 193*
Allen, J. R., 92, 100*
Alterauge, W., 94, 100*
Amano, K., 96, 100*
Amano, T., 24, 29*
American Meat Institute Center for continuing Education, 419, 420*
American Public Health Association, 122, 123, 125*, 410, 413*, 418, 420*
Andal, M., 336, 341*
Anderson, A. W., 318, 319, 324, 324*
Anderson, E. S., 48, 53*
Andrewes, C., 258, 266*
Anellis, A., 93, 100*
Angelotti, R., 94, 102*, 276, 282*, 283*, 318, 324*
Anon, 1972, 59, 60, 65*, 95, 100*, 396, 401, 402, 403*
Anrep, G. V., 335, 338*
Anusz, Z., 94, 100*
Aoyama, A., 19, 28*, 189, 194*
Appleton, P. J., 182, 193*
Archer, J. F., 200, 205*
Arguedas, J., 261, 267*

Arie, K., 376, 284*
Arledge, W. L., 79, 86*
Arseculeratne, S. N., 184, 194*
Artischeva, L. J., 97, 103*
Asakawa, Y., 377, 383*
Asatoor, A. M., 330, 338*
Ashby, B. S., 182, 193*
Asperger, H., 79, 85*, 99, 100*
Atkinson, N., 157, 161*
Ay, C., 81, 87*
Ayadi, M. S., 335, 338*
Aylward, F., 396, 399, 403*
Ayres, J. C., 91, 102*, 318, 325*
Azuma, Y., 85, 85*

B

Bach, R., 94, 104*
Badstue, P. B., 225, 226*
Baer, E. F., 234, 236*
Bailey, J., 138, 140*, 190, 194*
Baird-Parker, A. C., 94, 100*
Bajlozov, D., 94, 100*
Baker, M., 177, 178*
Bamford, V., 159, 161*
Baran, W. L., 90, 100*
Barber, M. A., 234, 235*
Barja, I., 78, 85*, 86*
Barker, W. H., 274, 282*
Barnes, E. M., 6, 6*
Barnett, H. W., 93, 104*
Baross, J., 190, 193*
Barrow, G. I., 27, 27*, 138, 140*, 189, 190, 193*, 194*
Barsoum, G. S., 333, 334, 335, 338*
Bartl, V., 92, 96, 99, 101*, 104*
Bartlett, D. I., 177, 178*
Bashford, T., 360, 367*, 448, 450*
Battey, Y. M., 26, 27*, 138, 140*, 160, 161*, 190, 193*
Batty, I., 317, 323, 325*
Bauman, H. E., 425, 426*
Baumgart, J., 190, 194*
Baumgarten, U., 332, 338*
Baur, H., 334, 339*
Bavdeau, H., 83, 86*
Bayard, H., 328, 338*

Subject Index

A

Abattoir,
  design and practices, 4, 7, 41–46
Aboriginals, Australian,
  salmonellosis among, 158, 159, 162
Acetic acid preserves, 97
Achromobacter,
  in semi- preserved fish, 96
    meats, 92
A. histaminum,
  histamine production by, 330
Acridine orange,
  "cure" of prophage infection, 32
Aerobic plate count,
  limitations of, 349, 350, 351
  of foods, 90, 91, 95–99
Aflatoxins, 296, 298, 299
Algal blooms, 186, 187, 188
Alkaligenes in preserved meats, 92
American Public Health Association,
  standard methods, 387–392
Animal feeds,
  chemical treatment of, 15
  growth of salmonellae in liquid, 13
  heat treatment of, 9, 15, 221, 225
  pelleting of, 251
  salmonella in, 10–15, 45, 173–178, 220,
    221, 226
  survival of *Salmonella senftenberg* in,
    206
Anisakis, 182, 384
Anisakis marina, 230
Antagonism,
  microbial, 348
Antibiotic resistance,
  of salmonellae, 179
  of staphylococci, 279
Antigens,
  of *Vibrio parahaemolyticus,* 21, 22, 191,
    382, 383, 385
Arizona, 177
Arothron hispidus, 184
Artemia salina,
  in mycotoxin assay, 296, 304
Aspergillus amstellodami,
  toxicity of, 300, 301
A. flavus,
  toxicity of, 300, 301
A. niger,
  toxicity of, 301

A. ochraceus,
  toxicity of, 301
A. oryzae var. *effusus,*
  toxicity of, 300, 301
A. terreus,
  toxicity of, 301
A_w, *see* Water activity

B

Bacillus,
  in semi-preserved fish, 95, 96, 97
B. anthracis, 230
B. cereus,
  control of, 135, 136
  causing 'bitty' cream, 58
    sweet curdling of milk, 58
  ecology of, 135, 136
  food poisoning, 48, 59, 60, 69, 130, 135,
    454
    incubation period of, 74
    symptoms of, 135, 136
  growth of, 132, 135, 136
  haemolysin of, 71, 72, 73
  in canned hams, 91
  in gelation of UHT milk, 59
  in milk and dairy products, 57–67
  in semi-preserved meats, 92
  lethicinase of, 71, 72, 73
  pathogenicity of, 69–75
  sources of, 58
  spores,
    germination of in milk, 61–66
    heat activation of, 62, 63, 64
    in milk cans, 58
  sporulation of, 135, 136
  survival of, 132, 135, 136
  toxic dose of, 66, 73
  toxin,
    production by, 70, 71, 73, 136
    properties of, 71, 72
B. coagulans,
  in canned hams, 91
  in semi-preserved meats, 92
B. licheniformis, 92
B. megaterium, 91
B. mesentericus, 96
B. pumilis, 91
B. subtilis,
  in canned hams, 91

Food poisoning–*cont.*
  by histamine, 327–343
    marine microorganisms, 181–196
    Salmonella, 12, 13, 94, 97, 130, 131,
        133, 144, 145, 146, 150, 151,
        154–160,   172–175,   180,
        197–201, 453
    Shigella sonnei, 53
    Staphylococcus aureus, 130, 131,
        145, 146, 150, 151, 155, 156,
        162, 163, 180, 273–*285, 453,
        454*
    streptococci, 146, 156, 162
    Vibrio parahaemolyticus, 19, 24–27,
        130, 148, 149, 156, 160, 454
  due to,
    baker's confectionery, 12
    beef, 146
    butter, 146, 291
    candy love beads, 148
    catered meals, 155
    cereal products, 135
    cheese, 144, 147, 155
    corn spread, 155, 156
    crab, 148, 160, 190
    cream pastries, 60, 95, 155
    dairy products, 133, 274, 275, 284,
        291
    fish, 97, 98, 182–184, 327, 336, 337
    frankfurters, 95
    ham, 94, 95, 151, 163, 274, 275, 281
    herring, 97
    izushi, 98
    liver paste, 147
    lobster, 160
    marinated fungi, 98
    meat, 60, 132, 133, 135, 154, 274,
        275, 281
    meat balls, 60
    meat loaf, 60
    meat pies, 132, 133
    milk, 133, 155
    mushrooms, 147, 150
    pork, 146, 162, 163, 167, 174, 180
    potato salad, 155
    potatoes, 59
    poultry, 132, 133, 135, 154, 156,
        197–201, 231, 274, 275, 281
    prawns, 156, 160, 275, 281, 285
    processed foods, 132
    puddings, 60
    red beets, 94
    rice, 60
    salami, 95, 151
    sausage, 60
    seafoods, 154, 155, 160, 181–196

Food poisoning–*cont.*
  due to,
    shellfish, 154, 155, 156, 160, 182,
        183, 186–190
    soups, 60
    tomato juice, 94
    tuna fish, 147
    tunny fish salad, 98
    vanilla sauce, 60
    vegetables, 60, 274, 275
  epidemiology of, 454
  incidence of,
    in Australia, 153–163
        England and Wales, 129–142
        U.S.A., 143–151
  reporting of, 129, 130, 153, 154
  statistics of, 130–133, 139, 153, 160
  surveillance of, 130, 132–134, 139,
        143–151, 156
Food protection, 125
  storage, temperatures for, 50, 52
Foods,
  dried,
    bacterial survival in, 107–119
  frozen,
    thawing and holding of, 50
  mycotoxins in, 87, 88
  non-acid, canned,
    code of practice for, 449
  semi-preserved, 89–106
    definition of, 89
  virus isolations from, 264–266
  weaning, 77–88
    microbiological standard for, 81, 82
Foot and mouth disease virus, 230, 259,
        262, 265, 270
  heat resistance of, 269
Frankfurters,
  food poisoning due to, 95
  spoilage of, 92
Fugu poison, 184
Fungi, marinated,
  botulism due to, 98

G

Geese,
  salmonellosis in, 212, 214
Gel diffusion,
  enterotoxin detection by, 276–278
Germination, *see* Spores
Glucono-delta-lactone, 368
Glucose-salt-teepol (GST) broth, 190, 378
Glycerol-mannitol-acetamide-cetrimide
    (GMAC) agar, 85